T0389355

An Introduction to Groups, Groupoids and Their Representations

Alberto Ibort
Department of Mathematics
University Carlos III of Madrid
Madrid, Spain

Miguel A Rodríguez
Department of Theoretical Physics
University Complutense of Madrid
Madrid, Spain

CRC Press is an imprint of the
Taylor & Francis Group, an **informa** business
A SCIENCE PUBLISHERS BOOK

CRC Press
Taylor & Francis Group
6000 Broken Sound Parkway NW, Suite 300
Boca Raton, FL 33487-2742

Printed on acid-free paper
Version Date: 20190919

International Standard Book Number-13: 978-1-138-03586-7 (Hardback)

Visit the Taylor & Francis Web site at
http://www.taylorandfrancis.com

and the CRC Press Web site at
http://www.crcpress.com

To our wives, Conchi and Teresa
and our sons, Alberto and Pedro

Preface

This project grew from lecture notes prepared for courses on group theory taught by the authors a long time ago. They were recovered and put in their present form because of the impetus provided by recent investigations on groupoids that convinced the authors of the interest in providing a way to graduate and undergraduate students as well to researchers without a strong background on geometry and functional analysis to access the subject.

The theory of groups and their representations has had an extraordinary success and reached a state of maturity in the XXth century. Motivated by problems in Physics, mainly Quantum Mechanics, the theory of group representations had a fast development in the first half of the century, boosted by the contributions and insight of such extraordinary people as H Weyl, E Wigner, etc. Later on, the definitive contributions of A Kirillov, B Kostant, GW Mackey and many others closed the circle by establishing the deep geometrical link between the theory of group representations and symplectic geometry and quantization. There are many books on the subject covering the full spectrum of rigor, depth and scope: From elementary to advanced, encyclopedic and highly specialized, etc.

At the same time, the notion of groupoid [7], the 'most natural' extension of the notion of group, sometimes even proposed to be called 'groups', has had a very different history. A Connes claimed that "it is fashionable among mathematicians to despise groupoids and to consider that only groups have an authentic mathematical status, probably because of the pejorative suffix oid", but that they are the relevant structures behind many notions in Physics [17, page 13].

Whatever the reason, the truth is that the theory of groupoids has not been developed to the extent that the theory of groups has. Only more recently, because of various facts, among them the insight coming from Operator algebras (see for instance the book by Renault [57]); the relevant role assigned to them in the mathematical foundations of classical and quantum mechanics in the work by N Landsman [45]; the combinatorial applications in graph theory [43]; the applications of groupoid theory to geometrical mechanics, control theory (see, for instance, [19] and references therein) and variational calculus [52]; the recent contributions to the spectral theory of aperiodic systems [4], the recent groupoids-based analysis of Schwinger's algebraic foundation for Quantum Mechanics [14], [15], or the appealing description by A Weinstein of groupoids as unifying internal and external symmetries [66] extending, in a natural way, the notion of symmetry groups and groups of transformations with the associated wealth of applications, groupoids are attracting more and more attention.

There are many treatments of groupoid theory in the literature, from the very abstract, like the seminal categorical treatment in [33], to the differential geometrical approach in [48], passing through the Harmonic analysis perspective in [67] or the C^*-algebraic setting [57] mentioned already. The elementary notions on groupoids can be explained, however, at the same level that those of groups, and actually, some of the ideas on group theory related to groups as transformations on spaces, gain in clarity and naturality when presented from the point of view of groupoids. We feel that there is a lack of a presentation, as elementary as possible, of the theory and structure of the simplest possible situation, that is, of finite groupoids, that will help researchers, in many areas where group theory is useful, to effectively use groupoids and their representations.

On the other hand, the theory of linear representations of finite groupoids is interesting enough to deserve a detailed study by itself. We will just mention here that the fundamental representation of the groupoid of pairs of n elements is just the matrix algebra $M_n(\mathbb{C})$. Hence, the theory of linear representations of finite groupoids provides a different perspective to the standard algebra of matrices. There are a number of relevant contributions to the theory of linear representations of groupoids that describe the notion of measurable and continuous representations (see, for instance, the aforementioned work by Landsman [45] or [6] and references therein), unitary representations of groupoids and the extension of Mackey's imprimitivity theorem [54], [55], [29] or the representation theory of Lie groupoids [30], but again, as before we feel that there is lack of an elementary approach to the theory of representations of groupoids discussing the theory for finite groupoids.

Hence the proposal to write a monograph developing the foundations of groups and groupoids simultaneously and at the same elementary level, enriching both approaches by explaining the basic concepts together and providing a wealth of examples. The first part of this work will describe the algebraic theory and try to dwell in the structure of finite groups and groupoids. The second part will develop the theory of linear representations of finite groups and groupoids.

There are many subjects that will not be covered in this work, like the theory of topological and Lie groups and groupoids (the latter being particularly interesting because of the recent proof of Lie's third theorem for groupoids [20], [21]) or the many applications mentioned earlier that will be left for further developments. We believe that the time is ripe for a project like this, so we will try to offer the reader a pleasant and smooth introduction to the subject.

Acknowledgements

Of course, this work would not have been possible without the help and complicity of many people, starting with the closest ones, our families, thanks for your infinite patience and support, to our colleagues, students and friends.

As mentioned above, this text has profited from old lecture notes on group theory. Thanks to JF Cariñena for even older notes that inspired many parts of this text.

We would also like to thank many of our teachers: LJ Boya, JF Cariñena, M Lorente, from whom we got infected since our Ph.D. studies with the virus of the harmony and inner beauty of group theory that has evolved into the present book.

During the period of preparation of the manuscript, various aspects of the book were discussed with many people in different contexts and presentations. In particular AI would like to thank the students of the Master course on *Abstract Control Theory* at the Univ. Carlos III de Madrid that have 'suffered' the exposition of parts of this work.

Finally we would like to thank the editor of this book for his patience and encouragement.

Contents

APPENDICES

Introduction

As mentioned in the previous paragraphs, in spite of the growing interest in the subject, we feel that there is a lack of expository works where the basic notions would be explained, allowing the interested reader to enter the subject. These notes pretend exactly so, that is, to offer the interested reader a gentle introduction to the theory of groupoids and their representations in a similar way as many books offer an introduction to the theory of groups and their representations.

Actually, we pretend to fuse the theory of groups within the theory of groupoids, offering at the same time the basis of them from this more general viewpoint. In doing so, we have to warn the reader that there are important difficulties, as many natural questions in the theory of groupoids are in a much less mature state than the corresponding ones in the theory of groups, for instance, the notions of normal subgroups and simple groups.

The impetus provided by Quantum Mechanics to the development of the theory of groups was paramount. Thus, it is of fundamental importance to orient our aim when laying the grounds of a theory with such potential for widespread influence. In these notes, we have chosen to follow the insight from the relationship between the Theory of Groups and Quantum Mechanics in order to obtain inspiration in the development of groupoids. In doing so, a guiding idea that will ignite the process is required. In the case of groups and Quantum Mechanics, it was Wigner's theorem that provided such departing point. In our case, the key idea from Quantum Mechanics will be the algebra of measurements introduced by Schwinger. For the interested reader, we will discuss all of this in Appendix C, where it will be shown that Schwinger's algebra is actually the algebra of a groupoid.

Conceptually, the book attempts to convince the reader that the notion of groupoid (and category) is natural. It appears from extremely simple problems, disguised in the text as 'games' or puzzles, either invented or existing already, introduced so that we can gain insight on them and explore their structure. At the same time, the notion of 'realization' or 'representation' of the abstract structure appears naturally and it shows the relevance of the abstract and powerful notion of functor that will serve us, among other things, to define homomorphisms of groupoids. The first part will end with the discussion of a structure theorem for groupoids and some elementary classification theorems for groupoids of small order.

The second main idea that permeates the book is that understanding the structure of groupoids is inseparable from understanding their representations. This is done by combining the 'geometrical' description of representations on fibered sets and the abstraction provided by constructing the algebra of a groupoid and the corresponding notion of modules. Both roads will converge to prove semisimplicity theorems that will simultaneously show that any finite-dimensional representation of a groupoid is a direct sum of irreducible ones and that the algebra of a groupoid is a direct sum of simple ones.

The book will be organized as follows: First to describe the foundations of the theory, that is, the notion of group and groupoid as well as the basic applications and structure theorems. Because of the intrinsic categorical nature of the notion of groupoids we would like to introduce the basic notions of categorical language in order to be able to use them in a smooth and natural way throughout the book. Thus, the first chapter of the work will be devoted to developing the basic language of the theory of categories that will be used throughout the text. The corresponding definitions of the fundamental notions will be introduced as needed in a language as simple and elementary as possible. Even if most of the categories, and definitely all groups and groupoids used in the exposition, are going to be small categories, i.e., they are going to be sets, we feel that a sensible use of categorical notions brings an important and meaningful light to the subject.

The second chapter will be devoted to developing the basic notions of the theory of groups, mostly finite groups, preparing the ground to discuss the notion of groupoids. Moreover, because of the fundamental role played by the notion of isotropy groups of a groupoid, it is convenient to state the basic notions and results of the theory first in the case of groups alone. Then, in chapter three, the notion of groupoids is developed from a set-theoretical perspective. A first structure theorem for groupoids is presented, showing that any groupoid is a direct union of connected (or transitive) groupoids. The basic ideas about direct union and products of groupoids are discussed and the chapter is ended by discussing a family of groupoids that appear by the hand of such well-known puzzles as Loyd's puzzle and Rubik's cube. Emphasis is put on showing that the theory of groupoids is very well adapted to discuss such combinatoric problems.

Chapter four will be devoted to the analysis of the relation between the notion of symmetry and groupoids. The discussion extends the well-known analysis of the relation between symmetries and groups (of transformations) and it is completed by reviewing Weinstein's observation on the relation between local symmetries and the theory of groupoids. The chapter ends by extending Cayley's theorem for groups to groupoids.

The ground is laid to discuss in-depth the notion of homomorphism of groupoids, a key notion in the theory of groups that we have avoided to introduce so far. The notion of homomorphism of groupoids is just the basic notion of functor in categorical terms. Thus, in Chapter 5, functors are introduced, offering even an information-like interpretation in terms of databases, and all basic properties of homomorphisms of groupoids are discussed. The chapter ends by discussing the categorical notion of natural transformation in order to understand the meaning of equivalence in the various uses of functors and homomorphisms of groupoids discussed so far.

Finally, after all previous preparations, the main structure theorem of groupoids is presented in Chapter 6: Any connected groupoid is an extension of a groupoid of pairs by a totally disconnected trivial groupoid. In arriving at this result, the notion of normal subgroupoid is thoroughly discussed usings the discussion held in Chapter 2, Sect. 2.1.3, for normal subgroups as a template. Combining all the information gathered so far, a classification theorem up to groupoid isomorphisms of groupoids of order smaller than or equal to 20 is presented. It is remarkable that there are very few groupoids of such small order and all of them are trivial, in the sense that the extension defining them (according to the main structure theorem) is trivial, that is, they are isomorphic to a product of a groupoid of pairs and a group. The proof of this fact is not completely trivial and requires elaboration on the basics of what would constitute a cohomology theory for groupoids.

The second part will be devoted to the study of the theory of linear representations of finite groupoids. In Chapter 7, the theory of linear representations of finite groups is discussed in detail. The basic notions of irreducibility and complete reducibility as well as Schur's Lemmas are introduced, laying the ground for a more advanced discussion, albeit elementary, of the theory in Chapter 8 with the introduction of the theory of characters. The main standard theorems of the theory are proved and various examples are discussed at length.

The notion of linear representations of groupoids is introduced in Chapter 9 and, to do that, we return to the spirit of the first chapters of this book and the notion of linear representation of quivers and categories are discussed first. The basic notions, like irreducibility and indecomposability of representations, are presented using a variety of examples and some classical structure theorems, for example, Krull-Schmidt or Gabriel's theorems are either explained or proved.

To proceed further with the analysis of the structure of representations of groupoids, the main idea of identifying them with A-modules is introduced and exploited in Chapter 10, which will be devoted to the study of the theory of finite-dimensional associative algebras and the corresponding modules over them. Many examples will be provided and, after the discussion of Schur's Lemma and the Jordan-Hölder theorem, the chapter ends by establishing the key relation between representations of groupoids and modules over the algebra of the given groupoid.

Before developing the theory of characters for representations of groupoids, we felt that it was necessary to devote a chapter to analysing the notion of semi-simplicity, instrumental in the theory of representations of groups and groupoids. In doing so, we have surrendered to the spell of the beautiful presentation of the subject in [26] that has profoundly inspired this chapter, and that ends with a proof of the semi-simplicity of the algebra of a finite groupoid.

The last chapter, Chap. 12, is devoted to the theory of characters. The presentation of the subject in [36] is followed and the corresponding extension of the orthogonality relations for characters of representations of groupoids is obtained. We will show that, again, the theory of groupoids not only extends, in a natural way, that of groups, but helps to enlighten many aspects of it. The extension of Burnside's theorem for groupoids and the analysis of the regular representation of a groupoid are good examples of this.

The book will be completed with a few Appendices added for the sake of the reader. The first one is a glossary/summary of Linear and Matrix Algebra containing a description of the basic results and notations used throughout the text. Appendix B provides a succinct description of the notion of algebraic structures constructed out of generators and relations, a basic technique used throughout the text.

Appendix C is devoted to offering a brief glimpse of how the theory of groupoids is instrumental in Schwinger's abstract description of Quantum Mechanics and could give the reader an extra motivation to work on the subject. Finally Appendix D provides a graphical solution of Rubik's cube discussed in the main text, Sect. 3.2.4.

I

WORKING WITH CATEGORIES AND GROUPOIDS

Chapter 1

Categories: Basic Notions and Examples

In this chapter, the basic notions of the theory of categories will be introduced using natural examples directly related to the theory of graphs.

A category is the natural algebraic notion that captures the associativity property of composition of maps. Usually, it is stated in a very abstract way that does not require the theory of sets, even though, in most instances, we will just be dealing with sets and the standard notions of the theory of sets will be applicable.

However, we will try to emphasise that the way of thinking in the realm of categories is not only natural but both simple and powerful as it provides a new insight into some basic manipulations with such simple mathematical objects as graphs and relations.

Both, the notion of groupoid, and its directly related notion of group, will be introduced as natural categorical ones.

1.1 Introducing the main characters

1.1.1 Connecting dots, graphs and quivers

Let us start our discussion by playing the children's game of connecting dots. The idea is to lay a family of points in a piece of paper and to provide the player with a set of instructions to connect them so that, if followed faithfully, a pattern or picture will emerge that will mesmerise the little child doing it. We will play the game expecting this time to raise the attention of the reader when appreciating the structure that will be uncovered.

Consider a finite family of dots, denoted formally as $V = \{v_1, v_2, \ldots, v_n\}$ and a finite list of rules $L = \{l_1, l_2, \ldots, l_N\}$ that indicates how to draw lines connecting them. Thus, each rule, also called a 'link', l_a will start at a dot, say v_i, that will be called its 'source' or 'tail' and it will end at another dot, say v_j, that will be called its 'target' or 'head'. We can make this explicit by writing $l_a(v_i, v_j)$, the head (target) of l_a will be denoted by $t(l_a)$ and the tail (source) as $s(l_a)$.

The lines that we will be drawing to connect the dots (or 'vertices', using the more technical term preferred by mathematicians) will be oriented, so that we may draw an arrow on them. For instance, we can play the game defined by the dots $V = \{v_1, v_2, v_3, v_4, v_5, v_6\}$ and instructions (or links):

$$\begin{aligned} L &= \{l_1(v_1,v_2), l_2(v_2,v_3), l_3(v_2,v_2), l_4(v_1,v_4), \\ &\quad l_5(v_5,v_6), l_6(v_1,v_2), l_7(v_6,v_5), l_8(v_4,v_4), l_9(v_4,v_4)\}, \end{aligned} \tag{1.1}$$

and after playing it, we will get a drawing like the one shown below (Fig. 1.1).

Notice that there could be more than one link (also called 'edge') joining two dots. If we allow, at most, one link joining any two dots, we will call the resulting figure a 'graph', or, to be more precise, a 'directed graph' or a 'digraph', as the links are directed. The slightly more general setting we are using here is called a 'multigraph' or more precisely, a 'directed multigraph' or a 'multi-digraph' (or, even fancier and shorter, a 'quiver').

A number of features are apparent when we look at the design we have got, however, before starting to discuss them, it is worth remarking that the actual 'rendering' of the instructions to draw the quiver $Q = (V, L)$ given in (1.1), is by no means unique and we could have arranged the dots and the links in ways that would seem to appear very different from what we have got. All of them, however, will share the same design or 'structure' though, and it is this structure that we would like to understand better[1].

Once we have the picture above, we can travel around it by following the links. That is, we will define a 'path' (or a 'walk') as a sequence of links, such that the head of each one is the tail of the next one, that is, $l_2 l_1$ is a path because the head of l_1 is the dot v_2 which is the tail of l_2. Notice that we have chosen to write the list of steps l_1, l_2 in the path before sinistrally, that is from right to left, contrary to the perhaps natural dextral convention inherent to the writing in English. However, this is not by chance, and it hides one of the first consequences of the abstract structures lurking below this little exercise. The same happens for $l_7 l_5$, $l_8 l_9 l_8$ or $l_2 l_3 l_1$, however, neither $l_4 l_8$, $l_1 l_6$ are paths, while $l_8 l_4$ is.

More formally, given a sequence $p = l_{a_r}(v_{i_r}, v_{j_r}) \ldots l_{a_2}(v_{i_2}, v_{j_2}) l_{a_1}(v_{i_1}, v_{j_1})$ of links in the quiver $Q = (V, L)$, it will be called a path if $t(l_{a_k}) = s(l_{a_{k+1}})$, $k = 1, \ldots, r-1$, or equivalently, $v_{j_k} = v_{i_{k+1}}$. Each one of the links l_{a_i} in a path will be called a step (or link) of the path. We can denote the path p by using a compact notation like $l_{a_r \cdots a_2 a_1}$ for the path before.

[1] There is a whole branch of research devoted to understanding how to actually draw graphs and to recognise their structure. See, for instance, [63], in particular Chapter 1 in relation to the introduction so far and Chapter 2 for the basic background on graphs.

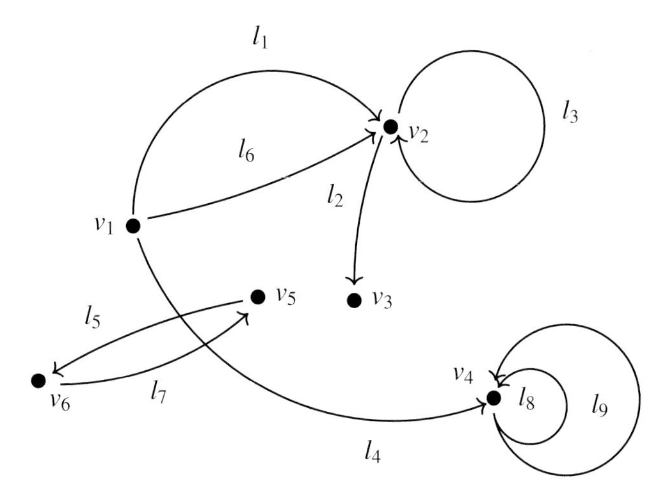

Figure 1.1: The output of a 'connecting dots' game.

The fundamental observation about quivers and paths is that, if we have two paths of the form $p = l_{a_r \cdots a_2 a_1}$ and $q = l_{a_s \cdots a_{r+1} a_r}$, we can form the new path $q \circ p = l_{a_s \cdots a_{r+1} a_r \cdots a_2 a_1}$. In other words, we can walk the quiver using the path p first and then continue the walk using the path q.

We will extend the terminology introduced before and call the initial point of a path its 'source' and the endpoint its 'target'. Two paths p, q, will be said to be 'composable' (or 'consistent') when the head of the first is the tail of the second, i.e., $t(p) = s(q)$. In such case, the new path $q \circ p$ obtained by the concatenation of both paths, that is $q \circ p = qp$, can be defined and it will be called the composition of p and q. Consistently, we may also think that each path is a new link that we are adding to the picture of (V, L) until all possible paths on the picture are included.

For any dot v we may also consider the path that simply stays there. We will use a special notation, 1_v, for these paths. Notice that $l \circ 1_v = l$ for l any path l starting at v and $1_v \circ l = l$ for any path with endpoint v.

Consider now the collection of all paths in a given quiver $Q = (V, L)$ and denote it by $\mathbf{P}(Q)$. There is a natural composition law in $\mathbf{P}(Q)$ given by the composition of consistent paths given above. The composition law defined in this way clearly satisfies the following identity:

$$p'' \circ (p' \circ p) = (p'' \circ p') \circ p,$$

whenever the paths p, p' and p'' can be composed. In other words, it is associative. Moreover the special paths 1_v act as units with respect to this composition law, that is:

$$1_v \circ p = p \qquad p' \circ 1_v = p',$$

whenever the source of p' or the target of p is v.

Once we have drawn all possible paths in our picture, we are almost done with the game, in the sense that all possible ways of travelling around the pattern provided by the quiver $Q = (V, L)$ have been explored, however, a number of natural questions begin to emerge: How many paths are available to travel from to given dots v to u? Which path is the shortest (less number of steps involved)? Can we draw a path that passes through all dots just once?, and through all links just once?, etc.

Answering these questions will not be the main aim of this book. Many of them belong to the branch of mathematics called graph theory[2], however we will encounter some ideas related to them along our trip. What we want to stress here is that in playing a simple game we have found a fundamental structure that is going to accompany and help us all along the book. The collection of paths $\mathbf{P}(Q)$ with the composition law above is actually a category, the basic algebraic structure we are going to use in this book. This particular instance of category will be called the category generated by the quiver $Q = (V, L)$.

1.1.2 Drawing simple quivers and categories

Apart from the fancy names we have been introducing ('graphs', 'quivers', 'categories', etc.) we have not yet started to grasp the properties of the structures that are showing up. As it often happens, it is convenient to think first about simple examples. Let us consider a few of them.

Consider first the extremely simple quiver $Q_{2,1}$ given by $V = \{v_1, v_2\}$, $L = \{l(v_1, v_2)\}$, with two dots and a single link (we will call it the 'singleton'). It can be drawn as (see Fig. 1.2):

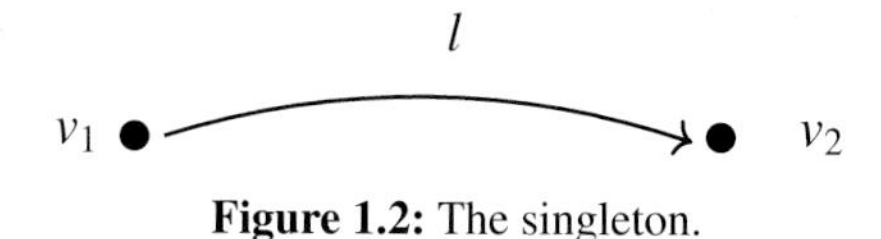

Figure 1.2: The singleton.

The category[3] $\mathbf{P}_{2,1}$ generated by the singleton consists of just two objects (the name that will be given to the vertices in the abstract setting) the set V, and three arrows (the name given to links in the abstract setting) 1_{v_1}, 1_{v_2} and l, with the obvious composition law. The only walk we can take on it is to go from v_1 to v_2; rather boring!

We may consider the (apparently) simpler example of the quiver $Q_{1,1}$ with only one dot $V = \{v\}$, and a single link $L = \{l(v, v)\}$. We call this quiver the 'loop' (see Fig. 1.3).

The path l takes us back to v, so we can keep going and we may can back to v as many times as we want, thus, the category generated by the singleton consists of all paths $p_n := l \circ l \circ \cdots (n-\text{times}) \cdots \circ l = l^n$, including the unit $1_v (= l^0)$. There is again

[2] And many of them are remarkably hard!

[3] Mathematicians have a natural tendency, shared by the authors, to introduce terminology like this, using various numbers as labels. Contrary to the custom in other areas where the use of acronyms becomes insufferable, this tradition has the ultimate goal of helping constructing classifications of the objects under study. We will have some opportunities along this work to check the usefulness of this, ending up with the classification theorem of groupoids of lower order, Thm. 6.5.

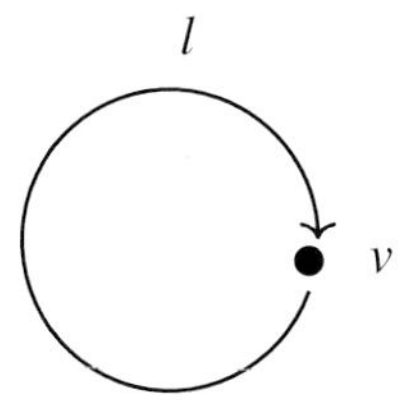

Figure 1.3: The loop.

a technical name invented by mathematicians for a category with just one object, they are called 'monoids' but we will not insist on this for the moment. What is noticeable about this example, is that the composition law of this category is remarkably simple, actually we all know it. The composition of the path p_n with the path p_m is the path p_{n+m}, with $n+m$ indicating the sum of the natural numbers n and m. Thus, we may conclude that the category generated by the loop is the 'same' as the natural numbers $\mathbb{N}$ (including zero) with the standard addition of natural numbers as composition law. We have quoted 'same' just to draw attention to the fact that the two categories, the category generated by the loop and $(\mathbb{N},+)$, have the same form and the same properties, once their objects and arrows are properly identified (in our case the dot v is identified with 0 and the link l with 1). We will say that they are 'isomorphic' (we will come back to discuss this notion in depth in Chapter 5) and we write $\mathbf{P}_{1,1} \cong \mathbb{N}$. For the moment, we will be content enough in being able to identify the single loop category with the natural numbers, our first significant achievement.

Encouraged by the last example, we jump to consider the two-loops quiver $Q_{1,2}$ (Fig. 1.4): $V = \{v\}, L = \{l_1(v,v), l_2(v,v)\}$.

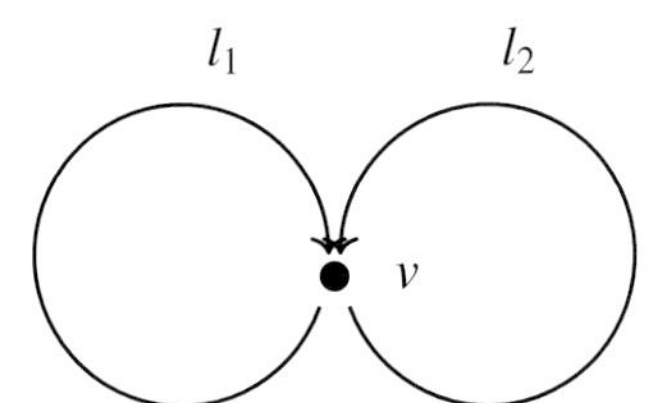

Figure 1.4: The two-loops.

Now, we may travel the following paths, like $l_1 \circ l_2$ or $l_2 \circ l_1$, however, both paths are different even if we always end up at v. To better appreciate the difference, it is clear that the path $l_1 \circ l_2 \circ l_1$ is different from $l_1 \circ l_1 \circ l_2$ and so on. Thus, the category generated by the two-loops quiver consists of all finite paths $l_{i_1 i_2 \cdots i_r}$ with the indices $i_1, \ldots, i_r$, taking the values $1,2$. The monoid $\mathbf{P}_{1,2}$ generated by this quiver is far from being as simple as $\mathbf{P}_{1,1}$, which we identified easily with the natural numbers. The monoid $\mathbf{P}_{1,2}$ (also called the free monoid generated by two elements) actually has some 'wild' features that we will discuss succinctly later on (see the section on free groups, Chap. B.1).

The difficulty with this example is that the two paths l_1l_2 and l_2l_1 are different. Suppose, however, that, for whatever reason, we want to dispose with the difference, that is, we decide that we want to identify l_1l_2 with l_2l_1 (this, of course, cannot be shown in the drawing), so we declare that both paths are the same (for us, that is). We will call such declaration a 'relation' and we will explore this notion and its consequences in the following section.

1.1.3 Relations, inverses, and some interesting problems

It is perhaps better to think first in the example provided by the three-dots quiver consisting of the quiver $V = \{v_1, v_2, v_3\}$, $L = \{l_1(v_1, v_2), l_2(v_2, v_3)\}$ drawn below (Fig. 1.5):

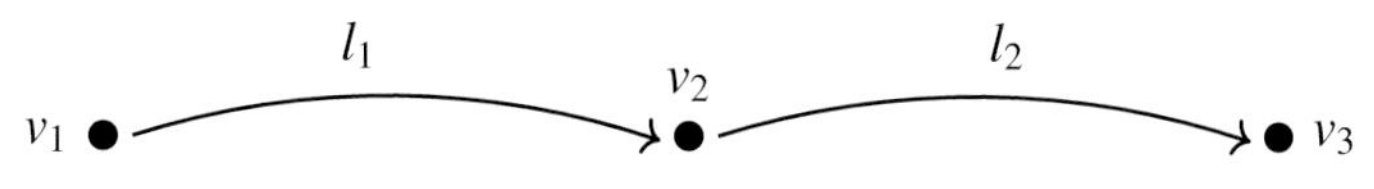

Figure 1.5: The three-dot quiver.

The category generated by it contains the dots and links above together with the path $l_{21} = l_2 \circ l_1$ (and the three units $1_{v_1}, 1_{v_2}, 1_{v_3}$). However, we may have considered the quiver with the same three dots and links as the three-dots quiver with one more link $l_3(v_1, v_3)$ shown in the Fig. 1.6.

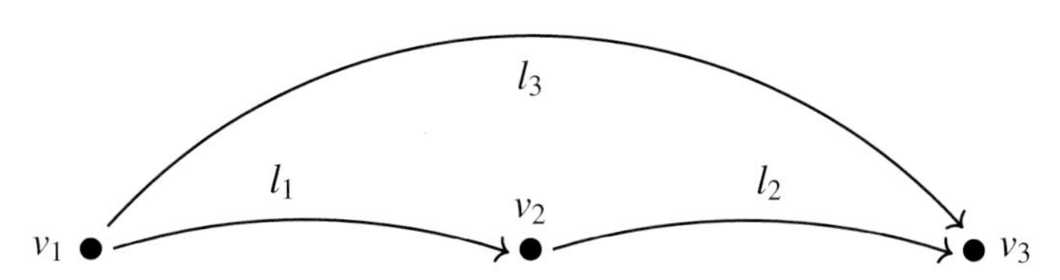

Figure 1.6: The extended three-dot quiver.

This extended three-dot quiver is not the same as the category generated by the three-dot quiver above as $l_3 \neq l_2 \circ l_1$. Actually, the category generated by it consists of the arrows l_1, l_2, l_3, l_{21} and the three units (notice that l_3 cannot be composed with neither l_1 or l_2. However, we may introduce a relation in this category by declaring that $l_3 = l_1 \circ l_2$. Once we do that, the drawing Fig. 1.6 is actually describing the category generated by the three-dot quiver (we will omit the units as we often do unless it is necessary to avoid confusions).

There are special relations which play an important role in what follows and they are those that allow us to introduce the inverse of a link, that is, consider the extension of the singleton quiver (Fig. 1.7), for example. This category will be obtained from the quiver with vertices and links: $V = \{v_1, v_2\}$, $L = \{l(v_1, v_2), l'(v_2, v_1)\}$, by

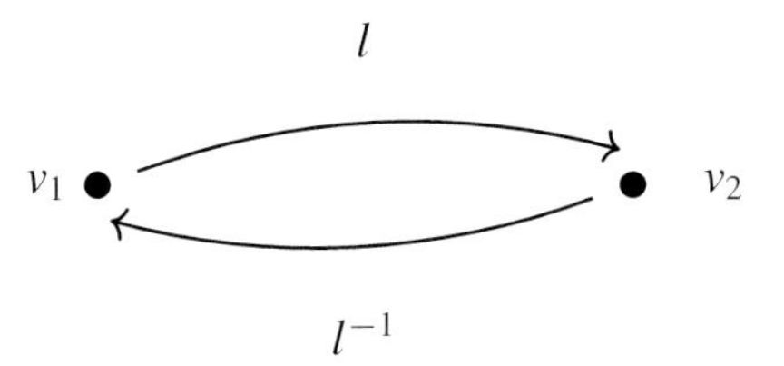

Figure 1.7: The extended singleton or groupoid $\mathbf{A}_2$.

introducing relations. Notice that the category generated by this quiver consists of all alternating paths $l \circ l' \circ l \circ \cdots \circ l' \circ l$, etc. We may, however, wish to consider that the path $l' \circ l$ that starts and ends at v_1 is the same as 1_{v_1}, that is, that l' traces back the step that takes you from v_1 to v_2 and, in the same way that the path $l \circ l'$ is the same as 1_{v_2}. Hence, we introduce the relations:

$$l' \circ l = 1_{v_1}, \qquad l \circ l' = 1_{v_2}.$$

In such case, we will say that l' and l are inverses of each other and we will write $l' = l^{-1}$. Notice that the resulting category contains four arrows: l, l^{-1} and the two units, 1_{v_1} and 1_{v_2} (contrary to what happens in the case of the single loop, we cannot compose l with itself). Again, there is a technical name for a category like this, where the arrows (links) are invertible, they are called "groupoids" and they constitute the main subject of this book. We have just found the simplest groupoid with two dots and two arrows that will be called the $\mathbf{A}_2$-groupoid.

Example 1.1 *The integers. We can do the same trick with the single loop category (that is, with the natural numbers $\mathbb{N}$). We can add the inverse l^{-1} to the single loop quiver. Alternatively, we can consider the two-loop quiver with loops l_1, l_2 and we introduce the relations $l_2 \circ l_1 = 1_v$ and $l_1 \circ l_2 = 1_v$. Then, the resulting category can be identified with the integer numbers $\mathbb{Z}$.*

Exercise 1.2 *Describe the category obtained from the single loop category when we introduce the relation given by $l^2 = 1_v$. Do the same when we consider the relations $l^3 = 1_v$ and, now in complete generality, when $l^n = 1_v$ for some given natural number $n \in \mathbb{N}$.*

We can proceed in the same way with the two-loops category, that is, we introduce the links l_1^{-1} and l_2^{-1} with relations:

$$l_a^{-1} \circ l_a = l_a \circ l_a^{-1} = 1_v, \quad l_a \circ l_b = l_b \circ l_a, \qquad l_a \circ l_b^{-1} = l_b^{-1} \circ l_a, \qquad a, b = 1, 2,$$

and we get the category $\mathbb{Z}^2 = \mathbb{Z} \times \mathbb{Z}$ consisting of pairs of integers with the standard additive composition law.

Exercise 1.3 *Describe the category obtained when we consider $\mathbb{Z}^2$ by introducing in addition the relation: $l_1^2 = l_2^3 = 1_v$. A bit harder: What will be the category obtained introducing the relation $l_1^2 = l_2^3$?*

In general, given a quiver $Q=(V,L)$ and a collection of relations $\mathcal{R}$, that is, a finite list $p_i = q_i$, $i = 1,\ldots,s$, of identifications between paths p_i and q_i in Q, we may consider the category generated by Q with relations $\mathcal{R}$. Such category will be defined as the category obtained by identifying paths in the category $\mathbf{P}(Q)$ generated by Q, that differ only in the identifications provided by the list of relations $\mathcal{R}$ (see Sect. B.1 where the construction of the category generated by Q with relations $\mathcal{R}$ is discussed in detail).

We can ask ourselves how can be recognised if two categories obtained from quivers Q_1 and Q_2 with relations $\mathcal{R}_1$ and $\mathcal{R}_2$ respectively are isomorphic. We will not try to discuss this problem here, we will just mention that it is a really hard problem!

1.2 Categories: Formal definitions

1.2.1 Finite categories

In this section, we will provide the formal definitions and precise setting for the discussion before. We will start with the definition of a finite category.

Definition 1.1 A finite category $\mathbf{C}$ consists of a finite set Ω, more formally denoted by $\mathbf{Ob}(\mathbf{C})$, whose elements are called objects and, for each pair x,y of objects, there is a finite set[4] $\mathbf{C}(x,y)$ whose elements are called morphisms or arrows. If $\alpha \in \mathbf{C}(x,y)$ is a morphism, we will say that x is the source of α and that y is the target of α. The set of objects and morphisms of a category satisfy the following axioms:

i.- (Composition law) Given two morphisms $\alpha \in \mathbf{C}(x,y)$ and $\beta \in \mathbf{C}(y,z)$ there exists a morphism $\beta \circ \alpha \in \mathbf{C}(x,z)$ called the composition of α and β.

ii.- (Existence of units) Given an object x, there exists a morphism denoted by $1_x \in \mathbf{C}(x,x)$, called the unit at x, such that for any morphism $\alpha \in \mathbf{C}(x,y)$: $\alpha \circ 1_x = \alpha$, and $1_y \circ \alpha = \alpha$.

iii.- (Associativity) Given three morphisms $\alpha \in \mathbf{C}(x,y)$, $\beta \in \mathbf{C}(y,z)$, $\gamma \in \mathbf{C}(z,u)$, then the associative law is satisfied:

$$\gamma \circ (\beta \circ \alpha) = (\gamma \circ \beta) \circ \alpha . \tag{1.2}$$

A morphism $\alpha \in \mathbf{C}(x,y)$ is said to 'transform the object x in the object y' and it will be conveniently denoted also by $\alpha \colon x \to y$ (sometimes we prefer the backwards notation $y \overset{\alpha}{\leftarrow} x$, which proves to be more convenient in many occasions)[5].

[4]Later on, when dealing with groupoids, Chapt. 3, the notation for $\mathbf{C}(x,y)$ will modified and a backwards convention will be adopted, see after Def. 3.1.

[5]The notation $\alpha \colon x \to y$ does not imply by any means that α is a map from the 'set' x to the 'set' y as the set theoretical notation would imply, this notation is just emphasizing the abstract relation existing between the objects x and y, whatever they are, and that is captured by the arrows α in $\mathbf{C}(x,y)$, recall the list of links and paths in the connecting dots games played in the previous section.

A morphism $\alpha\colon x\to y$ will be said to be an isomorphism (which is not the notion of isomorphism of categories mentioned earlier) if there exists another morphism $\beta\colon y\to x$ such that $\beta\circ\alpha = 1_x$ and $\alpha\circ\beta = 1_y$. We will also denote $x\overset{\alpha}{\cong} y$ and $\beta=\alpha^{-1}$.

We will denote the finite set of all morphisms of the category $\mathbf{C}$ with the same symbol[6] $\mathbf{C}$. Notice that the set of objects of the finite category $\mathbf{C}$ can be considered to be a subset of the set of morphisms of the category by means of the map $\iota\colon \Omega\to\mathbf{C}$, sending each object x into its associated unit 1_x, that is $\iota(x)=1_x$. We may also consider the source and target maps $s,t\colon \mathbf{C}\to\Omega$, defined respectively as $s(\alpha)=x$, $t(\alpha)=y$ if $\alpha\colon x\to y$.

Example 1.4 *In the previous section, Sect. 1.1.2, we have already found some examples of finite categories: The singleton category, the three-dots category. Notice however that the single loop category, i.e.,* $\mathbb{N}$*, or the two-loops category are not finite.*

Apart from these simple examples, there are examples of trivial categories associated to sets. For instance, any finite set S defines a finite category whose objects are the elements of S and whose morphisms are just the unit morphism, 1_x, $x\in S$. There is, however, a simple example of category associated to any set that will be of considerable interest in what follows and that we will describe in the following example.

Example 1.5 *The category* $\mathbf{G}(\Omega)$ *of pairs of a given set* Ω*.*

Let Ω *be a set (the category* $\mathbf{G}(\Omega)$ *will be finite if* Ω *is finite but the construction is for an arbitrary set) and consider the collection of all pairs* $(x,y)\in\Omega\times\Omega$*. The category of pairs* $\mathbf{G}(\Omega)$ *is the category whose objects are the elements of* Ω *and whose morphisms are the pairs* (y,x)*. Thus, given two objects* $x,y\in\Omega$*, there is just one morphism* $\alpha\colon x\to y$ *which is the pair* (y,x)*. The composition of the morphism* $\alpha: x\to y$ *and the morphism* $\beta: y\to z$ *is the morphism* $\beta\circ\alpha: x\to z$*, i.e., the pair* (z,x)*. Using the notation* $\alpha=(y,x)$ $(y\overset{\alpha}{\leftarrow}x)$*, then*

$$\beta\circ\alpha = (z,y)\circ(y,x) = (z,x)\,, \tag{1.3}$$

or more graphically: $z\overset{\beta}{\leftarrow}y\overset{\alpha}{\leftarrow}x$*. The morphism* 1_x *is just the pair* (x,x)*.*

In this category, any morphism is an isomorphism, actually $(y,x)^{-1}=(x,y)$ *since* $(x,y)\circ(y,x)=(x,x)=1_x$ *and* $(y,x)\circ(x,y)=(y,y)=1_y$*.*

Definition 1.2 A subcategory $\mathbf{C}'$ of a finite category $\mathbf{C}$ is a category whose objects are a subset of the set of objects of $\mathbf{C}$ and whose morphisms are a subset of the set of morphisms $\mathbf{C}$.

Notice that a subcategory of a finite category must be finite.

[6] The notation $\mathrm{Hom}\,(x,y)$ is often used to indicate the class of all morphisms $\alpha\colon x\to y$, then the class of all morphisms of the category $\mathbf{C}$ will be denoted by $\mathrm{Hom}\,(\mathbf{C})$, however we will rather use the convention indicated in the main text. Then using a different notation for the set of all morphisms is redundant.

Coming back to the example of the category of pairs, notice that the category generated by a finite directed graph (V,L) is a subcategory of the category of pairs of the set of vertices V. The category of pairs of the set of vertices V is called the complete graph of V and, in graph theory, it is commonly denoted by K_V. If V is finite, then K_V will be denoted by K_n with n the number of elements, or cardinal, of V, that is $n = |V|$. See below some illustrations of complete graphs of sets of vertices with $n = 10, 20, 40$. In what follows, we will use the notation $\mathbf{G}(n)$ or $\mathbf{A}_n$, for reasons that will be clear later on, to refer to the category of pairs of the set $\{1, 2, \ldots, n\}$.

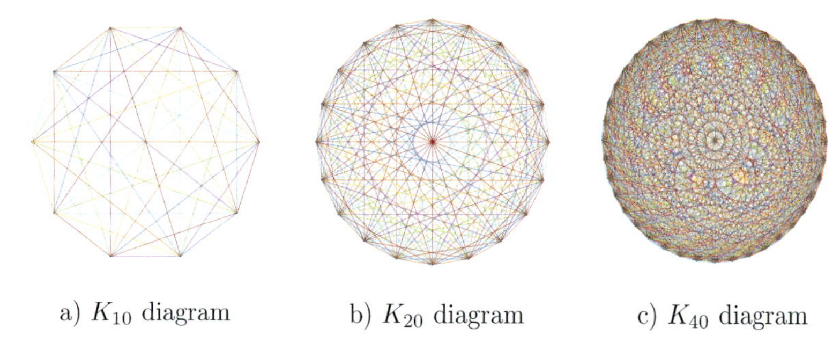

a) K_{10} diagram b) K_{20} diagram c) K_{40} diagram

Figure 1.8: The complete graphs (also the category of pairs) K_{10}, K_{20}, K_{40}.

1.2.2 Abstract categories

If we remove any mention of 'sets' and 'finite' in the previous definition of a finite category, Def. 1.1 above, we get the general definition of a category.

Definition 1.3 A category $\mathbf{C}$ consists of a class[7] of objects x, denoted $\mathbf{Ob}(\mathbf{C})$, and for each pair x, y of objects, a class of morphisms $\alpha \colon x \to y$, denoted $\mathbf{C}(x, y)$, satisfying the following axioms:

i.- (Composition law) If we have two morphisms $\alpha : x \to y$ and $\beta : y \to z$, then there exists a morphism, denoted by $\beta \circ \alpha$, from $x \to z$, $\beta \circ \alpha : x \to z$.

ii.- (Existence of units) Given an object x, there exists a morphism denoted by 1_x such that for any morphism $\alpha : x \to y$, then $\alpha \circ 1_x = \alpha$, and for any morphism $\beta : y \to x$, then $1_x \circ \beta = \beta$.

iii.- (Associativity) Given three morphisms $\alpha : x \to y$, $\beta : y \to z$, $\gamma : z \to u$, then it is satisfied:

$$\gamma \circ (\beta \circ \alpha) = (\gamma \circ \beta) \circ \alpha \tag{1.4}$$

[7]The word 'class' has a definite mathematical meaning that we will not need to use here; just use it as an alternative to the word 'set', so that we are free from the restrictions imposed by standard axioms (restrictions) of set theory.

Many structures commonly used in Mathematics (but also in other areas of Science and Technology) are categories (see, for instance, Sect. 5.2 for an example of the use of categories in Information Theory where categories are shown to be abstract databases).

We will offer now a few examples that go beyond the examples in the previous section in order to illustrate the abstract use of categories[8].

Example 1.6 *The category $\mathbb{N}$ generated by the single loop is a non-finite category. This category has only one object and such categories are called monoids. The category $\mathbb{N}$ is non-finite but it is still a set.*

Example 1.7 *The category of finite sets:* **FinSets**.

The objects of this category are all finite sets. Given two finite sets S_1 and S_2, the class of morphisms from S_1 to S_2 are all maps f from S_1 to S_2, $f\colon S_1 \to S_2$. In this case the class of morphisms between two objects is a set. The composition of morphisms is just the composition of maps, $g \circ f$ if $f\colon S_1 \to S_2$ and $g\colon S_2 \to S_3$. The identity morphism 1_S is just the identity map. A morphism f is an isomorphism if and only if (iff) f is a bijective map, Two objects are isomorphic iff they have the same cardinal, i.e., they are bijective.

The same example can be worked out but considering now all sets and all maps among them (not necessarily finite). Such category will be denoted by **Sets**. *Clearly* **FinSets** *is a subcategory of* **Sets**.

A small category is a category whose class of objects and morphisms are sets. For instance, the category $\mathbb{N}$ is a small category. Any category generated by a finite quiver is small (notice that we may enumerate its morphisms). However, the category of all sets **Sets** is not small[9].

If **C** is a small category, then we can consider the category whose objects are the objects of **C** but with only the identity morphisms. Such category is just the set of objects of **C**. We may consider it to be the "setification" of the small category **C**. This process of converting a small category in a set is not very interesting as we are just erasing the abstract relations implied by the category structure.

Two types of categories which are very common in Mathematics are:

Example 1.8 *Equivalence relations: Given a set Ω, an equivalence relation on Ω is a subset $\mathcal{R} \subset \Omega \times \Omega$, whose pairs (x,y) will be denoted as $x \sim y$ satisfying that:*

1. *(Reflexivity) $x \sim x$ for all $x \in S$ (i.e., $(x,x) \in \mathcal{R}$ for all $x \in \Omega$).*

2. *(Symmetry) If $x \sim y$, then $y \sim x$.*

[8] Sometimes called also 'abstract nonsense' to indicate both the higher abstract realm of these ideas and where they could lead if not handled with care.

[9] Because of Russell's paradox the class of all sets cannot be a set (if it were, call it $\mathcal{S}$, we could define a subset $\mathcal{A}$ (from 'absurd') of it as follows: A set $X \in \mathcal{S}$ is in $\mathcal{A}$ if X doesn't belong to X. Formally $\mathcal{A} = \{X \in \mathcal{S} \mid X \notin X\}$. Then we may ask if $\mathcal{A}$ belongs to $\mathcal{A}$. If $\mathcal{A} \in \mathcal{A}$ were true, by definition of $\mathcal{A}$, then $\mathcal{A} \notin \mathcal{A}$, which is a contradiction. Conversely if $\mathcal{A} \notin \mathcal{A}$, then again because of the definition of $\mathcal{A}$, we get that $\mathcal{A} \in \mathcal{A}$, which is again a contradiction. Thus, we have defined formally a subset, $\mathcal{A}$, that cannot be defined!).

3. *(Transitivity) If $x \sim y$ and $y \sim z$, then $x \sim z$.*

Clearly, an equivalence relation $\mathcal{R}$ on Ω is a subcategory of the category of pairs $\mathbf{G}(\Omega)$.

The equivalence classes defined by $\mathcal{R}$ are the sets of elements which are isomorphic among them.

Example 1.9 *Order relations: Given a set Ω, a partial order relation on Ω is a subset $\mathcal{O} \subset \Omega \times \Omega$, whose pairs (x,y) will be denoted as $x \leq y$ satisfying that:*

1. *(Reflexivity) $x \leq x$ for all $x \in \Omega$ (i.e., $(x,x) \in \mathcal{O}$ for all $x \in \Omega$).*

2. *(Antisymmetry) If $x \leq y$ and $y \leq x$, then $x = y$.*

3. *(Transitivity) If $x \leq y$ and $y \leq z$, then $x \leq z$.*

As it happens with equivalence relations, partial order relations on Ω are subcategories of the pair category $\mathbf{G}(\Omega)$.

An interesting construction for a given category is the following. Let $\mathbf{C}$ be a category. We will introduce an equivalence relation[10] among its objects as follows: $x \sim y$ if there exists an isomorphism $\alpha \colon x \to y$. Clearly $x \sim x$ as 1_x is an isomorphism. Notice that if $x \sim y$ then $y \sim x$ because if α is an isomorphism, α^{-1} is an isomorphism too, and, finally, if $x \sim y$ and $y \sim z$, then $x \sim z$ as $(\beta \circ \alpha)^{-1} = \alpha^{-1} \circ \beta^{-1}$.

Then, we may construct the class defined by the equivalence classes of objects in $\mathbf{C}$. We will denote by $[x]$ the equivalence class containing x. We will denote the class obtained by the previous process, $\mathbf{C}/\sim$, and it will be called the skeleton of the category $\mathbf{C}$. If $\mathbf{C}/\sim$ is a small category we will say that $\mathbf{C}/\sim$ is the set of isomorphism classes of $\mathbf{C}$.

Consider, for instance, the category **FinSets** discussed in Example 1.7. Then the skeleton **FinSets**$/\sim$ is just the natural numbers $\mathbb{N}$.

It is convenient to work out this example with more detail. It is clear that in the equivalence class of any finite set S we can choose the set $\{1,2,\ldots,n\}$ where n is the natural number denoting the cardinal of S. Thus, we can denote each equivalence class with the natural number $n \in \mathbb{N}$.

Moreover, if we have an injective map $\alpha : \{1,2\ldots,n\} \to \{1,2\ldots n,n+1\}$, there is always the representative defined by $\alpha(h) = h$, $h = 1,2,\ldots$ in the class $[\alpha]$. Such morphism will be denoted by $n \to n+1$ (or, in Peano's axiomatization of natural numbers, this is just the 'follow' or 'successor' map). Notice that the composition $n \to n+1 \to n+2$ gives the map $n \to n+2$, etc. The associative property of addition of natural numbers corresponds to the associativity of morphisms.

The category whose objects are linear spaces will be of particular importance along this book. We will introduce it briefly in the following example.

[10] Notice that the notion of equivalence relation above works as well for arbitrary categories.

Example 1.10 *The category* **FinVects***. Let us consider now the category of finite-dimensional complex (or real) linear spaces. The objects of the category are finite-dimensional complex vector spaces and, given two linear spaces E, F, the morphisms from E to F are the set of all linear maps $\alpha\colon E \to F$. The composition of morphisms is just the composition of the corresponding linear maps and the identity morphism is just the identity map. On this occasion, we will attach to the common use in category theory and we will denote the set of morphisms from E to F as* $\mathrm{Hom}(E,F)$ *(see Appendix A).*

Notice that two linear spaces are isomorphic as objects in the category **FinVects** *if they are isomorphic as linear spaces. Hence, the isomorphism classes of linear spaces in the category* **FinVects** *is characterized by a natural number n, the dimension of the linear space, as it is well known that two finite-dimensional linear spaces are isomorphic if and only if they have the same dimension (see A.1 for a glossary on linear algebra terminology and basic results and, in particular, Thm. A.4).*

The map $n \to n+1$, discussed before, corresponds to the linear map $L\colon \mathbb{C}^n \to \mathbb{C}^{n+1}$ sending the basis $\{e_1, e_2, \ldots, e_n\}$ of $\mathbb{C}^n$ into the first n vectors of the basis $\{e_1, e_2, \ldots, e_n, e_{n+1}\}$ of $\mathbb{C}^{n+1}$.

The last two examples show that the set $\mathbb{N}$ of natural numbers with the addition basis arises naturally as the isomorphism classes of different categories, **FinSets** and **FinVects**, for instance. Thus, we see that natural numbers constitute a 'rough' description of much finer structures. This process of passing from categories to isomorphic equivalence classes can be thought as a sort of 'setification' of categories by using the natural notion of 'sameness' provided by the class of isomorphisms within the category.

The converse procedure is much more interesting (and difficult), namely, given a notion in the standard set-theoretical setting, try to guess which is the 'categorification' of it, i.e., find out a category whose isomorphism classes should correspond to the given structure. Of course, such 'categorification' process is not well defined, as the previous examples show, and it should be understood as a research process that will unveil hidden structures beneath the set-theoretical disguise of the structures being studied.

The language of categories is extremely powerful as it allows us to encompass many situations often found in Mathematics. Thus, it is clear that categories wherein all morphisms are isomorphisms are called to play an important role. Actually, they are the main subject of this work.

1.3 A categorical definition of groupoids and groups

Definition 1.4 A groupoid is a category **G** such that all its morphisms are isomorphisms, i.e., such that all its arrows are invertible.

A groupoid set is a groupoid which is a small category. We will not use a different name for them. In most cases, we will be considering groupoids which are sets and, furthermore the main theme of this book consists of the study of finite groupoids.

Given a category $\mathbf{C}$, the class of all isomorphisms of $\mathbf{C}$ define a groupoid $\mathbf{G}(\mathbf{C})$. The objects of $\mathbf{G}(\mathbf{C})$ are the objects of $\mathbf{C}$ and its morphisms the isomorphisms of $\mathbf{C}$.

Given a category $\mathbf{C}$, the natural equivalence relation on it, discussed before, can be rephrased by saying that two objects $x, y \in \mathbf{Ob}(\mathbf{C})$ are equivalent if there exists $g \in \mathbf{G}(\mathbf{C})$ such that $g\colon x \to y$, and two morphisms $\alpha\colon x \to y$, $\alpha'\colon x' \to y'$ are equivalent if there exists $g\colon x \to x'$, $h\colon y \to y'$, $g, h \in \mathbf{G}(\mathbf{C})$) such that $\alpha' = h \circ \alpha \circ g^{-1}$. We will also say that the groupoid $\mathbf{G}(\mathbf{C})$ acts on the category $\mathbf{C}$ and x, x' and α, α' are equivalent if they are in the same "orbit" of $\mathbf{G}(\mathbf{C})$.

Example 1.11 *The groupoid of pairs of a set* $\mathbf{G}(\Omega)$*: Consider the category of pairs* $\mathbf{G}(\Omega)$. *It is clear that* $\mathbf{G}(\Omega)$ *is a groupoid because any morphism* (x,y) *has an inverse* (y,x). *Notice that in this case all objects are equivalent and all morphisms are equivalent:*

$$(y',x') = (y',y) \circ (y,x) \circ (x,x') . \tag{1.5}$$

In particular if the set Ω is the set of indexes $\{1,2,\ldots,n\}$, we will denote such groupoid as $\mathbf{A}_n$ or $\mathbf{G}(n)$.

Example 1.12 *Darboux coverings. Let* P *be a set and two projections* $s\colon P \to \Omega$, $t\colon P \to \Omega$. *We will call the category* $\mathbf{Darboux}(P,\Omega)$, *the category whose objects are the elements of* Ω *and, given two points* $x, y \in \Omega$, *the morphisms* $\mathbf{D}(x,y)$ *are all maps from* $s^{-1}(x)$ *to* $t^{-1}(y)$. *The composition law is the composition of maps.*

Notice that now the groupoid of the category $\mathbf{Darboux}(P,\Omega)$ *is the category whose objects are the elements of* Ω *and morphisms the bijections from* $s^{-1}(x)$ *to* $t^{-1}(y)$.

Definition 1.5 A group G is a groupoid with only one object $\{e\}$. The object is commonly called the neutral element of the group.

If the group is small, we recover the standard definition of groups. As the reader would possible know, there are a great variety of examples of groups, but, as the previous constructions and examples show, the structure of group is just a particular instance of the notion of groupoid. A lot is known about the structure of groups (most notably about the structure of finite groups). Their classification, properties, representations, etc. We will try to extend some of this knowledge to the much larger class of finite groupoids. This will be the content of the rest of this book. We will start by describing the fundamental algebraic notions of the theory of groups and how to extend them to finite groupoids. This will be the purpose of the next chapter.

1.4 Historical notes and additional comments

1.4.1 Groupoids: A short history

Although the introduction to groupoids used in this book has its roots in the theory of categories, the concept of groupoid is twenty years older than that of category. The first definition of groupoids can be found in the work of Brandt on composition of quadratic forms [7] (and, later on, in his work on multiplication of ideals in orders defined over Dedekind domains [8]). The notion of groupoid appears when extending to the noncommutative case the ideal class group in rings of algebraic integers [34]. A similar concept, compound groups, was introduced by Loewy [47] in his research on isomorphisms between conjugate field extensions, and used by Baer in his studies on Galois Theory [2]. Later on, Reidemeister used the concept of groupoid, citing Brandt, in his book on combinatorial topology, as the fundamental groupoid of a topological space [56].

However these two approaches were forgotten for many years, mainly because partial operations were not as appreciated at that time.

1.4.2 Categories

Categories were introduced by Eilenberg and MacLane in their work on natural equivalences [24]. Ten years after its publication, category theory was widely accepted, partial operations acquired a new significance and the concept of groupoids was developed, mainly through the work of Ehresmann [23] and applied in different fields (for instance in Galois theory [47] and group theory [32]).

Leading mathematicians in their fields have used the concept of groupoids extensively in their researches: A. Grothendieck (construction of moduli spaces in algebraic geometry, groupoids as pre-equivalence relations [31]), G.W. Mackey (virtual groups in his work on ergodic actions [49],[50],[51]), A. Connes (noncommutative geometry [17]), R. Brown (for instance, in his work on the van Kampen Theorem [9] [10] and his excellent survey [11]) and A. Weinstein (symplectic geometry, see the interesting review [66] where the groupoid concept is connected to the use of internal and external symmetries of the systems).

1.4.3 Groupoids and physics

Although it may not be a common opinion, one of the main applications in physics of groupoid theory could be to put together internal and external symmetries of physical systems [46].

However, a common ground for groupoids in Physics and Mathematics is their application in noncommutative geometry. In this last discipline, a groupoid $\mathbf{G}$ is associated to a C^*-algebra, $C^*(\mathbf{G})$. In Quantum Physics the algebra of observables has a set of symmetries which form a groupoid. In classical physics, a Poisson manifold can be described as the dual of the Lie algebroid of a Lie groupoid.

1.4.4 Groupoids and other points of view

The change of point of view, from global symmetries (groups) to local symmetries (groupoids) has attracted the interest of philosophers or, at least, has motivated the interest of mathematicians to introduce this concept into the realm of philosophy, see, for instance, the expository article of A. Vistoli in [64]. The paper of Corfield [18] uses the concept of groupoid and its applications as an example of the value of the research in mathematics. He studies, in his own words, *the debate surrounding the question as to whether groupoids are significantly more powerful than groups at capturing the symmetry of a mathematical situation.*

From another point of view, the fact that groupoids have been left aside in many occasions could be originated by the name used to describe this concept. As commented in the introduction, in words of A. Connes [17]: *It is fashionable among mathematicians to despise groupoids and to consider that only groups have an authentic mathematical status, probably because of the pejorative suffix oid. To remove this prejudice we start Chapter I with Heisenberg's discovery of quantum mechanics. The reader will, we hope, realise there how the experimental results of spectroscopy forced Heisenberg to replace the classical frequency group of the system by the groupoid of quantum transitions. Imitating for this groupoid the construction of a group convolution algebra, Heisenberg rediscovered matrix multiplication and invented quantum mechanics.*

Chapter 2

Groups

We cannot go further in the study of groupoids without introducing the most basic facts of the theory of groups. We will restrict this chapter to a brief exposition, albeit complete enough to provide significant results regarding the corresponding theory of finite groupoids and their representations. We will use the symmetric group or the group of permutations, the first instance of a group, as a fundamental example of the notions we will revise here. Note that, from a chronological point of view, when groupoids were introduced, groups had acquired a high degree of maturity and all the concepts, which could be also viewed as part of the theory of groupoids, are now regarded as the theory of groups.

Thus, in this chapter, an overall description of the standard set theoretical theory of groups will be offered, starting with the basic definitions of groups, subgroups, normal subgtroups, quotient groups, and homomorphisms of groups. The language of exact sequences and the various notions of direct and semidirect products of groups will be discussed as well. Cayley's theorem for groups will be proved, emphasizing that it is the first instance of a representation theorem found in the book. The group of automorphisms of a given group will be also briefly considered.

Throughout the text, many examples will be provided and, in addition to the symmetric group, the theory of Abelian groups will be developed as well as the structure of the dihedral groups. As it was pointed out before, the symmetric group will be treated with special care as a first example of a family of games that will lead to the introduction of further families of groupoids in subsequent chapters.

2.1 Groups, subgroups and normal subgroups: Basic notions

2.1.1 Groups: Definitions and examples

Recall from the previous chapter the categorical definition of a group: A group is a small category whose morphisms are isomorphisms and possess a single object called the neutral element, Def. 1.5. The composition law of morphisms satisfies the usual properties of associativity, existence of a neutral element and for each morphism there exists an inverse morphism. We can recapitulate the definition of a group in a set theoretical language as it is done in many elementary courses in Mathematics.

Definition 2.1 A group is a set G with a composition law $*\colon G \times G \to G$ called the product, also multiplication or composition law, of the group, satisfying the following properties:

1. Associative: $g * (g' * g'') = (g * g') * g''$, $g, g', g'' \in G$.
2. Neutral element: $e \in G$, $e * g = g * e = g$, $g \in G$.
3. Inverse element: $\forall g \in G$, $\exists g^{-1} \in G$ such that: $g * g^{-1} = g^{-1} * g = e$.

If $g * g' = g' * g$, $\forall g, g' \in G$ the group is called commutative (or Abelian). The cardinal of G is called the order of the group and will be denoted by $|G|$. A finite group G is a group whose order $|G|$ is finite.

In the following, we will omit the operation symbol $*$ if there is no risk of confusion. As a consequence of the above definitions, the neutral element e is unique and each element has a unique inverse.

Exercise 2.1 *Prove the previous statement.*

Example 2.2
1. *We have already encountered the integer numbers $\mathbb{Z}$ as a category, Sect. 1.1.3, Example 1.1. The set $\mathbb{Z}$ of integer numbers with the operation $*$ defined by the standard addition, that is $n * m = n + m$ is an instance of an Abelian group. The neutral element being* 0 *and the inverse of a given element n is $-n$. Notice that $\mathbb{Z}$ is countable infinite.*

2. *The group $\mathbb{Z}_n$ of congruence classes of integers module n. We have already described the group $\mathbb{Z}_n$ from a categorical viewpoint, Sect. 1.1.3, Exercise 1.2. It is convenient to present it from the more mundane perspective of Arithmetic. Consider the set of equivalence classes of integer numbers module n (n a fixed positive integer), that is $p = q \bmod (n)$ if n divides $p - q$ or, in other words, if there exists an integer c such that $p = q + cn$. Notice that given n, for any integer p there exists a unique integer $0 \leq q < n$ such that $p = cn + q$ (q is the remainder of the integer division of p by n). Integers p, q which are equivalent module n are also said to be congruent (module n). It is clear that the congruence relation is an equivalence relation and the class containing a given*

integer p will be denoted by $[p]$. Because we can select a unique representative $0 \leq q < n$ in each class, we will denote the corresponding classes by $[q]$ or $\mathbf{q}$. Thus, the set of congruence classes of integers module n is given by:

$$\mathbb{Z}_n = \{\mathbf{0}, \mathbf{1}, \ldots, \mathbf{n-1}\}.$$

The natural operation of addition induced by the addition of integers defines a group structure on $\mathbb{Z}_n$, that is we define:

$$[p] + [q] = [p+q], \qquad \forall p, q \in \mathbb{Z}.$$

It is a trivial exercise to check that the previous operation is well defined and it defines a group structure on $\mathbb{Z}_n$, the neutral element being $\mathbf{0}$ and the inverse of $[q]$ being $-[q] = [n-q]$. Notice that $|\mathbb{Z}_n| = n$. The group $\mathbb{Z}_n$ is the natural instance of an Abelian group and, in fact a classification theorem of finite Abelian groups will be discussed further on, Sect. 2.5.2, as well as their linear representations, Sect. 8.7.1.

3. *The group of symmetries of a regular polygon, the dihedral group D_n. Consider the nth regular polygon P_n, $n = 3, 4, \ldots$, in the plane $\mathbb{R}^2$. We will label its vertices by $1, 2, \ldots, n$ and we consider the set of rigid transformations (Euclidean transformations) of the plane that map P_n into P_n. We will call such transformations symmetries of the polygon P_n and it is clear that there are two types of symmetries: Rotations r_m around the centre of the polygon of angle $\theta_m = 2\pi m/n$, $m = 0, 1, \ldots, n-1$, and reflections τ with respect to the 'symmetry axes' of the polygon, i.e., the lines passing through the origin and the vertices or midpoints of the sides of P_n. Thus, for instance, P_3 has three symmetry axes (the mediatrices of the triangle) and P_4 has four symmetry axes, the lines passing through two opposite vertices and the lines passing through the mid-points of opposite sides.*

   We will denote by D_n the set of symmetries of P_n with the product defined by the composition of transformations. It is clear that D_n becomes a group, called the dihedral group, with neutral element the identity transformation. The dihedral group D_n, $n \geq 3$, is non Abelian with order $|D_n| = 2n$. A description of the dihedral group in terms of generators and relations is offered in App. B.2 and its linear representations are discussed in Sect. 8.7.3.

Exercise 2.3 *Describe explicitly the dihedral groups D_3 and D_4.*

The abstract notion of group has a fascinating and convoluted history (see the historical notes at the end of this chapter 2.6). Symmetries and groups are paired today following Herman Weyl's powerful insight (reading his masterful book "Symmetry" [70] is always a delight). We will take back this point of view in the coming chapter, Chap. 4, where groups (and groupoids) will be considered defining 'actions' on sets bringing more examples beyond the ones provided above. In the present chapter, we will keep pushing forward the idea, already presented in the previous chapter, of groups (and groupoids) being the most convenient way of describing certain games (see Sects. 2.2 and 3.2.1 later on), but before doing that, we need to explore some basic abstract properties of groups and their subgroups. This will be the task of the following sections.

2.1.2 Subgroups and cosets

If we are given a group G, the first thing we can do to understand its structure is to look for its subsets and, among them, those that reproduce the group structure. Certainly those subsets will 'capture' relevant aspects of the group (and, being smaller, will be easier to understand too).

Definition 2.2 A subset H of a group G is a subgroup of G if it is a group with the operation in G.

Proposition 2.1
If G is a group, $H \subset G$ is a subgroup if and only if $gg'^{-1} \in H$, for all $g, g' \in H$.

Exercise 2.4 *Prove the previous statement.*

Example 2.5

1. *Let $2\mathbb{Z} \subset \mathbb{Z}$ be the set of all even integers, then $2\mathbb{Z}$ is a subgroup of $\mathbb{Z}$. In general, given n a positive integer the set $n\mathbb{Z}$ of all integers which are multiples of n is a subgroup of $\mathbb{Z}$.*

2. *Consider the subset $p\mathbb{Z}_n = \{\mathbf{0}, \mathbf{p}, \mathbf{2p}, \ldots\}$ of the group $\mathbb{Z}_n$ where p divides n, then $p\mathbb{Z}_n$ is a subgroup of $\mathbb{Z}_n$.*

3. *The rotations r_m of angle $\theta_m = 2\pi m/n$, $m = 0, 1, \ldots, n-1$ define a subgroup H of order n of the dihedral group D_n. Consider the subset $\{e, \tau\}$ where τ is any reflection with respect to a symmetry axis of P_n, then $\tau^2 = e$, the element τ has order 2 and the subset $\{e, \tau\}$ defines a subgroup K of D_n.*

Definition 2.3 Let G be a group and $g \in G$ an element. The set $C_g = \{g^p \mid p \in \mathbb{Z}\}$ defines a subgroup of G called the cyclic subgroup of G generated by the element g. If C_g is finite with order $r = |C_g|$ we will say that the element g has finite order r and we will denote it by $|g|$ or $o(g) = r$, otherwise we will say that g has infinite order.

Notice that the order of the element g can be defined alternatively as the smallest positive integer r, such that $g^r = e$ (note that in such case $g^k \neq e$, for all $k = 1, 2, \ldots, r-1$.

Example 2.6

1. *Consider the Abelian group $\mathbb{Z}_n$. The element $\mathbf{1}$ has order n. If p is prime, any element in $\mathbb{Z}_p$ has order p.*

2. *Consider the group consisting of the reflections along the coordinate axes of a square centered at the origin and parallel to them. Such group has four elements* $e, \tau_x, \tau_y, \tau_y \tau_x = \tau_x \tau_y$. *Notice that all elements different from the neutral element have order 2. This group is called the Klein group of order 4 and it is sometimes denoted as* V_4.

3. *Consider the dihedral group* D_n. *The element r, the rotation of angle* $\theta = 2\pi/n$ *is an element of order n, while the reflection* τ *has order 2. They generate cyclic subgroups* $H = \{e, r, r^2, \ldots, r^{n-1}\}$ *and* $K = \{e, \tau\}$, *of orders n and 2, respectively.*

Exercise 2.7 *Prove that if* $H, H' \subset G$ *are subgroups of G, then* $H \cap H'$ *is a subgroup of G.*

Exercise 2.8 *Let* $\mathbb{Z}_r$, $\mathbb{Z}_s$ *be two subgroups of* $\mathbb{Z}_n$. *Show that* $\mathbb{Z}_r \cap \mathbb{Z}_s = \mathbb{Z}_{\text{g.c.d.}(r,s)}$, *with* $g.c.d.(r,s)$ *denoting the greatest common divisor of r and s.*

Contrary to what the previous exercise shows, the union of subgroups is, in general, not a subgroup (consider for instance the group of all integer numbers and the union of the subgroups $2\mathbb{Z} \cup 3\mathbb{Z}$). Given two subgroups we may consider however the smallest subgroup containing them.

Definition 2.4 Let $H, H' \subset G$, two subgroups of G, then we call the subgroup generated by H and H' the smallest subgroup of G that contains both H and H'. It will be denoted[1] by $\langle H, H' \rangle$.

Exercise 2.9 *Let* $\mathbb{Z}_r$ *and* $\mathbb{Z}_s$ *as in Exercise 2.8. Prove that* $\langle \mathbb{Z}_r, \mathbb{Z}_s \rangle = \mathbb{Z}_{\text{g.c.d}(r,s)}$, *with* $\text{g.c.d}(r,s)$ *the greatest common divisor of r and s.*

Clearly the definition can be extended to a finite family[2] of subgroups H_i, and we will denote by $\langle H_1, \ldots, H_r \rangle$ the smallest subgroup containing all of them.

Any element g in a finite group G defines its cyclic subgroup C_g, then if we are given a family of elements $g_1, \ldots, g_r$, we may consider the corresponding family of subgroups $C_{g_1}, \ldots, C_{g_r}$.

Definition 2.5 Let $g_1, \ldots, g_r$ be a subset of G. We call the subgroup generated by $\{g_1, \ldots, g_r\}$ the subgroup $\langle C_{g_1}, \ldots, C_{g_r} \rangle$ generated by the cyclic subgroups $C_{g_1}, \ldots, G_{g_r}$ and we will denote it by $\langle g_1, \ldots, g_r \rangle$ (also $\langle g_1, \ldots, g_r \mid g_1^{n_1} = \cdots = g_r^{r_r} = e \rangle$ with n_1,..., n_r the order of the elements g_1,..., g_r respectively).

We will say that the group G is generated by the set of elements $g_1, \ldots, g_r$ if $\langle g_1, \ldots, g_r \rangle = G$. We will not pursue, for the moment, the discussion of how to describe groups by using generators and relations (see Appendix B).

[1]This notation is consistent with the notation chosen to denote a group generated by a family of generators and relations, App. B.2.

[2]The reader may have noticed that the definition applies to arbitrary families.

Exercise 2.10 *Consider the dihedral group D_n again. Prove that the elements r and τ generate D_n. Prove that $\tau r \tau^{-1} = r^{n-1} = r^{-1}$. Using the relations $r^n = \tau^2 = e$ and the previously proved formula, show that all elements of D_n can be listed as*

$$e, r, r^2, \ldots, r^{n-1}, \tau, r\tau, r^2\tau, \ldots, r^{n-1}\tau .$$

Given two subgroups $H, H' \subset G$, we may also define the set $HH' = \{hh' \mid h \in H, h' \in H'\}$, however, the set HH' is not, in general, a subgroup.

Exercise 2.11 *Prove that if the group G is Abelian, then $HH' = H'H = \langle H, H' \rangle$.*

Exercise 2.12 *Provide an example of two subgroups H, H' of a group G, such that HH' is not a subgroup.*

Exercise 2.13 *Prove that $D_n = HK$ with H the subgroup generated by the rotation r of angle $2\pi/n$ and K is the subgroup generated by τ.*

One way to address the problem of understanding the structure of a group would be to ask wether it is possible or not to write it in the form $G = HK$ where H and K are two subgroups. Because of Exercise 2.11, if G is Abelian, then if $G = HK$, then G is generated by $H \cup K$ and the two subgroups commute, that is $HK = KH$. Later on, see Prop. 2.16, we will describe the exact meaning of such factorization.

Thus given a group G, we may consider the family of all its subgroups. Actually, we can organize them using a *Hasse diagram*, that is, we partially order them by using inclusion. That is, we draw a diagram starting with the trivial subgroup $\{e\}$ lying at the bottom and staking on top of it the subgroups H of G linked by lines if they are contained on the ones on top.

Eventually, we would like to see if it is possible to 'decompose' the group into a bunch of 'simple' pieces and to get in this way an explicit description of the structure of the group. In the following paragraphs, we will start to take the first timid steps towards this program[3]. For that, we need to introduce a notion that complements that of a subgroup, and is the process of quotienting groups.

Definition 2.6 Given a group G and a subgroup H, we can define an equivalence relation in G, called the right equivalence relation associated to the subgroup H as: $g \sim_R g'$ if there exists $h \in H$ such that $g' = gh$.

Exercise 2.14 *Prove that the relation $\sim_R$ defined by the subgroup H is an equivalence relation.*

This relation classifies the elements of G into classes of equivalence, called right cosets. Each coset is determined by a representative g and all its elements have the form gh, $h \in H$, and because of that, it is often denoted by gH.

[3]This is a fascinating topic that we only would be able to cover partially, in the realm of groupoids, in Chap. 6. The classification of finite simple groups is a staggeringly hard problem whose solution is arguably one of humankind's greatest achievements.

Similarly, there is another equivalence relation, the left equivalence relation $\sim_L$ associated to the subgroup H.

Definition 2.7 Let $H \subset G$ be a subgroup, then, given $g, g' \in G$, we will say that g is left equivalent to g' and it will be denoted by $g \sim_L g'$, if there exists $h \in H$ such that $g' = hg$.

The equivalence classes determined by the left equivalence relation associated to the subgroup H are called left cosets and are denoted by Hg.

Notice that, in general (for non-Abelian groups), both equivalence relations $\sim_L$, $\sim_R$ are different.

Exercise 2.15 *Even if we are are still short of examples, we can test these ideas with the examples 2.5. Describe the equivalence relations $\sim_L$, $\sim_R$ defined by the subgroups $R_n \subset D_n$ and $\{e, \tau\} \subset D_n$.*

The left and right equivalence relations associated to a subgroup H allow us to define the corresponding quotient spaces $H \backslash G$ and G/H, respectively. Thus, for instance, G/H consists of the set of right cosets gH:

$$G/H = \{gH \mid g \in G\}. \tag{2.1}$$

Theorem 2.1
(Lagrange's theorem). Let G be a finite group. The number of elements in each coset (left or right) is equal to the order of the subgroup and is a divisor of the order of the group. The quotient $|G|/|H|$ is called the index of the subgroup H and is denoted by $[G:H]$.

Proof 2.1 Given two cosets gH and $g'H$, it is easy to check that the map $\varphi\colon gH \to g'H$, given by $\varphi(gh) = g'h$, is well-defined and bijective, in particular, the coset containing the neutral element is just H. Then, all cosets have the same number of elements, which is equal to the order of H, that is $|gH| = |H|$ for all $g \in G$. On the other hand, the equivalence classes defined by an equivalence relation partition the set, that is $|G| = r|H|$ where r is the number of equivalence classes, thus, $r = [G:H]$ and the order of H divides G.

Example 2.16

1. *The subgroup $p\mathbb{Z}_n$ of $\mathbb{Z}_n$ has order n/p if p divides n.*
2. *The subgroup $R_n \subset D_n$ has order n and index 2. The subgroup $\{e, \tau\} \subset D_n$ has order 2 and index n.*

Exercise 2.17 *Prove that if G is a group such that its order is prime then it is the cyclic group generated by any of its elements. Moreover it is Abelian.*

Exercise 2.18 *Show that, if a group has order 4, then its elements have order 2 or 4. Show that, in the former case, the group must be Abelian and conclude that it must be the Klein group V_4. Show that all groups of order less than six are commutative.*

2.1.3 Normal subgroups

The following discussion on a particular class of subgroups, normal subgroups, will be of outmost importance in the setting of groupoids (see Sect. 6.1). We can introduce the notion of normal groups in various ways. For the purposes of this work, it is convenient to consider the situation when the two canonical equivalence relations associated to a subgroup $H \subset G$, that is the right and left equivalence relations $\sim_R$, $\sim_L$ (recall Defs. 2.6 and 2.7), coincide.

Definition 2.8 Let G be a group. A subgroup $H \subset G$ is a normal subgroup if the canonical equivalence relations $\sim_R$, $\sim_L$, defined by H, coincide, or, in other words, if given $g, g' \in G$, there exists $h \in H$ such that $g' = gh$, then there exists $h' \in H$ such that $g' = h'g$ and conversely.

This leads immediately to the standard definition of normal group

Proposition 2.2
Let H be a subgroup of the group G, then H is a normal subgroup if and only if $gHg^{-1} = H$ for all $g \in G$ or, more explicitly, $ghg^{-1} \in H$ for all $h \in H$ and $g \in G$.

Exercise 2.19 *Prove the previous proposition.*

Notice that, if H is a normal subgroup, the left and right cosets coincide, that is, $gH = Hg$ for all $g \in G$. Sometimes a normal subgroup H is denoted as $H \lhd G$.

Proposition 2.3
The subgroup $H \subset G$ is a normal subgroup if and only if $gH = Hg = HgH$ for all $g \in G$. Then the quotient spaces G/H, $H\backslash G$ and the double coset space $H\backslash G/H$ (defined as the family of sets of the form gHg', $g, g' \in G$), coincide and will be called the quotient space of G by H.

Proof 2.2 Note that gH consists of all elements equivalent to g with respect to $\sim_R$, but that the equivalence relation $\sim_R$ coincides with $\sim_L$ if and only if $gH = Hg$ because Hg is the equivalence class of all elements equivalent to g with respect to $\sim_L$.

Finally notice that not only does $G/H = H\backslash G$ (because $gH = Hg$) but also $gH = HgH$ (because $HgH = gHH = gH$) and then $G/H = H\backslash G/H$.

If $H \subset G$ is a normal subgroup, we will just denote the quotient space in the proposition before by G/H. The trivial subgroups $\{e\}$ and G of the group G are always normal. A normal subgroup $H \triangleleft G$ is called proper it is not trivial, that is, $H \neq \{e\}, G$. Notice that if G is Abelian, any subgroup is normal.

Exercise 2.20 *Show that all subgroups of index 2 are normal subgroups.*

Example 2.21 *As a consequence of the previous exercise, we get that the subgroup $R_n \subset D_n$ is normal.*

Sometimes, we can introduce a group structure in the set of cosets G/H associated to a subgroup. The following theorem shows that not only normal subgroups are subgroups such that this is true but that if $\sim$ is an equivalence relation such that $G/\sim$ inherits a group structure, then $\sim$ is the canonical equivalence relation defined by a normal subgroup.

Theorem 2.2

Let G be a group and $\sim$ an equivalence relation. Then, the space of equivalence classes $G/\sim$ has a group structure such that the canonical projection map $\pi\colon G \to G/\sim$ satisfies $\pi(gg') = \pi(g)\cdot\pi(g')$, $g, g' \in G$ if and only if $\sim$ is the canonical equivalence relation defined by a normal subgroup H. In such case, $G/\sim = G/H$ and H is the equivalence class containing the neutral element $e \in G$.

Proof 2.3 Let us denote by $[g]$ the equivalence class containing g, that is $[g] = \{g' \in G \mid g' \sim g\}$. Assume that $G/\sim$ has a group structure with composition law $[g]\cdot[g']$ such that the canonical map $\pi\colon G \to G/\sim$, $\pi(g) = [g]$, verifies $\pi(gg') = \pi(g)\cdot\pi(g')$, that is, π is a group homomorphism (see later on Def. 2.16). We may also use the brackets notation: $[gg'] = [g]\cdot[g']$.

We will show first that the equivalence class $[e] \subset G$ is a subgroup. Clearly, if $g \in [e]$, $[g^{-1}]\cdot[g] = [g^{-1}g] = [e]$, then $[g^{-1}] = [g]^{-1}$. Then if $g, g' \in [e]$, $[g] = [e] = [g']$, and we get $[g^{-1}g'] = [g]^{-1}\cdot[g'] = [e]^{-1}\cdot[e] = [e]$ and because Prop. 2.1, $[e]$ is a subgroup.

We denote the subgroup $[e]$ by H and it is clear that H is normal. Note the following tautological formula: $[g] = \{g' \in G \mid g \sim g'\} = \pi^{-1}([g])$. That is the symbol $[g]$ on the left hand side denotes the subset of G defined by all elements equivalent to g and the symbol $[g]$ on the right hand side denotes the abstract element in $G/\sim$ defined by the equivalence class containing g. The previous formula can also be written as $\pi([g]) = [g]$. Then $\pi(g[e]) = \pi(g)\cdot\pi([e]) = [g]\cdot[e] = [g]$, then $g[e] = [g]$, but this shows that $gH = [g]$, or, in other words, the elements g' which are equivalent to g are the elements in the coset gH. Then, we have shown that the equivalence relation $\sim$ is the canonical equivalence relation $\sim_R$ associated to the subgroup H.

The same argument shows that $Hg = [g]$, then $gH = Hg$ and the subgroup H is normal.

Definition 2.9 Let $H \triangleleft G$ be a normal subgroup. The group G/H is called the quotient group of G by the normal subgroup H.

The previous discussion shows the preeminent role played by normal subgroups. If a group G has a normal subgroup, then G/H is also a group and Thm. 2.12 will show that G can be reconstructed from H and G/H. Thus, groups without normal subgroups are somehow impossible to 'decompose into smaller' pieces. Hence, the following definition.

Definition 2.10 A group is called simple if it has no proper normal subgroups.

Example 2.22 *The group $\mathbb{Z}_p$, p prime, is simple. In fact, since $\mathbb{Z}_p$ is Abelian any subgroup is normal, but there are no proper subgroups of $\mathbb{Z}_p$ (their order should divide p because of Lagrange's theorem, Thm. 2.1).*

Exercise 2.23 *Show that, conversely, the only Abelian groups which are simple are the cyclic groups of prime order.*

We introduce now a natural equivalence relation in groups that will play an important role in the discussion on their structure (and in the theory of linear representations).

Definition 2.11 Two elements g, g' of a group G are conjugate if there exists $g_0 \in G$ such that $g' = g_0 g g_0^{-1}$.

The conjugation is an equivalence relation and places the elements of G into conjugacy classes. For instance, the neutral element of any group is the only element in its conjugacy class. Clearly, in an Abelian group, all conjugacy classes have only one element. Notice that H is a normal subgroup of G if and only if H is a conjugacy class by itself. The set of conjugacy classes does not have the structure of a group although the product of two classes is a union of classes.

We will come back to the discussion of quotient group structures later on (see Sect. 6.1) in connection with the corresponding notions for groupoids. For the moment, we just want to introduce these fundamental notions so that we have the proper vocabulary when analising the important example to follow: the Symmetric group. We will describe it by introducing another family of simple games that will be of interest later on when discussing groupoids.

2.2 A family of board games: The symmetric group

In order to extend our collection of examples, we will discuss another instance of board games, more sophisticated than the 'connecting dots' games described in the previous chapter. As we will see later (Sect. 3.2.1), these games will lead the discussion of some famous puzzles, such as Loyd's "15 puzzle", one the many puzzles created by Samuel Loyd[4].

[4]The greatest puzzle maker [28].

The first game will take place on a board (see Fig. 2.1 (a)) with n positions that we may imagine numbered from 1 to n. There will be n square pieces or tiles, also marked with numbers from 1 to n, placed on the board and that can be moved by exchanging two nearby ones. Each exchange of pieces at each step of the game will be called a 'move'. The move that will exchange the pieces i and j (recall that they must be sitting side by side) will be denoted as m_{ij}. We will denote by m_0 the 'trivial' move, that is, the move that does not exchange tiles. At each step of the game, the actual disposition of the pieces in the board will be called a configuration of the game. The objective of the game will be to obtain a given configuration α by starting from an arbitrary random one. The game with a board with n positions will be denoted by S_n (see, for instance, the configuration on the game S_7 in Fig. 2.1 (b)). Given two solutions of the game, the winner is the one that used fewer moves. The game can be complicated by imposing restrictions on the moves that can be used at each step. For instance, it can be accorded that only moves obtained from a given list of allowed moves at each step of the game can be used, etc.

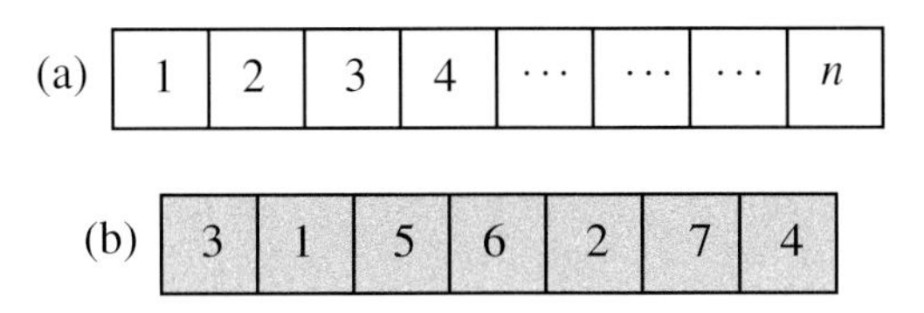

Figure 2.1: (a) A numbered board for the game S_n. (b) How many moves are needed to reach this position starting from $1, 2, \cdots, 7$?

The history of the game will be provided by listing from right to left the sequence of moves thus, or instance $h = m_{23}m_{13}m_{45}m_{25}m_{46}m_{26}m_{74}$ is a history producing the standard configuration starting from Fig. 2.1 (b). Notice that a 'move' is reversible and it can be undone in the next step.

There is however a different way of describing the configurations of the game and its evolution. We may describe the configuration as the bijective map associating the positions i in the board with the tile occupying such position $\alpha(i)$, then each configuration corresponds to the bijective map $\alpha\colon \{1,2,\ldots,n\} \to \{1,2,\ldots,n\}$. The traditional name for such bijective maps is 'permutations' (of the numbers $1,2,\ldots,n$) and then a configuration or permutation α will be written as:

$$\begin{pmatrix} 1 & 2 & \cdots & n \\ \alpha(1) & \alpha(2) & \cdots & \alpha(n) \end{pmatrix},$$

indicating that the position i on the board is occupied by the piece marked $\alpha(i)$, $i = 1,\ldots,n$. Now it is clear that the set of all configurations, i.e., the set of all permutations, form a group with the product given by composition of maps. We will study this group in order to better understand our game.

Definition 2.12 We will denote by S_n the group of all bijective maps of the set $\{1,\ldots,n\}$, whose elements will be called indices, onto itself with composition law the standard composition of maps. The group S_n is called the symmetric (or permutation) group of n indices and its elements α are called permutations. The composition law will be denoted by juxtaposition of the permutations, that is $\alpha\beta = \alpha \circ \beta$, where the expression on the right hand side is the standard composition of maps.

Proposition 2.4
The order of the group S_n is $n!$.

Proof 2.4 The image of the first index under a bijective map α could be any other index (n choices), the image of the second index could be any among the remaining $n-1$ indices, etc. Thus, the order of the group S_n will be $|S_n| = n(n-1)\cdots 1 = n!$.

Definition 2.13 We will say that the permutation $\alpha \in S_n$ fixes the index x if $\alpha(x) = x$; moreover, if $\alpha(x) \neq x$, it will be said that α changes the index x.

Let $i_1,\ldots,i_r$ be a subset of different indices in the set $\{1,\ldots,n\}$. If the permutation $\alpha \in S_n$ fixes all the indices x different from $i_1,\ldots,i_r$, but $\alpha(i_k) = i_{k+1}$, for all $1 \leq k \leq r-1$ and $\alpha(i_r) = i_1$, we will say that α is an r-cycle or that α is a cycle of length r.

The identity map is by convention a cycle of length one.

The cycle described in Def. 2.13, determined by the indices $i_1,\ldots,i_r$, will be denoted by $(i_1,i_2,\ldots,i_r)$ or just $(i_1\, i_2\, \ldots\, i_r)$.

Example 2.24 1. *If $\alpha,\beta \in S_3$, are given by:*

$$\alpha = \begin{pmatrix} 1 & 2 & 3 \\ 3 & 2 & 1 \end{pmatrix}, \qquad \beta = \begin{pmatrix} 1 & 2 & 3 \\ 2 & 3 & 1 \end{pmatrix},$$

then

$$\alpha\beta = \begin{pmatrix} 1 & 2 & 3 \\ 2 & 1 & 3 \end{pmatrix} \qquad \beta\alpha = \begin{pmatrix} 1 & 2 & 3 \\ 1 & 3 & 2 \end{pmatrix}.$$

The permutation α is a 2-cycle, $\alpha = (1\,3)$. The permutation β is a 3-cycle, $\beta = (1\,2\,3)$. The product $\alpha\beta$ is a 2-cycle, $\alpha\beta = (1\,2)$, whereas $\beta\alpha$ is a different 2-cycle, $\beta\alpha = (2\,3)$.

2. *The permutation $\alpha \in S_4$ given by*

$$\alpha = \begin{pmatrix} 1 & 2 & 3 & 4 \\ 2 & 3 & 4 & 1 \end{pmatrix},$$

is a 4-cicle, $\alpha = (1\,2\,3\,4)$.

The permutatiom $\beta \in S_5$ defined by

$$\beta = \begin{pmatrix} 1 & 2 & 3 & 4 & 5 \\ 5 & 1 & 4 & 2 & 3 \end{pmatrix},$$

is a 5-cycle, $\beta = (15342)$, however, the permutation

$$\gamma = \begin{pmatrix} 1 & 2 & 3 & 4 \\ 4 & 3 & 2 & 1 \end{pmatrix},$$

is not a cycle, although it can be written as a product of disjoint cycles $(14)(23)$ (see Def. 2.14 and Thm. 2.4 below).

Remark 2.1 Notice that S_1 is the trivial group (it only contains the identity) and that S_2 is a group with two elements (and then Abelian). However, S_3 which has six elements, is a non-Abelian group (the unique non Abelian group of order 6):

$$(12)(23) = (123), \quad (23)(12) = (132). \quad \blacksquare \tag{2.2}$$

Proposition 2.5
The order of a r-cycle is r, that is, if α is an r-cycle, then $\alpha^r = e$ and $\alpha^k = e$ for any $1 < k < r$.

Proof 2.5 If α is the cycle[5] $\alpha = (i_1, \ldots, i_r)$, then α^2 is the permutation

$$\alpha^2(i_1) = i_3, \alpha^2(i_2) = i_4, \ldots, \alpha^2(i_{r-1}) = i_1, \alpha^2(i_r) = i_2,$$

and, repeating this process, we get $\alpha^r(i_k) = i_k$, in other words, α^r is the identity.

Definition 2.14 We will say that two permutations $\alpha, \beta \in S_n$ are disjoint if each index that is changed by one is fixed by the other. We will say that a set of permutations is disjoint if all of them are pairwise disjoint.

Proposition 2.6
Two cycles $\alpha = (i_1, \ldots, i_r)$ and $\beta = (j_1, \ldots, j_s)$ are disjoint iff $\{i_1, \ldots, i_r\} \cap \{j_1, \ldots, j_s\} = \emptyset$. If two cycles are disjoint, they commute.

Proof 2.6 Is a direct consequence of the definition that if two cycles are disjoint then $\{i_1, \ldots, i_r\} \cap \{j_1, \ldots, j_s\} = \emptyset$. Then, because α and β move different indices, the order in which such permutations are performed is irrelevant.

---

[5]In what follows, we will keep the notation (123) for specific cycles and we will insert commas if needed to make the expressions more clear, like in: $(i_1, i_2, \ldots, i_r)$.

It is also an immediate consequence of the previous definitions that:

Proposition 2.7
Let α and β be two r-cycles in S_n. If there exists an index $j \in \{1,\dots,n\}$ that is moved by both, α and β, and such that $\alpha^k(j) = \beta^k(j)$ for any integer k, then $\alpha = \beta$.

Proof 2.7 If j is in the cycle α, then $(j,\alpha(j),\dots\alpha^m(j))$ is this cycle for some m, but then both cycles (α and β) are the same.

Theorem 2.3
Let $\alpha \in S_n$ be a permutation. There exists a partition of the set of indices $\Omega = \{1,\dots,n\}$ in disjoint subsets $\Omega = \bigcup_{i=1}^r Y_r$, with $Y_i \cap Y_j = \emptyset$, $i \neq j$, such that the restriction of α to each Y_i is a cycle whose order is the number of indices in Y_i.

Proof 2.8 Consider the equivalence relation defined on Ω as follows: j is equivalent to i if there is k such that $\alpha^k(i) = j$. It is trivial to verify that this relation is an equivalence relation. Notice that if $\alpha = (i_1,i_2,\dots,i_r)$ is an r-cycle, then the indices $\{i_1,i_2,\dots,i_r\}$ form an equivalence class and the remaining indices form equivalence classes containing just one element each.

Now, consider the decomposition of Ω in the equivalence classes $Y_1,\dots,Y_r$ defined before. It is clear that these subsets Y_i provide the sought after decomposition.

Theorem 2.4
Any permutation $\alpha \in S_n$, $\alpha \neq e$, is the product of disjoint cycles of length greater or equal than 2; this factorisation is unique up a reordering of the cycles.

Proof 2.9 Let $Y_1,\dots,Y_r$ be the decomposition of indices defined in Thm. 2.3 associated to α. If β_i is the permutation that coincides with α on Y_i and fixes all other indices, then α can be written as: $\alpha = \beta_1\cdots\beta_r$. Finally, we discard all factors that reduce to the identity.

To prove the uniqueness, we suppose that we have two decompositions: $\alpha = \beta_1\cdots\beta_r = \gamma_1\cdots\gamma_r$. If α changes the index x, there must be β_i and γ_j that should change x. Let us assume that they are β_1 and γ_1, respectively (recall that the β_i commute among themselves as well as the γ_j). But $\alpha(x) = \beta_1(x) = \gamma_1(x)$, hence, $\beta_1^k(x) = \gamma_1^k(x)$ for any integer k, implying that $\beta_1 = \gamma_1$ (Prop. 2.7). Then, we cancel these two factors and we proceed with another index which is moved by α. After a finite number of steps, there will be no more indices changing under α, hence, there will be no more factors β_i, γ_l and the decompositions will be identical (up to reordering the factors).

Definition 2.15 A 2-cycle will be called a transposition. Because a 2-cycle is of order 2, the inverse of a transposition is the transposition itself.

Theorem 2.5
Each permutation $\alpha \in S_n$ can be expressed as the product of transpositions.

Proof 2.10 The cycle $(i_1,\ldots,i_r)$ can written as $(i_1,\ldots,i_r) = (i_1,i_r)\cdots(i_3,i_1)(i_1,i_2)$. As any permutation can be written as a product of disjoint cycles, Thm. 2.4, and any one of them can be written as a product of transpositions because of the previous observation, then an arbitrary permutation can be written as a product of transpositions.

Remark 2.2 Notice that the transpositions appearing in the factorization of an arbitrary permutation do not have to commute among each other, i.e., they do not have to be disjoint. Moreover the factors are not uniquely determined. For instance: $(123) = (13)(12) = (23)(13) = (13)(42)(12)(14)$. ■

Remark 2.3 Because of Thm. 2.5, the set of all transpositions defines a system of generators for the symmetric group (see the following exercise). ■

Exercise 2.25 *Prove that the transpositions of the form $(k,k+1)$ define a system of generators of S_n, that is, that any permutation α can be written as a composition of transpositions of the form $(k,k+1)$.*

Exercise 2.26 *Prove that the transposition $\tau = (12)$ and the n-cycle $\gamma = (1,\ldots,n)$ define a system of generators for S_n. (Hint, show that $\gamma^k\tau\gamma^{-k} = (k+1,k+2)$ for all integer k).*

What is the relation between moves and permutations? The answer is simple. Suppose that we have the tile $\alpha(i)$ sitting in position i and $\alpha(i+1)$ in the adjacent position $i+1$, then the move $m_{\alpha(i),\alpha(i+1)}$ on the configuration α will correspond exactly to the transposition $(i, i+1)$ acting on the permutation α describing the configuration of the game, that is $(i,i+1)\alpha$. Now, if we consider the move $m_{\alpha(i),\alpha(i+2)}$, the consecutive moves $m_{\alpha(i),\alpha(i+1)}m_{\alpha(i),\alpha(i+2)}$ will amount to the transpositions $(i,i+1)$, $(i,i+2)$ acting on α, that is $(i,i+1)(i,i+2)\alpha$. Hence moving a tile r steps to the right of the board will be written as $(i,i+1)(i,i+2)\ldots(i,i+r)\alpha$, that is, acting on α with the inverse of the cycle $(i,i+1,i+2,\ldots,i+r)$. Then, we observe that, because of Remark 2.3, the cycles $(k,k+1)$ form a generating system for S_n and the board game S_n always has a solution.

Exercise 2.27 *Work out an algorithm to solve the board game S_n and compute the number of steps needed.*

We will continue discussing the structure of the symmetric group in Sect. 2.4 after introducing more tools that will help to deal with it.

2.3 Group homomorphisms and Cayley's theorem

2.3.1 Group homomorphisms: First properties

We will now briefly discuss maps between groups. Most relevant among them are those that respect their group structures. This notion has a strong categorical flavour but its full meaning and implications will not be uncovered until we have developed more tools and examples (see Chapt. 5).

Definition 2.16 Let G, G' be groups and $f : G \to G'$ a map. We say that f is a group homomorphism if:

$$f(g_1 g_2) = f(g_1) f(g_2), \qquad g_1, g_2 \in G.$$

If f is surjective, it is sometimes called an epimorphism, and a monomorphism if it is injective. If f is a one-to one map onto G', f is called an isomorphism (an automorphism if $G = G'$). Hence, an isomorphism is a homomorphism which is both an epimorphism and a monomorphism.

Example 2.28 *1. Consider the group $\mathbb{Z}_n$, given any natural number m, the map $f_m \colon \mathbb{Z}_n \to \mathbb{Z}_n$ given by $f_m(\mathbf{k}) = m\mathbf{k}$, $\mathbf{k} \in \mathbb{Z}_n$, is a group homomorphism.*

2. The map $f \colon D_3 \to S_3$ given by $f(r_0) = e$, $f(r_1) = (123)$, $f(r_2) = (132)$, $f(\tau_1) = (12)$, $f(\tau_2) = (23)$, $f(\tau_3) = (13)$ is a group homomorphism.

Exercise 2.29 *Show that the homomorphism f_m in Example 2.28.1 is an isomorphism iff n and m are relatively prime, i.e.,* g.c.d.$(n,m) = 1$.

Definition 2.17 There are two trivial homomorphisms: The identity $\mathrm{id}_G \colon G \to G$, that maps any element g on itself, $\mathrm{id}_G(g) = g$ and which is an isomorphism, and the map that maps any element into the neutral element, that is $g \mapsto e$, for all $g \in G$, which is an epimorphism from the group G onto the group $\{e\}$. We will denote such homomorphism simply by $G \to \{e\}$ or $G \to 1$.

If $H \subset G$ is a subgroup the canonical immersion $i \colon H \to G$, $i(h) = h$, for all $h \in H$ is a homomorphism, which is a monomorphism. In particular, because $\{e\} \subset G$ is a subgroup the canonical immersion map $\{e\} \to G$ is a monomorphism (denoted also by $1 \to G$).

Exercise 2.30 *Prove that the image of the neutral element e in G under a homomorphism $f \colon G \to G'$ is the neutral element e' in G', i.e., $f(e) = e'$.*

Exercise 2.31 *Let $f \colon G \to G$ be a group homomorphism. Show that $f(g)^{-1} = f(g^{-1})$ for any $g \in G$.*

Definition 2.18 Define the kernel of the homomorphism $f \colon G \to G'$, as:

$$\ker f = \{g \in G \colon f(g) = e'\}, \tag{2.3}$$

It is easy to show that it is a normal subgroup of G. Then we can construct the quotient group $G/\ker f$.

Exercise 2.32 *Prove the last statement, that is* $\ker f \triangleleft G$ *for any group homomorphism* f.

Proposition 2.8
The homomorphism of groups $f\colon G \to G'$ *is a monomorphism iff* $\ker f = \{e\}$.

Proof 2.11 If f is a monomorphism it is injective, then $\ker f = \{e\}$ because $f(e) = e'$ (Exercise 2.30). Conversely, if $\ker f = \{e\}$, then f is injective because if $f(g) = f(g')$ for some $g, g' \in G$, then $e' = f(g)^{-1} f(g') = f(g^{-1}g')$, then $g^{-1}g' \in \ker f$, hence, $g^{-1}g' = e$, and we conclude $g = g'$.

Finally notice that the range of f

$$\operatorname{ran} f = \{g' \in G' : \exists g \in G,\ f(g) = g'\} \equiv f(G), \tag{2.4}$$

is a subgroup of G' (not necessarily normal though).

Exercise 2.33 *Consider the map* $f\colon S_n \to S_m$, $m > n$, *given by* $f(\sigma)(k) = \sigma(k)$ *if* k *is an index in* $\{1, 2, \ldots, n\}$ *and* $f(\sigma)(k) = k$ *if* $k > n$, *for any* $\sigma \in S_n$. *Show that* f *is a group homomorphism. Compute its range* $\operatorname{ran} f = f(S_n)$ *and show that it is not a normal subgroup of* S_m.

Using these groups associated to a homomorphism f, we can easily prove the following isomorphism theorem.

Theorem 2.6
Let G, G' *be groups and* $f : G \to G'$ *a group homomorphism. Then the following diagram is commutative:*

$$\begin{array}{ccc} G & \xrightarrow{\ f\ } & G' \\ {\scriptstyle \pi}\downarrow & & \uparrow{\scriptstyle i} \\ G/\ker f & \xrightarrow[\ \tilde{f}\ ]{} & \operatorname{ran} f \end{array}$$

that is, $f = i \circ \tilde{f} \circ \pi$, *where* π *is the projection map,* i *is the inclusion map, and* $\tilde{f}$, *defined as* $\tilde{f}(g \ker f) = f(g)$, *is an isomorphism.*

Exercise 2.34 *Prove the statements in Thm. 2.6.*

2.3.2 Cayley's theorem for groups

Isomorphisms allow us to identify groups. Thus, if a group is isomorphic to a subgroup of another group, we may consider that it is realised as a subgroup of the second group by means of that isomorphism. This way of thinking will find its first application in the following statement.

Theorem 2.7

Cayley's theorem. Any finite group G is isomorphic to a subgroup of the symmetric group S_N, with $N = |G|$.

Proof 2.12 The proof is rather easy and constructive. Suppose that we label the elements of the finite group G as $\{g_1 = e, g_2, \ldots, g_N\}$ with $N = |G|$ the order of the group. Now to any element g_k we associate the permutation $\sigma_k \in S_N$ that takes the sequence $(1,2,\ldots,N)$ into $(\sigma_k(1), \sigma_k(2), \ldots, \sigma_k(N))$ with $\sigma_k(j)$ being the index defined by $g_k g_j = g_{\sigma_k(j)}$, that is, in plain words, the permutation σ_k sends j to the index labelling the element obtained by multiplying g_j by g_k on the left. Let us denote this map by σ, that is:

$$\sigma \colon G \to S_N\,, \qquad \sigma(g_k) = \sigma_k\,, \quad g_{\sigma_k(j)} = g_k g_j\,, \forall k, j = 1, \ldots, N\,.$$

Checking that σ is an homomorphim is routine. In fact, if $g_k g_l = g_m$, then $\sigma(g_k g_l) = \sigma(g_m) = \sigma_m$ with $g_{\sigma_m(j)} = g_m g_j = g_k g_l g_j$. But $\sigma(g_k)\sigma(g_l)(j) = \sigma(g_k)\sigma_l(j) = \sigma_k(\sigma_l(j))$ which means that $g_{\sigma_k(\sigma_l(j))} = g_k g_{\sigma_l(j)} = g_k g_l g_j$, then $\sigma_m = \sigma_k \sigma_l$, proving that σ is a homomorphism.

It is also easy to check that σ is injective. But this is trivially so because if $\sigma(g_k) = \sigma(g_l)$ then $\sigma_k(j) = \sigma_l(j)$, that means that $g_k g_j = g_l g_j$, then $g_k = g_l$. Then the group G is isomorphic to the subgroup of S_N defined by the image of σ.

Cayley's theorem is the first major representation theorem we are finding in this book. That means that we are 'identifying' (by means of the map σ) elements of the abstract group G (have we said anything on how the group G is obtained or constructed?) with permutations of the numbers $1, 2, \ldots, N$ in such a way that the relations among the elements of the group (given by the composition law among them) are preserved (because the map σ is a homomorphism). Only a word of caution here, the map σ is not canonical[6]. Actually, any other labelling of the elements of the group will do the trick.

Exercise 2.35 *For the adventurous reader, we may suggest to prove the following extension of the previous theorem: Any group G is isomorphic to a subgroup of the group of bijective maps of G onto itself, that accordingly could be called the symmetric group of G and denoted by S_G.*

[6]The word "canonical" is a mathematicians favorite. It should be understood as meaning "natural", "unique in a certain sense", etc. Even if it comes from latin "canon", meaning that appartains to a list of sacred books, its mathematical use has obviously nothing to do with the principles of religious doctrine, and is only used to emphasize is privileged status among a family of choices.

Thus, we have shown that there is a monomorphism $\sigma\colon G \to S_{|G|}$ that represents our group. This monomorphism, in disguise, will be exploited throughout this work under the name of the 'regular representation' (see, for instance, Sect. 7.1.4) and will prove to be instrumental in studying the linear representations of both groups and groupoids.

2.4 The alternating group

We return to study the structure of the symmetric group that will be used all along the rest of this work (the linear representations of S_3 will be discussed in Sect. 8.7.4).

Let $p = p(x_1, \ldots, x_n)$ be the polynomial in n-variables:

$$p(x_1, \ldots, x_n) = \prod_{j>i} (x_j - x_i)\,.$$

Given a permutation $\alpha \in S_n$ we denote by p^α the polynomial:

$$(p^\alpha)(x_1, \ldots, x_n) = \prod_{j>i} (x_{\alpha(j)} - x_{\alpha(i)})\,.$$

Definition 2.19 We say that the permutation α is even if $p^\alpha = p$; we will say that α is odd if $p^\alpha = -p$. If the permutation α is even, we say that its signature is $+1$ and it will be denoted as $\sigma(\alpha) = +1$. If it is odd, we say that its signature is -1, then, with these conventions, $p^\alpha = \sigma(\alpha)p$ (notice that any permutation is either even or odd because in p^α there appear the same factors as in p up to the sign).

Proposition 2.9
If α and β are two permutations in S_n, then $\sigma(\alpha\beta) = \sigma(\alpha)\sigma(\beta)$.

Proof 2.13 Notice that, by definition, $p^{\alpha\beta} = (p^\beta)^\alpha$. Then, on one side we get $p^{\alpha\beta} = \sigma(\alpha\beta)p$ and, on the other: $(p^\beta)^\alpha = \sigma(\alpha)p^\beta = \sigma(\alpha)\sigma(\beta)p$.

Then we obtain the following:

Corollary 2.1
The map $\sigma\colon S_n \to C_2$ defined by $\sigma : \alpha \mapsto \sigma(\alpha)$, where C_2 is the multiplicative group $\{1, -1\}$, is a group homomorphism which is surjective, i.e., σ is an epimorphism of groups.

Definition 2.20 We will denote by A_n the subgroup of S_n whose elements are all even permutations, that is, $A_n = \ker\sigma$, and A_n is a normal subgroup of S_n. The subgroup A_n is called the alternating group of n indices.

Proposition 2.10
$|A_n| = n!/2$.

Proof 2.14 The first isomorphism theorem for groups, Thm. 2.6, applied to the epimorphism $\sigma\colon S_n \to C_2$, gives us $S_n/A_n \cong C_2$. Then, because of Lagrange's theorem, Thm. 2.1, $|S_n| = 2|A_n|$.

Theorem 2.8
A permutation $\alpha \in S_n$ is even if and only if it is the product of an even number of transpositions and it is odd if and only if it is the product of an odd number of transpositions.

Proof 2.15 Let $\alpha = \tau_1 \cdots \tau_r$ be a factorization of α as a product of transpositions. Then, because of Prop. 2.9, we have that $\sigma(\alpha) = \sigma(\tau_1)\cdots\sigma(\tau_r) = (-1)^r$ and the conclusion follows.

Corollary 2.2
The number of factors that appear in the factorisation of a permutation as a product of transpositions is always even or odd, depending on its signature.

Remark 2.4 The symmetric groups S_n are not simple for $n > 2$ as they always have a proper normal subgroup. Hovewer, the alternating groups A_n are simple for $n = 2,3$ and $n > 4$. However, A_4 is not simple. ∎

2.4.1 Conjugacy classes: Young diagrams

Lemma 2.1
Let α and β be two permutations, then the permutation $\alpha\beta\alpha^{-1}$ has the same cyclic structure as β. The cycles in $\alpha\beta\alpha^{-1}$ are obtained by applying α to the indices in β.

Proof 2.16 Let $\beta = \gamma_1 \cdots \gamma_s$ the (unique up to reordering) decomposition in disjoint cycles of β. Then we get:

$$\alpha\beta\alpha^{-1} = \alpha\gamma_1\alpha^{-1}\alpha\gamma_2\alpha^{-1}\cdots\alpha\gamma_s\alpha^{-1}.$$

Thus all we have to do is to check the structure of the conjugate $\alpha\gamma\alpha^{-1}$ of a cycle γ by a permutation α. Let $\gamma = (i_1,\ldots,i_r)$ be an r-cycle and let $\{i_{r+1},\ldots,i_n\}$ be the set of fixed indices of γ. If α is the permutation $\alpha(i_k) = j_k$, with $k = 1,\ldots,n$, then the indices j_k with $k = r+1,\ldots,n$, are fixed by $\alpha\gamma\alpha^{-1}$, while $\alpha\gamma\alpha^{-1}$ move the indices j_k with $k = 1,\ldots,r-1$ to j_{k+1} and j_r is moved to j_1. Then we get: $\alpha\gamma\alpha^{-1} = (j_1,\ldots,j_r)$.

The following theorem will allow us to identify the conjugacy classes of S_n.

Theorem 2.9

Two permutations α and β are conjugate if and only if they have the same cyclic structure.

Proof 2.17 Because of the previous Lemma 2.1, if α and β are conjugate they have the same cyclic structure. Conversely, if α and β have the same cyclic structure, then:

$$\begin{aligned} \alpha &= (i_1,\ldots,i_{r_1})(i_{r_1+1},\ldots,i_{r_2})\cdots(i_{r_s+1},\ldots,i_n), \\ \beta &= (j_1,\ldots,j_{r_1})(j_{r_1+1},\ldots,j_{r_2})\cdots(j_{r_s+1},\ldots,j_n), \end{aligned}$$

and the permutation μ defined by $\mu(i_k) = j_k$ is such that $\mu\alpha\mu^{-1} = \beta$.

Proposition 2.11

Each conjugacy class in S_n is in one-to-one correspondence with a set of non-negative integer numbers $\{\nu_1,\ldots,\nu_n\}$, such that $\nu_1 + 2\nu_2 + \cdots + n\nu_n = n$.

Proof 2.18 Let α be a permutation in S_n then, because Thm. 2.9, any permutation in its conjugacy class has the same cycle structure, thus, if α consists of ν_1 1-cycles, ν_2 2-cycles,...., ν_n n-cycles, then $\nu_1 + 2\nu_2 + \cdots + n\nu_n = n$.

Conversely, if we are given a set of non-negative integers $\nu_1,\ldots,\nu_n$, such that $\sum_{k=1}^n k\nu_k = n$, then we may consider the permutation with ν_1 1-cycles, ν_2 2-cycles, etc.:

$$\begin{aligned} \alpha = {} & (1)\cdots(\nu_1)(\nu_1+1,\nu_1+2)\cdots \\ & \cdots(\nu_1+2\nu_2-1,\nu_1+2\nu_2)(\nu_1+2\nu_2+1,\nu_1+2\nu_2+2,\nu_1+2\nu_2+3)\cdots \\ & \cdots(\nu_1+2\nu_2+3\nu_3-2,\nu_1+2\nu_2+3\nu_3-1,\nu_1+2\nu_2+3\nu_3)\cdots. \end{aligned} \tag{2.5}$$

Any other permutation in the conjugacy class of α has the same cyclic structure.

Definition 2.21 Given a natural number n, a partition of n is a set of non-negative integers $\lambda_i \geq 0$, $i = 1,\ldots,n$, such that $\lambda_1 \geq \lambda_2 \geq \cdots \geq \lambda_n$ and $\lambda_1 + \lambda_2 + \cdots + \lambda_n = n$.

Partitions can be described graphically by using Young diagrams. A Young diagram of order n consists of a pile of boxes with rows of non-increasing size, such that the first row contains λ_1 boxes, the second one λ_2, etc. (empty rows will not be displayed). Thus the total number of boxes is n and the number of boxes on each row give the factors of a partition of n.

$$\begin{array}{ll} \lambda_1 & \square\square\square\square\square\cdots\square \\ \lambda_2 & \square\square\square\square\cdots\square \\ \vdots & \vdots \\ \lambda_{r-1} & \square\square\square\cdots\square \\ \lambda_r & \square\cdots\square \end{array} \tag{2.6}$$

Theorem 2.10
There is a one-to-one correspondence between the set of conjugacy classes of S_n and partitions of n.

Proof 2.19 Let the conjugacy class of S_n be characterised by the numbers $\{\nu_1, \ldots, \nu_n\}$, such that $\nu_i \geq 0$ and $\nu_1 + 2\nu_2 + \cdots + n\nu_n = n$ given by Prop. 2.11. Define

$$\lambda_1 = \nu_1 + \nu_2 + \cdots + \nu_n, \quad \lambda_2 = \nu_2 + \cdots + \nu_n, \quad \ldots, \quad \lambda_n = \nu_n \,.$$

Then, it is clear that $\lambda_1 \geq \lambda_2 \geq \cdots \geq \lambda_n \geq 0$ and $\lambda_1 + \lambda_2 + \cdots + \lambda_n = n$, hence, each conjugacy class determines a partition of n.

Conversely, let $\{\lambda_1, \ldots, \lambda_n\}$ be a partition of n. Then,

$$\nu_1 = \lambda_1 - \lambda_2 \,, \nu_2 = \lambda_2 - \lambda_3 \,, \ldots, \nu_i = \lambda_i - \lambda_{i+1} \,, \ldots, \nu_n = \lambda_n \,.$$

is a set of non-negative integers such that: $\nu_1 + 2\nu_2 + \cdots + n\nu_n = n$.

Theorem 2.11
The number of elements in the conjugacy class characterized by the numbers $\{\nu_1, \ldots, \nu_n\}$ is given by:

$$n_{\{\nu_1,\ldots,\nu_n\}} = \frac{n!}{\nu_1! \cdots \nu_n! 1^{\nu_1} 2^{\nu_2} \cdots n^{\nu_n}} \,.$$

Proof 2.20 There are $n!$ ways of putting n different objects in n cells. However, because the cycles are disjoint, they commute and the ν_k k-cicles can be rearranged in $\nu_k!$ different forms. Moreover the k-cycle $(i_1, \ldots, i_k)$ coincides with $(i_2, \ldots, i_k, i_1)$ and all its cyclic permutations, that is, the same k-cycle can be written in k different forms.

Example 2.36 *Let us consider S_4. The partitions of $n = 4$ are given by $1+1+1+1$, $2+1+1$, $2+2$, $3+1$ and 4 (see the corresponding Young diagrams below). The first one corresponds to one 4-cycle as $\nu_1 = \nu_2 = \nu_3 = 0$ and $\nu_4 = 1$, the second describes a cycle structure with $\nu_1 = 1$, $\nu_2 = 0$ and $\nu_3 = 1$, that is one 3-cycle. The third describes a cycle structure with two transpositions, the fourth a single transposition and the fifth the neutral element because $\nu_1 = 4$. Denoting each class as $\mathcal{C}_{\nu_1,\ldots,\nu_n}$, we get that $|\mathcal{C}_{0,0,0,1}| = 6$, $|\mathcal{C}_{1,0,1,0}| = 24/3 = 8$, $|\mathcal{C}_{0,2,0,0}| = 24/(2 \times 4) = 3$, $|\mathcal{C}_{2,1,0,0}| = 24/(2 \times 2) = 6$, $|\mathcal{C}_{4,0,0,0}| = 24/4! = 1$. Then $|\mathcal{C}_{0,0,0,1}| + |\mathcal{C}_{1,0,1,0}| + |\mathcal{C}_{0,2,0,0}| + |\mathcal{C}_{2,1,0,0}| + |\mathcal{C}_{4,0,0,0}| = 6 + 8 + 3 + 6 + 1 = 24$.*

Example 2.37 *Consider the conjugacy class in S_{13} corresponding to the partition $\{5,3,3,2\}$. Such class can be represented by the Young diagram:*

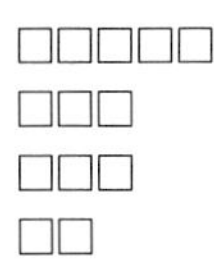

The cyclic structure corresponding to this class will be: $\nu_1 = \lambda_1 - \lambda_2 = 2$, $\nu_2 = \lambda_2 - \lambda_3 = 0$, $\nu_3 = \lambda_3 - \lambda_4 = 1$, $\nu_4 = \lambda_4 - \lambda_5 = 2$, $\nu_5 = \lambda_5 = 0$. *In other words, any permutation in the class will be the product of a 3-cycle and two 4-cycles. The number of elements on this class will be:*

$$n_{\{2,0,1,2\}} = \frac{13!}{2! \times 1! \times 2! \times 3 \times 4^2} = 32432400.$$

2.4.2 Homomorphisms and exact sequences

The use of homomorphisms becomes ubiquitous in the study and applications of groups. Some particular combinations of homomorphisms appear frequently and it is convenient to introduce some special notations and names for them.

Definition 2.22 Given two homomorphisms of the groups G_1, G_2 and G_3, $f_1 \colon G_1 \to G_2$ and $f_2 \colon G_2 \to G_3$, we will say that the sequence of homomorphisms f_1, f_2 is an exact sequence of group homomorphisms if $\operatorname{ran} f_1 = \ker f_2$. In such case, we will write them as $G_1 \xrightarrow{f_1} G_2 \xrightarrow{f_2} G_3$.

The definition extends naturally to any sequence of homomorphisms $f_i \colon G_i \to G_{i+1}$, $i = 1, \ldots, r+1$ and we will say that the sequence of homomorphisms $f_1, f_2, \cdots, f_r$ is exact if it is exact at each node, i.e., $\operatorname{ran} f_i = \ker f_{i+1}$, $i = 1, \ldots, r$.

Proposition 2.12
Let G be a group and $f \colon G \to G'$ a group homomorphism, then:

1. *The sequence of group homomorphisms $1 \to G \xrightarrow{f} G'$, where the first homomorphism is the trivial homomorphism sending the neutral element into the neutral element of G (Def. 2.17), is exact iff f is a monomorphism.*

2. *The sequence of group homomorphisms $G \xrightarrow{f} G' \to 1$ is exact iff f is an epimorphism.*

3. *The sequence of group homomorphisms $1 \to G \xrightarrow{f} G' \to 1$ is exact iff f is an isomorphism.*

Proof 2.21 1. Notice that if $1 \to G \to G'$ is exact then $\ker f = \{e\}$, because the range of the first map is just $\{e\}$. Then f is a monomorphism because of Prop. 2.8.

2. Clearly, if $G \to G' \to 1$ is exact, then $\operatorname{ran} f = G'$ as the kernel of the trivial map $G' \to 1$ is all G'.

3. Combining 1 and 2, we get the conclusion.

Example 2.38 *1. Consider the symmetric group S_n. The sequence of homomorphisms $i\colon A_n \to S_n$ and the signature homomorphism $\sigma\colon S_n \to C_2$ form an exact sequence of homomorphisms $1 \to A_n \to S_n \xrightarrow{\sigma} C_2 \to 1$.*

2. The sequence of homomophisms $n\mathbb{Z} \to \mathbb{Z} \to \mathbb{Z}_n$ is exact.

Let $f\colon G \to G'$ be a homomorphism, then the sequence $1 \to \ker f \to G \to G/\ker f \to 1$ is exact. More generally:

Theorem 2.12

Let $H \lhd G$ be a normal subgroup, then we have the following exact sequence:

$$1 \to H \xrightarrow{i} G \xrightarrow{\pi} G/H \to 1\,,$$

where $i\colon H \to G$ is the canonical inclusion and $\pi\colon G \to G/H$ is the canonical projection.

Conversely, if $1 \to H \xrightarrow{f} G \xrightarrow{f'} N \to 1$ is an exact sequence of homomorphisms, then f is a monomorphism, $f(H)$ is a normal subgroup of G, and H is isomorphic to a normal subgroup of G. In addition, f' is an epimorphism and N is isomorphic to the quotient group $G/f(H)$.

Proof 2.22 Because Prop. 2.12.1, f is a monomorphism. Moreover $\operatorname{ran} f = f(H)$ is a subgroup of G. It is normal because $f'(gf(h)g^{-1}) = f'(g)f'(f(h))f'(g^{-1}) = f'(gg^{-1}) = e$ (note that the sequence being exact implies that $f'(f(h)) = e$ for all $h \in H$), then $gf(h)g^{-1} \in \ker f' = \operatorname{ran} f$ and there exists $f(h') \in f(H)$ such that $gf(h)g^{-1} = f(h')$.

Then, we conclude that H is isomorphic to a normal subgroup of G.

Because Prop. 2.12.2, f' is an epimorphism of groups. Consider the map $\varphi\colon G/f(H) \to N$ given by $\varphi(gf(H)) = f'(g)$. We check that φ is an isomorphism of groups.

First, φ is a homomorphism:

$$\varphi(gf(H)g'f(H)) = \varphi(gg'f(H)) = f'(gg') = f'(g)f'(g') = \varphi(gf(H))\varphi(g'f(H))\,.$$

φ is a monomorphism: If $gf(H) \in \ker \varphi$, then $f'(g) - e$, and $g \in \ker f' = \operatorname{ran} f$. Then $g = f(h)$ and $gf(H) = f(H)$.

φ is an epimorphism: We construct a right inverse $\psi\colon N \to G/f(H)$ for φ. Let $n \in N$, then we define $\psi(n) = gf(H)$ where $g \in G$ is such that $f'(g) = n$. Clearly, such g exists because f' is surjective. Moreover, ψ is well defined because if $g' \in G$ is another element, such that $f'(g') = n$, then $f'(g^{-1}g') = e$ and $g^{-1}g' \in \ker f' =$

$f(H)$, then $g' \in gf(H)$ and $g'f(H) = gf(H)$. Finally we check that $\varphi \circ \psi = \mathrm{id}_N$: $\varphi \circ \psi(n) = \varphi(gf(H)) = f'(g) = n$, for all $n \in N$.

Definition 2.23 Given two groups H and N and an exact sequence of homomorphisms $1 \to H \to G \to N \to 1$, the group G will be called an extension of the group N by H and the sequence will be called a short exact sequence.

Given a normal subgroup $H \lhd G$, the short exact sequence $1 \to H \to G \to G/H \to 1$ will be called the canonical short exact sequence defined by H.

Example 2.39 *1. The sequence of homomorphisms $1 \to A_n \to S_n \to C_2 \to 1$ is exact and S_n is an extension of C_2 by A_n.*

2. *The sequence of homomorphisms $1 \to n\mathbb{Z} \to \mathbb{Z} \to \mathbb{Z}_n \to 1$ is exact, and $\mathbb{Z}$ is an extension of $\mathbb{Z}_n$ by $n\mathbb{Z}$.*

3. *Let $n = pq$ with p, q relative primes, then the sequence of homomorphisms $1 \to \mathbb{Z}_p \to \mathbb{Z}_n \to \mathbb{Z}_q \to 1$ is exact, where the first monomorphism $\mathbb{Z}_p \to \mathbb{Z}_n$ is defined as $\mathbf{k} \mapsto q\mathbf{k}$, $\mathbf{k} = \mathbf{0}, \mathbf{1}, \ldots, \mathbf{p-1}$; and the second $\mathbb{Z}_n \to \mathbb{Z}_q$ is given by $\mathbf{j} \mapsto [j]_q$, $\mathbf{j} = \mathbf{0}, \mathbf{1}, \ldots, \mathbf{n-1}$, where $[j]_q$ denotes the congruence class of j module q. Then $\mathbb{Z}_n$ is an extension of $\mathbb{Z}_q$ by $\mathbb{Z}_p$.*

2.4.3 The group of automorphisms of a group

The conjugation equivalence relation in a group G was introduced in Def. 2.11 and it was shown to play a relevant role in the discussion of the symmetric group. We are going to present it from the perspective of group isomorphisms.

Definition 2.24 Given a group G and an element $g_0 \in G$, we define the automorphism:

$$\mathrm{Int}\,(g_0)\colon G \to G, \qquad \mathrm{Int}(g_0)(g) = g_0 g g_0^{-1}. \tag{2.7}$$

Two elements g, g' of the group G will be said to be conjugate if there exists g_0 such that $g' = g_0 g g_0^{-1}$, that is, if $g' = \mathrm{Int}\,(g_0)(g)$. The conjugation automorphisms $\mathrm{Int}(g)$ are called inner automophisms of the group and they will be denoted as $\mathrm{Int}\,(G)$.

Proposition 2.13
The conjugation $\mathrm{Int}\,(g)$ transforms subgroups into isomorphic subgroups. Moreover, a normal subgroup is invariant under conjugation by any element of the group.

Proof 2.23 The first statement follows from conjugation being an automorphism. The second statement is a direct consequence of the definition of normal subgroups.

Definition 2.25 The set of all automorphisms $\varphi\colon G \to G$ form a group under the standard composition of maps. This group will be called the group of automorphisms of G and will be denoted as $\mathrm{Aut}(G)$.

Proposition 2.14
The set of inner automorphisms of the group G form a normal subgroup of the group of all automorphisms of G, i.e., $\mathrm{Int}(G) \lhd \mathrm{Aut}(G)$.

Proof 2.24 A simple computation shows that $\mathrm{Int}(g) \circ \mathrm{Int}(g') = \mathrm{Int}(gg')$, and it is immediate to prove that $\mathrm{Int}(G)$ is a subgroup of $\mathrm{Aut}(G)$.

To prove that $\mathrm{Int}(G) \lhd \mathrm{Aut}(G)$, let $\varphi\colon G \to G$ be an automorphism of G, then

$$\begin{aligned} \varphi \circ \mathrm{Int}(g) \circ \varphi^{-1}(h) &= \varphi(g\varphi^{-1}(h)g^{-1}) \\ &= \varphi(g)\varphi(\varphi^{-1}(h))\varphi(g^{-1}) = \varphi(g)h\varphi(g)^{-1} = \mathrm{Int}(\varphi(g))(h), \end{aligned}$$

that shows that $\mathrm{Int}(G)$ is a normal subgroup.

Definition 2.26 Let G be a group. The quotient group $\mathrm{Out}(G) = \mathrm{Aut}(G)/\mathrm{Int}(G)$ is called the group of outer automorphisms of the group G and its elements are called outer automorphisms.

Example 2.40 1. *It is clear that if a group A is Abelian then* $\mathrm{Int}(a) = \mathrm{id}_A$, *for all $a \in A$. Then,* $\mathrm{Int}(A) = \{\mathrm{id}_A\}$. *In such case* $\mathrm{Out}(A) = \mathrm{Aut}(A)$.

2. *Let us consider the Abelian group $\mathbb{Z}_n$. Example 2.28 and Exercise 2.29 show that if m and n are relatively prime then the homomorphism f_m ('multiplying by m') is an automorphism. Then, we conclude that if p is prime* $\mathrm{Aut}(\mathbb{Z}_p) = \{f_1, f_2, \ldots, f_{p-1}\}$ *and* $|\mathrm{Aut}(\mathbb{Z}_p)| = p-1$. *We will often denote the multiplicative group of $\mathbb{Z}_p$, that is, its invertible elements, as $\mathbb{Z}_p^\times$. Note that $|\mathbb{Z}_p^\times| = p-1$, but $|\mathbb{Z}_m^\times| = \phi(m)$ where $\phi(m)$ is Euler's function of m, that is, the number of $n \leq m$ such that n, m are relatively primes.*

It is interesting to observe that the group S_n, $n \neq 6$ is such that $\mathrm{Out}(S_n) = 1$ and $\mathrm{Aut}(S_n) = \mathrm{Int}(S_n)$, i.e., any automorphism is inner (in the case of $n = 6$, $\mathrm{Out}(S_6) = C_2$).

2.5 Products of groups

Before turning our attention to the study of groupoids properly, we will end this excursion on the theory of groups by providing some strokes on the different ways of combining groups to produce new ones. We will call product of groups a generic operation that creates a new group out of two. In the coming sections we will study two of such products: The direct product and the semidirect one. In doing that, we will realize that the language of exact sequences fits perfectly into this analysis. Actually, we will see that the fact that a group is a semidirect product of groups is equivalent to the splitting of an exact sequence.

2.5.1 Direct product of groups

The first and simplest product of group is the direct product that simply uses the Cartesian product operation among sets in order to construct a new group. Note that the disjoint union of groups can not be a group (it would have two neutral elements).

Definition 2.27 Let G_1, G_2 be two groups. The Cartesian product $G_1 \times G_2$ becomes a group with the composition law: $(g_1, g_2) \star (g_1', g_2') = (g_1 g_1', g_2 g_2')$, $g_1, g_1' \in G_1$ and $g_2, g_2' \in G_2$. The resulting group is called the direct product of the groups G_1, G_2 and it is commonly denoted by $G_1 \times G_2$. The order of $G_1 \times G_2$ is $|G_1 \times G_2| = |G_1||G_2|$.

Exercise 2.41 *Prove that* $(G_1 \times G_2, \star)$ *is a group.*

The product of groups allows us to construct a large number of groups out of the examples that we have already found. They are rather trivial though. Notice that we can consider G_1 and G_2 as subgroups of $G_1 \times G_2$ by using the natural inclusions: $i_1 \colon G_1 \to G_1 \times G_2$, $g_1 \mapsto i_1(g_1) = (g_1, e)$ and $i_2 \colon G_2 \to G_1 \times G_2$, $g_2 \mapsto i_2(g_2) = (e, g_2)$.

Example 2.42

1. *Consider one of the simplest non-trivial examples:* $\mathbb{Z}_2 \times \mathbb{Z}_2$. *This group has order 4 and all its elements have order 2.*
2. *Given* n, m *positive integers we can form the direct product* $\mathbb{Z}_n \times \mathbb{Z}_m$, *which is an Abelian group of order* nm.
3. *We may form various direct products with the examples we already know, like* $S_n \times S_m$, $A_n \times \mathbb{Z}_m$, $A_n \times \mathbb{Z}_2$.

Exercise 2.43

1. *Show that* $\mathbb{Z}_2 \times \mathbb{Z}_2$ *is isomorphic to the Klein group of four elements* V_4 *(recall Example 2.6.2).*
2. *Determine under what conditions the group* $\mathbb{Z}_n \times \mathbb{Z}_m$ *is isomorphic to* $\mathbb{Z}_{nm}$.

The definition of direct product can be extended trivially to a family of groups G_i, $i = 1, \dots, n$, then we define $G = \prod_{i=1}^n G_i$ as the set of n-tuples $(g_1, \dots, g_n)$, $g_i \in G_i$ and the component-wise composition law: $(g_1, \dots, g_n) \star (g_1', \dots, g_n') = (g_1 g_1', \dots, g_n g_n')$. It is trivial to check that the projection maps $\pi_i \colon \prod_{k=1}^n G_k \to G_i$, $\pi_i(g_1, \dots g_n) = g_i$ are group homomorphisms (as well as the canonical injective maps $i_k \colon G_k \to \prod_{k=1}^n G_k$, $i_k(g_k) = (e, \dots, e, g_k, e, \dots, e)$).

A direct check shows that the direct product has the properties listed in the following proposition.

Proposition 2.15
Let $G = G_1 \times G_2$ be the direct product of the groups G_1 and G_2, then both G_1, G_2 are commuting normal subgroups of G (we identify G_k with the corresponding images $i_k(G_k)$ inside the direct product), that is $(g_1,e) \star (e,g_2) = (e,g_2) \star (g_1,e)$ for all $g_1 \in G_1$, $g_2 \in G_2$, such that $G_1 \star G_2 = G$, and $G_1 \cap G_2 = \{(e,e)\}$. Moreover, $G/G_1 = G_2$ and $G/G_2 = G_1$.

The nice thing is that the properties described in the previous proposition actually characterize direct products, that is:

Proposition 2.16
Let G be a group and H, K two commuting normal subgroups such that $HK = G$ (the set HK is defined as $\{hk \mid h \in H, k \in K\}$), i.e., any $g \in G$ can be written as $g = hk$, $h \in H$, $k \in K$, and $H \cap K = \{e\}$, then G is isomorphic to the direct product $H \times K$.

Proof 2.25 We construct explicitly an isomorphism $\varphi \colon G \to H \times K$. It is defined as follows: $\varphi(g) = (h,k)$ where $g = hk$, $h \in H$, $k \in K$. First we must show that φ is well defined, that is if $g = hk = h'k'$, $h,h' \in H$, $k,k' \in K$, then $h = h'$, $k = k'$. This is obvious because $g = hk = h'k'$ implies that $h^{-1}h' = k(k')^{-1}$ (we use the fact that $hk = kh$ for any $h \in H$, $k \in K$), but because $H \cap K = \{e\}$, then $h^{-1}h' = e$, $k(k')^{-1} = e$ and the conclusion follows.

It is trivial to check that φ is a group homomorphism (using the fact that $HK = KH$ again). Clearly $\ker \varphi = e$, then φ is a monomorphism. Finally, because $G = HK$, φ is an epimorphism.

We may rewrite the previous results by using exact sequences. This will prove to be worthwhile when advancing in the study of the problem. First, we will introduce some terminology related to short exact sequences.

Definition 2.28 We will say that the short exact sequence of groups $1 \to H \to G \stackrel{\pi}{\to} K \to 1$ splits if there exists a section[7] σ of the epimorphism $\pi \colon G \to K$ which is a group homomorphism, i.e, there exists a group homomorphism $\sigma \colon K \to G$ such that $\pi \circ \sigma = \mathrm{id}_K$.

Then, we can rewrite Proposition 2.16 as:

Proposition 2.17
Let $H \lhd G$ be a normal subgroup and $1 \to H \to G \to K = G/H \to 1$ the canonical short exact sequence defined by H. Let $\pi \colon G \to K$ be the canonical projection map. Then G is isomorphic to the direct product of H and K if and only if there exists

[7]A section of a surjective map $\pi \colon X \to Y$ is a right-inverse of it, that is, a map $\sigma \colon Y \to X$ such that $\pi \circ \sigma = \mathrm{id}_Y$, sometimes a section is also called a "cross-section".

a section $\sigma: K \to G$ of π which is a group homomorphism and $\sigma(K)$ is a normal subgroup, i.e., if the short exact sequence $1 \to H \to G \to K \to 1$ splits and K is normal subgroup of G.

2.5.2 Classification of finite Abelian groups

Before commencing the discussion of groupoids, we will end up in the discussion on groups by providing a classification of Abelian groups.

In the case of Abelian groups, we will change the notation introduced before and we will denote the direct product of the Abelian groups A and B as $A \oplus B$ instead of $A \times B$. This could seem bizarre at first, but there are very good reasons to do that. The rationale behind this notation is that any Abelian group is a $\mathbb{Z}$-module, a notion that will become relevant in the second part of this work (see Sect. 10.4.1). A module is the natural extension of the notion of linear space, but where the role of the field of scalars, tipically the complex numbers $\mathbb{C}$, is replaced by a ring, in our case the integer numbers $\mathbb{Z}$. In fact, given an Abelian group A, it is naturally defined $na = a + \overset{n}{\cdots} + a$, $a \in A$, $n \in \mathbb{Z}$. Then, in the same way that we define the direct sum of linear spaces $V \oplus W$, we can define the direct sum of $\mathbb{Z}$-modules $A \oplus B$. Notice that an element in $A \oplus B$ is just a pair (a,b), $a \in A$, $b \in B$, that is our old definition of the direct product group $A \times B$. The emphasis of the notation $A \oplus B$ is that there is a natural operation by the ring of integers (the scalars now). Such structure is instrumental in proving the classification theorem of 'finitely generated' Abelian groups (that we will not address here). In what remains, we will stick to this notation, but recall that $|\mathbb{Z}_n \oplus \mathbb{Z}_m| = nm$ (and not $n + m!$)[8].

Finite Abelian groups are completely characterized[9] and the following theorem summarizes part of it.

Theorem 2.13
Let A be a finite Abelian group of order n, then A is isomorphic to $\mathbb{Z}_{p_1^{r_1}} \oplus \cdots \oplus \mathbb{Z}_{p_s^{r_s}}$ where $n = p_1^{r_1} \cdots p_s^{r_s}$ and p_i are the prime factors of n. The factors $\mathbb{Z}_{p_k^{r_k}}$ appearing in the factorization are unique.

Proof 2.26 We will only prove here the easy part of the theorem (see [41, Sect. 7.5] for a detailed proof of the theorem). Let us consider an Abelian group A of order nm with $(n,m) = 1$, i.e., n, m are relatively prime. Then we define $A(n) = \{a \in A \mid a^n = e\}$. Then $A(n) \cap A(m) = \{e\}$ (note that if $a \in A(n) \cap A(m)$, then $a^n = a^m = e$, but $(n,m) = 1$, and there are integers x, y such that $xn + ym = 1$, $a = a^{xn+ym} = (a^n)^x (a^m)^y = e$. Again, because of the previous formula, we get that for any $a \in A$, $a =$

[8]The previous notation is also misleading in a more subtle and insidious way. There are natural categorical definitions of products and the so called coproducts. The direct product is the product in the category of groups and the coproduct becomes the direct sum in the category of Abelian groups.

[9]As well as finitely generated Abelian groups. However, there is not yet a complete classification of general Abelian groups!

$(a^n)^x(a^m)^y$ for some integers x, y, but $(a^m)^n = (a^n)^m = a^{mn} = e$ (recall that the order of an element divides the order of the group), then $(a^m)^y \in A(n)$ and $(a^n)^x \in A(m)$ and $A = A(n)A(m) = A(m)A(n)$ (because the group is Abelian). Then, because of the characterization of direct product of groups, Prop. 2.17, we get that A is isomorphic to $A(n) \oplus A(m)$.

If $|A| = p_1^{m_1} \cdots p_s^{m_s}$ is the decomposition of the order of A in product of primes, the pairs $p_i^{m_i}$, $p_j^{m_j}$ are relatively prime ($i \neq j$), then $A \cong A(p_1^{m_1}) \oplus \cdots \oplus A(p_s^{m_s})$. It remains to prove that any one of the factors is itself a direct product of powers of the corresponding primes, that is $A(p_i^{m_i}) = \mathbb{Z}_{p_i^{m_{i_1}}} \oplus \cdots \oplus \mathbb{Z}_{p_i^{m_{i_l}}}$ and $m_{i_1} + \cdots + m_{i_l} = m_i$.

Example 2.44 *If $n = 4$, the Abelian group A is isomorphic to either $\mathbb{Z}_4$ or $\mathbb{Z}_2 \oplus \mathbb{Z}_2$ (notice that both groups are not isomorphic and they are the only two different groups of order 4).*

If $n = 8$, the Abelian group A can be isomorphic to $\mathbb{Z}_8$, $\mathbb{Z}_4 \oplus \mathbb{Z}_2$, or $\mathbb{Z}_2 \oplus \mathbb{Z}_2 \oplus \mathbb{Z}_2$. Note that not all groups of order 8 are Abelian; D_4, for instance, is not Abelian.

2.5.3 Semidirect product of groups

From the results characterizing direct products of groups, Props. 2.16 and 2.17, it is clear that a slightly weaker notion of product of groups can be obtained by lifting the requirement that both subgroups H and K of G must be normal. That is, we may consider that one of the groups is normal, for instance H and that the canonical short exact sequence $1 \to H \to G \to K = G/H \to 1$ splits.

Looking at the examples at the end of the previos section, Example 2.44, we realize, for instance that the group $\mathbb{Z}_4 \oplus \mathbb{Z}_2$ is Abelian, while D_4, that has the same order $|D_4| = 8$, has a normal subgroup $\mathbb{Z}_4$ and a subgroup $\mathbb{Z}_2$ (which is not normal, recall Examples 2.16.2 and 2.21). Then, we may ask what the reason is for the difference between $\mathbb{Z}_4 \oplus \mathbb{Z}_2$ and D_4.

In order to answer this question, we will start with an observation that would prove to be instrumental not only at this point but later on when trying to extend some of these ideas to groupoids. Let us assume that we have a group G and two subgroups H, K, such that one of them, say H, is normal, and such that $G = HK$. We are not assuming normality of the second subgroup. If we ask that the two subgroups intersect only at $\{e\}$, we will say that they cut transversaly, and we could expect that the group G is some sort of 'product' of H and K, that is, that we can reconstruct the composition law of G in terms of the ones in H and K. Actually, if H and K cut transversaly the decomposition $g = hk$ is unique (we may use the same argument as in the proof of Prop. 2.16 to prove it). Then we will get:

$$gg' = (hk)(h'k') = (h(kh'k^{-1}))(kk') = (h\,\mathrm{Int}\,(k)(h'))(kk'), \tag{2.8}$$

for any two elements $g, g' \in G$ and $g = hk$, $g' = h'k'$ their corresponding decompositions. From the right hand side of Eq. (2.8), we observe that the composition law of two elements of the group G can be recovered from the product of elements in K and H, but the later product is twisted by an inner automorphism of G, i.e., we have $h\,\text{Int}(k)(h')$ instead of hh' that would be the direct product case. Note that the automorphism $\omega_k\colon H \to H$, given by $\omega_k(h) = khk^{-1}$ is the restriction of an inner automorphism of G but is just an automorphism of H (because k is not in H).

In fact, the map $\omega\colon K \to \text{Aut}(H)$ defined by $\omega(k) = \omega_k = \text{Int}(k)$ is a group homomorphism: A simple computation shows that $\omega(kk') = \omega_k \circ \omega_{k'}$, for all $k,k' \in K$.

Then we prove:

Proposition 2.18
Let G be a group, H,K two subgroups, such that $G = HK$, $H \cap K = \{e\}$ and H is normal. Let $\omega\colon K \to \text{Aut}(H)$ be the group homomorphism defined by $\omega_k = \text{Int}(k)$ and consider the composition law defined on the Cartesian product $H \times K$:

$$(h,k) \star (h',k') = (h\,\omega_k(h'), kk'), \qquad h,h' \in H, \quad k,k' \in K.$$

Then $(H \times K, \star)$ is a group isomorphic to G.

Proof 2.27 It is a simple exercise to check that $(H \times K, \star)$ is a group: Its neutral element is (e,e) and $(h,k)^{-1} = (\omega_{k^{-1}}(h^{-1}), k^{-1})$.

Then we define $\varphi\colon H \times K \to G$, as $\varphi(h,k) = hk$. First we check that φ is a homomorphism: $\varphi((h,k) \star (h',k')) = \varphi(h\,\omega_k(h'), kk') = (h\,\omega_k(h'))(kk') = (hk)(h'k') = \varphi(h,k)\varphi(h',k')$. Clearly $\ker\varphi = (e,e)$ and φ is onto.

The previous proposition allows us to define the semidirect product of the groups H and K associated to a group homomorphism $\omega\colon K \to \text{Aut}(H)$ as:

Definition 2.29 Let H, K be two groups and a group homomorphism $\omega\colon K \to \text{Aut}(H)$. Then, we define the semidirect product of the groups H and K, and denote it by $K \ltimes_\omega H$ as the group structure defined on the product $H \times K$ by the formula:

$$(h,k) \star (h',k') = (h\omega_k(h'), kk'), \qquad k,k' \in K \quad h,h' \in H.$$

Notice that both H,K are natural subgroups of $K \ltimes_\omega H$ with respect to the natural maps $i\colon H \to K \ltimes_\omega H$, $h \mapsto (h,e)$ and $\sigma\colon K \to K \ltimes_\omega H$, $k \mapsto \sigma(k) = (e,k)$. The name 'semidirect product' is justified by the following proposition:

Proposition 2.19
Let $K \ltimes_\omega H$ be the semidirect product of H and K with respect to ω, then H is a normal subgroup of $K \ltimes_\omega H$, $HK = K \ltimes_\omega H$ and the short exact sequence $1 \to H \to K \ltimes_\omega H \xrightarrow{\pi} K \to 1$ splits.

Proof 2.28 First, notice that $K \ltimes_\omega H/H$ is canonically isomorphic to K. This is trivial because the cosets kH can be identified with $k \in K$.

Then observe that the canonical monomorphism $\sigma\colon K \to K \ltimes_\omega H$, $k \mapsto \sigma(k) = (e,k)$ is a section of the canonical projection $\pi\colon K \ltimes_\omega H \to K$.

Note that, contrary to the direct product, the subgroup K is not normal in general and the two subgroups H, K do not commute.

Proposition 2.20
Let G be a group and H a normal subgroup, then G is the semidirect produt of H and $K = G/H$ iff the short exact sequence $1 \to H \to G \to K = G/H \to 1$ splits.

Example 2.45 *Perhaps the simplest non-trivial example of a semidirect product is provided by the dihedral group D_n. Remember that D_n is generated by two elements r, τ satisfying $r^n = \tau^2 = e$ and $\tau r \tau^{-1} = r^{-1}$ (recall Exercises 2.10, 2.13). Moreover, the subgroup H generated by r is normal while the subgroup K generated by τ is not. The homomorphism $\omega \colon K \to \mathrm{Aut}(H)$, given by $\omega_\tau(r) = \tau r \tau^{-1} = r^{-1}$, is then the homomorphism, sending τ into the inversion of H (which is an automorphism because H is Abelian, actually isomorphic to $\mathbb{Z}_n$). Then, we conclude that the dihedral group D_n is isomorphic to the semidirect product $D_n \cong \mathbb{Z}_2 \ltimes \mathbb{Z}_n$.*

Example 2.46 *The holomorph $\mathrm{Hol}(G)$ of a group. Given a group G, its holomorph $\mathrm{Hol}(G)$ is a group defined as the semidirect product of the group G and its group of automorphisms $\mathrm{Aut}(G)$, with the tautological action of $\mathrm{Aut}(G)$ in G, that is $(\varphi, g) \mapsto \varphi(g)$, $\varphi \in \mathrm{Aut}(G)$, $g \in G$. Then $\mathrm{Hol}(G) = G \rtimes \mathrm{Aut}(G)$, and we have the following short exact sequence: $1 \to G \to \mathrm{Hol}(G) \to \mathrm{Aut}(G) \to 1$.*

In the particular instance of G being the Abelian group $\mathbb{Z}_p$ with p prime, we get $\mathrm{Hol}(\mathbb{Z}_p) = \mathbb{Z}_p \ltimes \mathbb{Z}_p^\times$ (recall Example 2.40.2).

2.6 Historical notes and additional comments

In this work we have tried to offer a 'human face'[10] approach to the, in principle, highly abstract theory of groups and groupoids and their representations. In spite of that, in the present chapter, we have deliberately chosen to introduce today's popular notion of group (and the basic associated properties) in the rather formal and abstract way preferred by W. Burnside[11] (who published the first English textbook on the subject: *Theory of Groups of Finite Order* [12]). This is not contradictory with the spirit of the work as plenty of insight on the abstract notions are provided by exhibiting many examples related with games and further discussion on the role of symmetries discussed in Chap. 4.

[10]Close to the 'mathematics with a human face' V. Arnold's spirit: 'Algebraists usually define groups as sets with operations that satisfy a long list of hard-to-remember axioms. I think one cannot understand such a definition. I believe the algebraists set up obstacles in the path of students to make it harder for uninitiated to penetrate their field. Perhaps their goal, if only subconscious, is to boost the reputation of their field' [1, p. 118].

[11]Burnside writes in the preface of his book: 'The present treatise is intended to introduce to the reader the main outlines of the theory of groups of finite order apart from any applications' [12, p. vi]. We recommend the historical interlude "William Burnside and intellectual harmony in mathematics" on the role played by Burnside in the modern development of group theory [26, Sect. 5.5, p. 100].

It is relevant at this point to mention that group theory has a long and convoluted history. We will refer the reader to the article by I. Kleiner [40] where the major sources in the evolution of group theory are discussed. In accordance with Kleiner we will just mention them with the names of their originators:

1. Classical Algebra (J.L. Lagrange, 1770).

2. Number Theory (K. Gauss, 1801).

3. Geometry (F. Klein, 1874).

4. Analysis (S. Lie, 1874; H. Poincaré and F. Klein, 1876).

A detailed discussion of the origin of the abstract notion of a group can also be found in the book by Wussing [72], a travel through the three fields, according to the author, which provided a complete background for the development of the concept of group, the first structure in Mathematics: Number theory, algebraic equations and geometry. Among all the great mathematicians of the XIX Century, Evariste Galois played a fundamental role in the field under study, providing the first application of the still undefined group concept to a long standing unsolved problem.

Chapter 3

Groupoids

Once the basic theory of groups has been discussed in the previous chapter, the notion of groupoid from a set theoretical viewpoint will be analyzed in this one, building upon the notions already developed for groups. Thus, subgroupoid will be discussed as well as the basic notions which are genuine of groupoids, like the fundamental isotropy subgroupoid, connectedness and the disjoint union and direct product operations of groupoids. All these will lead to the first structure theorem of groupoids that will allow us to describe a groupoid as a disjoint union of connected groupoids.

The example of the groupoid of pairs will be exhaustively discussed, emphasizing the relationship between subgroupoids and equivalence relations. The family of cyclic groupoids, a natural generalization of cyclic groups, will be introduced and discussed too.

A family of examples related to Loyd's puzzles will be discussed, with particular attention to the 2×2 Loyd's puzzle as a beautiful example of a finite groupoid with 48 elements. Rubik's pocket puzzle and its corresponding groupoid will be the final example discussed in this chapter.

3.1 Groupoids: Basic concepts

3.1.1 Groupoids and subgroupoids

We are now ready to address the notion of groupoid from a set theoretical point of view, similar to the one developed for groups in the previous chapter, in contrast with the initial abstract categorical definition given in Chapt. 1, Def. 1.4.

A group is a particular instance of a groupoid, one having only one object. Then, any morphism can be composed with any other morphism (since the sources and

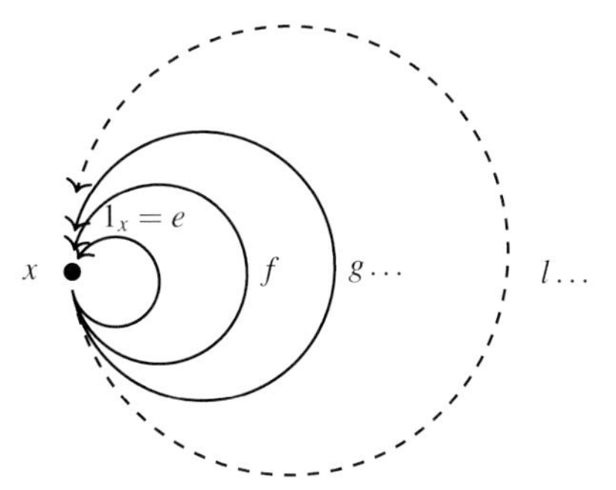

Figure 3.1: The (not very descriptive) diagram of a group.

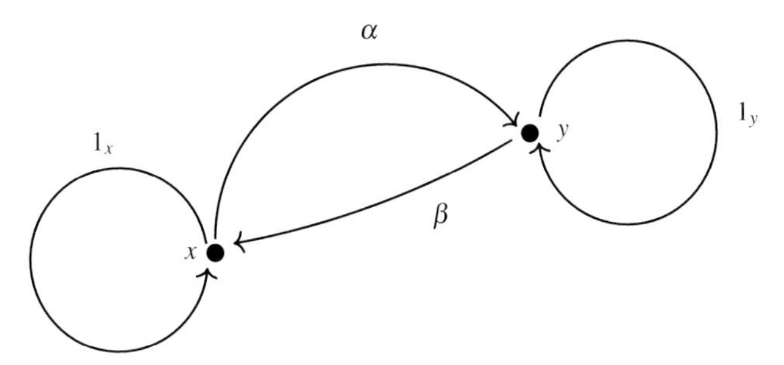

Figure 3.2: The diagram of a groupoid with two objects and four morphisms: The groupoid $\mathbf{A}_2$.

targets of all the morphisms are the same, the only object of the category) and there is only one identity element (see Fig.[1] 3.1).

When considering a groupoid which is not a group, a richer scenario comes into the field. We have more than one object in the category and a plethora of possibilities appears.

To get a flavor of these new possibilities consider a category with two objects, x, y and only four morphisms (see Fig. 3.2 for a diagram of this example). This is our extended singleton $\mathbf{A}_2$ groupoid (see Sect. 1.1.3):

$$1_x : x \to x, \quad 1_y : y \to y, \quad \alpha : x \to y, \quad \beta : y \to x \tag{3.1}$$

such that α, β are invertible and

$$\alpha \circ \beta = 1_y, \quad \beta \circ \alpha = 1_x \tag{3.2}$$

These four morphisms do not form a group. The composition is associative and α and β have inverses, but we have two units, 1_x, 1_y, and $\alpha \circ \alpha$ is not defined.

Thus we can summarize all properties characteristic of a groupoid again:

[1]As the attemptive reader would have noticed, diagram 3.1 is not very convenient to visualize the structure of a group as the relations between its elements are not displayed. For that, there is a much better representation, called Cayley diagrams, that have a deep relation with groupoids.

Definition 3.1 A groupoid consists of a set $\mathbf{G}$, a set Ω, and two maps $s, t\colon \mathbf{G} \to \Omega$, called the source and target map, respectively. The set Ω is called the space of objects of the groupoid whose elements will be denoted by x, y, etc. The elements in $\mathbf{G}$ will be called the morphisms or arrows of the groupoid and they will be denoted by using a convenient diagramatic notation $\alpha\colon x \to y$, where x is the source of the morphism α, that is $s(\alpha) = x$, and y its target, i.e., $t(\alpha) = y$. Two morphisms $\beta, \alpha \in \mathbf{G}$ will be said to be composable if $t(\alpha) = s(\beta)$. The set of pairs (β, α) of composable elements will be denoted by $\mathbf{G}_2$.

The axioms of a groupoid are completed by stating that there is a composition law defined only on pairs of composable morphisms, $\circ\colon \mathbf{G}_2 \to \mathbf{G}$, such that:

1. (Associativity) $(\gamma \circ \beta) \circ \alpha = \gamma \circ (\beta \circ \alpha)$, whenever (β, α) and (γ, β) are in $\mathbf{G}_2$.

2. (Units) For any object $x \in \Omega$, there exists a morphism $1_x\colon x \to x$, called the unit at x, such that if $\alpha\colon x \to y$, then $\alpha \circ 1_x = \alpha$ and $1_y \circ \alpha = \alpha$.

3. (Inverse) For any morphism $\alpha\colon x \to y$, there exists another morphism $\beta\colon y \to x$, such that $\beta \circ \alpha = 1_x$ and $\alpha \circ \beta = 1_y$. Such morphism will be denoted as α^{-1}.

A groupoid $\mathbf{G}$ over the space of objects Ω will be sometimes denoted as $\mathbf{G} \rightrightarrows \Omega$ in order to emphasize the source and the target maps s, t.

The set of morphisms from x to y will be denoted by[2] $\mathbf{G}(y,x)$ (while the standard notation in category theory would be $\mathrm{Hom}(x,y)$). It is clear that $\mathbf{G}(y,x) = s^{-1}(x) \cap t^{-1}(y)$. With these notations the composition law $\circ$ can be defined in a most conspicuous way as a family of maps $\circ\colon \mathbf{G}(z,y) \times \mathbf{G}(y,x) \to \mathbf{G}(z,x)$, for all $x, y, z \in \Omega$, satisfying the axioms (1), (2) and (3) in Def. 3.1.

The number of morphisms of $\mathbf{G}$ will be called the order of $\mathbf{G}$ and denoted by $|\mathbf{G}|$. Similarly the number of objects will be denoted by $|\Omega|$.

Groups are particular instances of groupoids whose space of objects consists of a single element and their morphisms are the elements of the group. The unit associated to it is the neutral element of the group.

Notice that units are unique, that is, if there exists a morphism $\delta\colon x \to x$ such that $\alpha \circ \delta = \alpha$ for all $\alpha\colon x \to y$, then $\delta = 1_x$ (because $1_x \circ \delta = 1_x$ by the defining property of δ and $1_x \circ \delta = \delta$ by the definition of 1_x). The family of units 1_x provides a canonical inclusion map $i\colon \Omega \to \mathbf{G}$, $i(x) = 1_x$, such that $s \circ i = t \circ i = \mathrm{id}_\Omega$.

In this sense note that the simplest example of a groupoid with more than one object is to consider two objects and the two identity morphisms (see Fig. 3.3). Similarly, any set Ω can be considered to be a groupoid:

Definition 3.2 Let Ω be a set. Consider the groupoid whose objects are the elements of Ω and its only morphisms the unit elements 1_x. Denoting the set of its morphisms by $\Omega = \{1_x \mid x \in \Omega\}$, its source and target maps $s, t\colon \Omega \to \Omega$ are just the map $s(1_x) = t(1_x) = x$, $x \in \Omega$. We will denote such groupoid by $\Omega \rightrightarrows \Omega$, by Ω if we

[2]Notice the backwards notation for the source and the target of morphisms $\alpha\colon x \to y$ in the sets $\mathbf{G}(y,x)$.

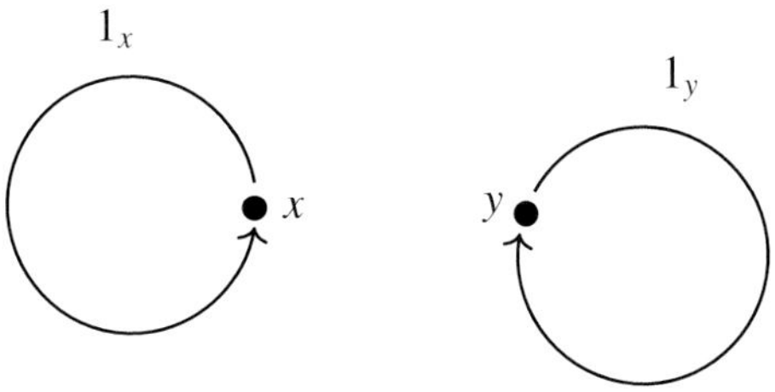

Figure 3.3: The diagram of a groupoid with two objects and two morphisms.

were to emphasize that we are considering the set Ω as a groupoid or by $\mathbf{1}_\Omega$, if we want to emhasize the role of the units 1_x (or just by Ω if there is no risk of confusion). We will call it the groupoid of objects of the set Ω.

It is a direct consequence of the definition that given an object x of the groupoid $\mathbf{G}$, the family of all morphisms with source and target x, $\alpha\colon x \to x$, form a group.

Definition 3.3 Let $\mathbf{G} \rightrightarrows \Omega$ be a groupoid and $x \in \Omega$. We call the isotropy group at x, and we denote it by G_x, all morphisms $\alpha\colon x \to x$, that is, $G_x = \mathbf{G}(x,x)$.

Exercise 3.1 *Prove that the isotropy groups G_x, $x \in \Omega$, of the groupoid $\mathbf{G} \rightrightarrows \Omega$ are actually groups.*

We will denote by $\mathbf{G}_+(x)$ the family of all morphisms whose source is x, that is:

$$\mathbf{G}_+(x) = \{\alpha \in \mathbf{G} \mid s(\alpha) = x\}.$$

We will sometimes call it the spray at x. Notice that $\mathbf{G}_+(x) = s^{-1}(x)$ and if $x \neq y$, then $\mathbf{G}_+(x) \cap \mathbf{G}_+(y) = \emptyset$. Similarly $\mathbf{G}_-(x)$ is the family of all morphisms whose target is x, that is $\mathbf{G}_-(x) = t^{-1}(x)$ and we will call it the sink at x. As in the case of sprays, sinks are also pairwise disjoint.

Moreover $\mathbf{G}(y,x) = \mathbf{G}_+(x) \cap \mathbf{G}_-(y)$, and $G_x = \mathbf{G}_+(x) \cap \mathbf{G}_-(x)$.

Clearly the groupoid is the disjoint union of the sprays $\mathbf{G}_+(x)$ (or the sinks $\mathbf{G}_-(x)$), that is:

$$\mathbf{G} = \bigcup_{x\in\Omega} \mathbf{G}_+(x) = \bigcup_{x\in\Omega} \mathbf{G}_-(x).$$

If $\mathbf{G}$ is finite, we conclude that both sprays and sinks are finite too, and:

$$|\mathbf{G}| = \sum_{x\in\Omega} |\mathbf{G}_+(x)| = \sum_{x\in\Omega} |\mathbf{G}_-(x)|. \tag{3.3}$$

Proposition 3.1

If $\mathbf{G}(y,x)$ is non-void, then the isotropy groups G_y and G_x are isomorphic and $|\mathbf{G}(y,x)| = |G_x| = |G_y|$.

Proof 3.1 It suffices to check that if $\alpha: x \to y \in \mathbf{G}(y,x)$, then the map $\phi_\alpha: G_x \to G_y$ given by $\phi_\alpha(\gamma_x) = \alpha \circ \gamma_x \circ \alpha^{-1}$ is a group isomorphism.

Example 3.2 *We will now review the old example of the groupoid of pairs $\mathbf{G}(\Omega)$ of a set Ω. The elements of the groupoid $\alpha: x \to y$ are pairs (y,x). Units 1_x are the pairs (x,x). Note that $\mathbf{G}(x,y)$ consists of just one element $\alpha = (y,x)$ and the isotropy groups G_x are trivial (only contain the unit 1_x). The set $\mathbf{G}_+(x)$ consists of all pairs of the form (y,x), $y \in \Omega$ and $\mathbf{G}_-(x)$ consists of all pairs of the form (x,y), $y \in \Omega$.*

As in the particular case of groups, any attempt to understand the structure of a groupoid will start by looking at its subgroupoids, recall Def. 2.2.

Definition 3.4 A subgroupoid $\mathbf{H}$ of the groupoid $\mathbf{G}$ is a subset of $\mathbf{G}$, that is a family of morphisms in $\mathbf{G}$, such that it is a groupoid with the induced source and target maps s,t and composition law $\circ$.

If $\mathbf{G}$ is a finite groupoid, a subgroupoid $\mathbf{H}$ must be finite too. The inclusion map $j: \mathbf{H} \to \mathbf{G}$ that describes $\mathbf{H}$ as a subset of $\mathbf{G}$, $j(\alpha) = \alpha$, $\alpha \in \mathbf{H} \subset \mathbf{G}$, is injective.

If $\mathbf{H}$ is a subgroupoid with respect to the restriction of the source and target maps, then $s(\mathbf{H}) = t(\mathbf{H}) = \Omega'$ will be its space of objects. Then, inclusion j induces also an injective map (denoted with the same symbol) between the objects Ω' of $\mathbf{H}$ and the objects Ω of $\mathbf{G}$, that is $j(x') = x'$, $x' \in \Omega' \subset \Omega$ because $j(1_{x'}) = 1_{j(x')} = 1_{x'}$. If we denote now by s' and t' the source and target maps of $\mathbf{H}$, respectively, $s', t': \mathbf{H} \to \Omega'$, then the compatibility conditions: $s \circ j = j \circ s'$, $t \circ j = j \circ t'$, are automatically satisfied.

As in the particular case of groups, Prop. 2.1, subgroupoids are easily characterized.

Proposition 3.2

Let $\mathbf{G}$ be a groupoid and $\mathbf{H} \subset \mathbf{G}$ a subset. Then $\mathbf{H}$ is a subgroupoid of $\mathbf{G}$ if and only for any $\alpha: x \to y$ and $\beta: z \to y$ morphisms in $\mathbf{H}$, then $\beta^{-1} \circ \alpha \in \mathbf{H}$.

In particular if $\mathbf{H} \subset \mathbf{G}$ is a subgroupoid then the isotropy groups $H_x = \{\alpha \in \mathbf{H} \mid s(\alpha) = t(\alpha) = x\}$ of $\mathbf{H}$ are subgroups of the isotropy groups G_x of $\mathbf{G}$ for all objects of $\mathbf{H}$, $x \in \Omega'$.

Proof 3.2 Clearly, if $\mathbf{H} \subset \mathbf{G}$ is a subgroupoid, then $\beta^{-1} \circ \alpha \in \mathbf{H}$ for any $\alpha: x \to y$ and $\beta: y \to z$ morphisms in $\mathbf{H}$.

Conversely, if given $\alpha: x \to y$ and $\beta: z \to y$ morphisms in $\mathbf{H}$, then $\beta^{-1} \circ \alpha \in \mathbf{H}$. By choosing $\beta = \alpha$, we conclude that $1_x \in \mathbf{H}$ for any $x \in \Omega'$, the object space of $\mathbf{H}$. Now if $\beta: y \to z \in \mathbf{H}$, then because $1_z \in \mathbf{H}$, then $\beta^{-1} \circ 1_z = \beta^{-1} \in \mathbf{H}$.

Then, finally, if $\alpha: x \to y$ and $\beta: y \to z$ are morphisms in $\mathbf{H}$, then $\beta^{-1}: z \to y$ and $(\beta^{-1})^{-1} \circ \alpha \in \mathbf{H}$.

Given a subgroupoid $\mathbf{H} \rightrightarrows \Omega'$ of $\mathbf{G} \rightrightarrows \Omega$ with space of objects Ω' smaller than Ω, we may always consider another subgroupoid $\widetilde{\mathbf{H}}$ of $\mathbf{G}$ whose space of objets is Ω, and that extends $\mathbf{H}$ naturally as a subgroupoid of $\mathbf{G}$. The groupoid $\widetilde{\mathbf{H}}$ is defined by simply adding the units 1_x, $x \notin \Omega'$, to $\mathbf{H}$.

Definition 3.5 The groupoid $\widetilde{\mathbf{H}}$ defined by adding the units 1_x, $x \notin \Omega'$, to $\mathbf{H}$ will be called the natural extension of the subgroupoid $\mathbf{H}$ to $\mathbf{G}$.

Hence, in what follows, when needed, a subgroupoid $\mathbf{H}$ of the finite groupoid $\mathbf{G}$ will be assumed to have the same space of objects Ω as $\mathbf{G}$.

Definition 3.6 Let $\mathbf{G} \rightrightarrows \Omega$ be a groupoid and $\mathbf{H} \subset \mathbf{G}$. We will say that the subgroupoid $\mathbf{H}$ is full if its space of objects is the same as that of $\mathbf{G}$, that is $s(\mathbf{H}) = \Omega$. The natural extension $\widetilde{\mathbf{H}}$ of any subgroupoid is full.

Consider the following examples:

Example 3.3 *1. Let $H \subset G$ a subgroup of the group G. Considered as groupoids with only one object, then H is a subgroupoid of G.*

2. *Consider a groupoid $\mathbf{G} \rightrightarrows \Omega$ and $x \in \Omega$, then the isotropy group $G_x \subset \mathbf{G}$ is a subgroupoid. The object space of G_x as a groupoid consists just of the object x. The extension $\widetilde{G_x}$ as a subgroupoid of $\mathbf{G}$ consists of all morphisms $\alpha \colon x \to x$ and all units 1_y, $y \neq x$.*

   Exercise 3.4 *Prove that the isotropy group G_x, $x \in \Omega$ is a subgroupoid of $\mathbf{G} \rightrightarrows \Omega$.*

3. *Given the groupoid of pairs $\mathbf{G}(\Omega)$ of the set Ω, and given any subset $\Omega' \subset \Omega$, the groupoid of pairs of Ω' is a subgroupoid of $\mathbf{G}(\Omega)$, $\mathbf{G}(\Omega') \subset \mathbf{G}(\Omega)$. The extension $\widetilde{\mathbf{G}(\Omega')}$ will consist of all pairs (y,x), $x,y \in \Omega'$ and all units (z,z), $z \notin \Omega'$. In other words, $\widetilde{\mathbf{G}(\Omega')}$ will be obtained by erasing from the complete graph defined by Ω all links from vertices in Ω' to vertices out of Ω' and all links in $\Omega \backslash \Omega'$.*

4. *Given a groupoid $\mathbf{G} \rightrightarrows \Omega$, the groupoid of objects Ω (recall Def. 3.2) of Ω can be considered as a subgroupoid of $\mathbf{G}$ by identifying it with the image of the canonical map $i \colon \Omega \to \mathbf{G}$, $i(x) = 1_x$. Notice that the extension $\widetilde{\Omega}$ coincides with itself. When considering the subgroupoid $\Omega \subset \mathbf{G}$, we will often denote it as $\mathbf{1}_\Omega$, emphasizing that it represents the set of units of $\mathbf{G}$.*

   Exercise 3.5 *Let $\mathbf{G} \rightrightarrows \Omega$ be a groupoid, show that the subset $\{1_x \mid x \in \Omega\}$ is a subgroupoid.*

5. *Given any set Ω, we can consider it as a subgroupoid of the groupoid of pairs $\mathbf{G}(\Omega)$, i.e., the groupoid of objects Ω is a subgroupoid of the groupoid of pairs $\mathbf{G}(\Omega)$.*

6. *Any subset $S \subset \Omega$, considered as a groupoid of objects, is a subgroupoid of the groupoid of objects Ω and consequently of $\mathbf{G}(\Omega)$.*

   The previous examples show that the theory of groupoids subsumes the theory of sets in a natural way.

Let $\mathbf{G} \rightrightarrows \Omega$ be a groupoid and $S \subset \Omega$ a subset. We denote by $\mathbf{G}_S$ the restriction of $\mathbf{G}$ to S, that is:

$$\mathbf{G}_S = \{\alpha \colon x \to y \in \mathbf{G} \mid x, y \in S\}. \tag{3.4}$$

Exercise 3.6 *Prove that $\mathbf{G}_S$ is a subgroupoid of $\mathbf{G}$ over the space of objects S and that its isotropy groups are the same as those of $\mathbf{G}$.*

Definition 3.7 Given a subset $S \colon \Omega$ and a groupoid $\mathbf{G} \rightrightarrows \Omega$, the subgroupoid $\mathbf{G}_S$ defined above, Eq. (3.4), will be called the restriction of the groupoid $\mathbf{G}$ to the subset S. Its isotropy groups will be denoted G_x, the same as those of $\mathbf{G}$.

Example 3.7 *Consider the groupoid $\mathbf{G}(\Omega)$ of pairs of Ω. Let $S \subset \Omega$ be any subset of Ω, then $\mathbf{G}(\Omega)_S = \mathbf{G}(S)$.*

It is an immediate consequence of the definition that the intersection of subgroupoids is a subgroupoid.

Proposition 3.3
Let $\mathbf{G} \rightrightarrows \Omega$ be a groupoid and $\mathbf{H}, \mathbf{H}' \subset \mathbf{G}$ two subgroupoids with object spaces Ω_H and Ω'_H, respectively, then $\mathbf{H} \cap \mathbf{H}'$ is a subgroupoid of $\mathbf{G}$ with space of objects $\Omega_H \cap \Omega'_H$.

As in the case of groups, the union of subgroupoids is not, in general, a subgroupoid, but contrary to the situation with groups, there is a natural operation that consists of considering the disjoint union of groupoids. Analyzing it will be the subject of the coming section.

3.1.2 Disjoint union of groupoids

As we were indicating before, a fundamental difference, in sharp contrast with the theory of groups, is that now we can form groupoids which are the disjoint union of groupoids.

Definition 3.8 Given two groupoids $\mathbf{G}_a$, with object spaces Ω_a, $a = 1, 2$, we define the coproduct (or disjoint union) of the groupoids $\mathbf{G}_1$ and $\mathbf{G}_2$ as the groupoid, denoted by $\mathbf{G}_1 \sqcup \mathbf{G}_2$, whose morphisms are the disjoint union of the morphisms in $\mathbf{G}_1$ and $\mathbf{G}_2$ and whose objects are the disjoint union of the objects Ω_1 and Ω_2, the composition law and source and target maps being the obvious ones. Then, the set of composable elements of $\mathbf{G}_1 \sqcup \mathbf{G}_2$ is the disjoint union of the set of composable elements of $\mathbf{G}_1$

and $\mathbf{G}_2$. It is clear that both $\mathbf{G}_1$ and $\mathbf{G}_2$ are subgroupoids of $\mathbf{G}_1 \sqcup \mathbf{G}_2$ with the obvious inclusions $j_a\colon \mathbf{G}_a \to \mathbf{G}_1 \sqcup \mathbf{G}_2$, $a = 1,2$.

Exercise 3.8 *Given the disjoint union of two groupoids $\mathbf{G}_1$ and $\mathbf{G}_2$, prove that both $\mathbf{G}_1$ and $\mathbf{G}_2$ are subgroupoids of $\mathbf{G}_1 \sqcup \mathbf{G}_2$.*

It is also obvious that:

$$|\mathbf{G}_1 \sqcup \mathbf{G}_2| = |\mathbf{G}_1| + |\mathbf{G}_2|. \tag{3.5}$$

Example 3.9
1. *Let G_1, G_2 be two groups with neutral elements e_1 and e_2 respectively. Then the disjoint union $G_1 \sqcup G_2$ is a groupoid over the space of objects $\Omega = \{e_1, e_2\}$.*
2. *Let Ω_1 and Ω_2 be two sets. We can consider them as groupoids of objects Ω_1 and Ω_2, recall Def. 3.2. Then, their direct union as groupoids is just the groupoid of objects of the disjoint union $\Omega_1 \sqcup \Omega_2$ of the sets Ω_1 and Ω_2.*
3. *Let $\mathbf{G}(\Omega)$ be the groupoid of pairs of Ω and G a group with unit e, then $\mathbf{G}(\Omega) \sqcup G$ is a groupoid with space of objects $\Omega \sqcup \{e\}$.*
4. *Let $\mathbf{G}(\Omega_1)$, $\mathbf{G}(\Omega_2)$, be the groupoids of pairs of Ω_1 and Ω_2 repectively. Then $\mathbf{G}(\Omega_1) \sqcup \mathbf{G}(\Omega_2)$ is a groupoid with object space $\Omega_1 \sqcup \Omega_2$. Note that $\mathbf{G}(\Omega_1) \sqcup \mathbf{G}(\Omega_2) \subset \mathbf{G}(\Omega_1 \sqcup \Omega_2)$ is a subgroupoid of the groupoid of pairs of $\Omega_1 \sqcup \Omega_2$ and is strictly contained on it (note that $|\mathbf{G}(\Omega_1) \sqcup \mathbf{G}(\Omega_2)| = |\Omega_1|^2 + |\Omega_2|^2$ while $|\mathbf{G}(\Omega_1 \sqcup \Omega_2)| = |\Omega_1 \sqcup \Omega_2|^2 = (|\Omega_1| + |\Omega_2|)^2 > |\Omega_1|^2 + |\Omega_2|^2$).*

The notion of disjoint union of groupoids will be complemented in the coming section, Sect. 3.1.4, with the notion of direct product of groupoids that extends the already familiar notion of direct product of groups. Before doing that, we will devote some attention to extracting some relevant consequences of the previous notions that will allow us to have a glimpse of the general structure of groupoids.

For this, we will now introduce a fundamental notion in the theory of groupoids that measures how the objects x of a given groupoid are related among themselves.

Definition 3.9 Given an object $x \in \Omega$, the orbit $\mathcal{O}_x$ of the groupoid $\mathbf{G}$ through x is the collection of objects corresponding to the targets of morphisms in $\mathbf{G}_+(x)$, that is, $y \in \mathcal{O}_x$ if there exists $\alpha\colon x \to y$ or, in other words, $\mathcal{O}_x = t(\mathbf{G}_+(x))$. The orbit $\mathcal{O}_x$ will also be denoted as $\mathbf{G}x$.

The orbits $\mathcal{O}_x$ define a partition of Ω in disjoint sets. We will denote by $\Omega/\mathbf{G}$ the space of orbits $\mathcal{O}_x$ of Ω.

Remark 3.1 The notion of orbit, introduced in the previous definition, is closely related to the orbit of the action of a group on a set as it will be discussed in Sect. 4.2. ▮

The following examples show two extreme situations.

Example 3.10 *1. Let $\mathbf{G}(\Omega)$ be the groupoid of pairs of Ω. The orbit $\mathcal{O}_x$ is the set Ω.*

2. Let Ω be the groupoid of objects of Ω. The orbit $\mathcal{O}_x$ is just the element x.

Exercise 3.11 *Show that the orbits of the disjoint union of two groupoids are the same as the orbits of the original groupoids.*

Definition 3.10 We will say that the groupoid $\mathbf{G}$ is connected (or transitive) if it has just one orbit, in other words, $\mathbf{G}$ is connected if for any x, y objects, there is a morphism $\alpha : x \to y$.

Corollary 3.1

The isotropy groups G_x, G_y corresponding to objects x, y in the same orbit are isomorphic, although not canonically isomorphic.

Proof 3.3 If x, y are in the same orbit, then there exists $\alpha : x \to y$, and because of Prop. 3.1, G_x and G_y are isomorphic (the isomorphism depends on α).

Example 3.12 *The groupoid of pairs $\mathbf{G}$ is connected while the disjoint union of two groupoids is never connected. For instance, the groupoid $\mathbf{G}(\Omega) \sqcup G$ has two orbits Ω and $\{e\}$. The isotropy group of elements in the orbit Ω is trivial while the isotropy group corresponding to the object e is G.*

In contraposition to conectedness, we will call totally disconnected those groupoids whose orbits are merely points.

Definition 3.11 We will say that a groupoid $\mathbf{G} \rightrightarrows \Omega$ is totally disconected if any connected component has only one element, that is $\mathcal{O}_x = \{x\}$. If a groupoid is totally disconnected, given any two different objects x, y, there are no arrows $\alpha : x \to y$.

The only arrows allowed in a totally disconnected groupoid are $\alpha : x \to x$, that is, a totally disconnected groupoid is the collection of elements in the isotropy groups G_x. For instance, if we label the objects in Ω as $\{x_1, x_2, \ldots, x_n\}$, and we consider a family of groups $G_1, \ldots, G_n$, then the totally disconnected groupoid

$$\mathbf{G} = G_1 \sqcup G_2 \sqcup \cdots \sqcup G_n = \bigsqcup_{i=1}^{n} G_i , \tag{3.6}$$

over Ω with isotropy groups $G_{x_k} = G_k$, is the disjoint union of the groups G_k (we will also say that $\mathbf{G}$ is a direct union of groups).

There is a natural projection[3] $\pi: \mathbf{G} \to \Omega$, $\pi(g_k) = x_k$, for any $g_k \in G_k$ corresponding to both the source and target map of the groupoid. The groups G_k are the inverse images of the object x_k with respect to the map π, for this reason, we will also call them the fibres of π. Then $\mathbf{G}$ is a groupoid with source and target maps $s = t = \pi$ and composition law the corresponding group multiplication on each fibre of π, that is, on each group G_k. The fibres of π are just the isotropy groups of the groupoid $\mathbf{G}$. The groupoid $\mathbf{G} = \sqcup_{i=1}^n G_i$ is totally disconnected as there are no arrows $g: x \to y$, $x \neq y$. Each connected component is just a element $\{x\}$. Notice that the isotropy groups of $\mathbf{G}$ do not have to be isomorphic as the groupoid is totally disconnected.

However, in general, a groupoid $\mathbf{G}$ has a richer structure than a direct union of groups, isomorphic or not, because there are additional relations among the isotropy groups provided by the arrows $\alpha: x \to y$. Later on, we will discuss a general structure theorem for groupoids (see Sect. 6.3 and the discussion there). Nevertherless, we must stress at this point that, given a groupoid $\mathbf{G}$, it always contains a totally disconnected groupoid $\mathbf{G}_0$ formed by the disjoint union of the isotropy groups G_x of $\mathbf{G}$, that is:

$$\mathbf{G}_0 = \bigsqcup_{x \in \Omega} G_x \,. \tag{3.7}$$

Clearly, $\mathbf{G}_0$ is totally disconnected. If $\mathbf{G}$ is connected, because Cor. 3.1, all the fibres G_x of $\mathbf{G}_0$ are isomorphic. However, and this is a relevant observation, even if the isotropy groups G_x are isomorphic, they are not canonically isomorphic, that is, there is not a preferred or special way of identifying them (see for instance, Sect. 4.2, where it is shown that the isotropy groups of the action groupoid associated to a group action corresponding to points in the same orbit, are conjugate but not canonically isomorphic).

Definition 3.12 The totally disconnected groupoid $\mathbf{G}_0$ associated to the groupoid $\mathbf{G}$ will be called its isotropy groupoid (or fundamental subgroupoid) and will play a fundamental role in the structure theorem and the theory of representations of groupoids.

Exercise 3.13 *Prove that the fundamental groupoid* $\mathbf{G}_0$ *is a subgroupoid.*

Example 3.14
1. *Consider the groupoid of pairs of* Ω. *Then* $\mathbf{G}(\Omega)_0 = \{1_x \mid x \in \Omega\} = \Omega$ *(recall Exercise 3.5 and Def. 3.2). In this sense, the set* Ω *itself, considered as the groupoid of objects* Ω, *is the fundamental subgroupoid of the groupoid of pairs* $\mathbf{G}(\Omega)$.

2. *Consider* G_x *as a subgroupoid of a groupoid* $\mathbf{G} \rightrightarrows \Omega$ *with only one object* $\{x\}$, *Example 3.3.2, then the natural extension* $\widetilde{G}_x$ *as a subgroupoid of* $\mathbf{G}$ *with space of objects* Ω *is totally disconnected.*

[3] Compare this with the direct product of groups, Sect. 2.5.1. In that case elements of $G_1 \times \cdots \times G_n$ are n-tuples of elements $(g_1, \ldots, g_n)$ with g_k in G_k. There are canonical projections $\pi_k: \prod_k G_k \to \Omega$, each π_k mapping the n-tuple $(g_1, \ldots, g_n)$ into the object of the kth factor g_k, that is, $\pi_k(g_1, \ldots, g_n) = x_k$; however there is no canonical projection $\pi: \prod_k G_k \to \Omega$.

3. *Let G be a a group and Ω a set. The fundamental subgroupoid of the disjoint union $\mathbf{G}(\Omega) \sqcup G$ is $\Omega \sqcup G$ with object space $\Omega \cup \{e\}$ (e the neutral element of G). The groupoid $\mathbf{G}(\Omega) \sqcup G$ has two orbits: Ω and $\{e\}$. Thus, it is neither connected nor totally disconnected.*

4. *Consider as in the Example 3 above a set Ω and a group G. Apart from the disjoint union, we can construct another groupoid: the Cartesian product $\Omega \times G$, with source and target maps the projection on the first factor, that is $s(x,g) = t(x,g) = x$, for any $(x,g) \in \Omega \times G$. The composition law is the group composition law on the group element, that is $(x,g) \circ (x,g') = (x,gg')$ (notice that only pairs (x,g), (x,g') with the same x-component, can be composed). This groupoid is totally disconnected.*

Exercise 3.15 *Prove that $(\mathbf{G} \sqcup \mathbf{G}')_0 = \mathbf{G}_0 \sqcup \mathbf{G}'_0$.*

Consider the restriction $\mathbf{G}_{\mathcal{O}}$ of the groupoid $\mathbf{G}$ to the orbit $\mathcal{O}$ (recall Def. 3.7). The source and target maps of the groupoid $\mathbf{G}_{\mathcal{O}}$ are just the restrictions of the source and target maps s,t of $\mathbf{G}$ to $\mathbf{G}_{\mathcal{O}}$.

Exercise 3.16 *Prove that $\mathbf{G}_{\mathcal{O}} \rightrightarrows \mathcal{O}$ is a connected subgroupoid of $\mathbf{G}$ over $\mathcal{O}$.*

Exercise 3.17 *Let $\mathcal{O}$ and $\mathcal{O}'$ be two different orbits of $\mathbf{G}$. Prove that $\mathbf{G}_{\mathcal{O}\cup\mathcal{O}'} = \mathbf{G}_{\mathcal{O}} \sqcup \mathbf{G}'_{\mathcal{O}}$, in other words, the restriction of the groupoid to the union of orbits is the direct union of the corresponding restriction to each orbit.*

Then we have:

Theorem 3.1

First structure theorem for groupoids. Given a groupoid $\mathbf{G} \rightrightarrows \Omega$ with orbits $\mathcal{O}$, then $\mathbf{G}$ is the direct union of the connected groupoids $\mathbf{G}_{\mathcal{O}}$:

$$\mathbf{G} = \bigsqcup_{\mathcal{O}\in\Omega/\mathbf{G}} \mathbf{G}_{\mathcal{O}} . \tag{3.8}$$

The previous decomposition will be called the connected components decomposition of the groupoid $\mathbf{G}$.

Proof 3.4 The proof is a direct consequence of the observation that if $\alpha : x \to y$, then $\alpha \in \mathbf{G}_{\mathcal{O}}$ with $x \in \mathcal{O}$, Exercises 3.16 and 3.17, and that $\Omega = \cup_{\mathcal{O}\in\Omega/\mathbf{G}} \mathcal{O}$.

Thus any groupoid $\mathbf{G}$ is a disjoint union of connected groupoids and its structure will be described by determining the structure of the corresponding connected subgroupoids.

Exercise 3.18 *Work out the decomposition on connected components of the groupoid $\sqcup_{i=1}^{n} \mathbf{G}(\Omega_i)$ and the extensions $\widetilde{\mathbf{G}(\Omega_i)}$ of the subgroupoids $\mathbf{G}(\Omega_i)$.*

3.1.3 *The groupoid of pairs revisited: Equivalence relations and subgroupoids*

We will review the groupoid of pairs of a set discussed in Examples 1.5 and 1.11 from the point of view of the ideas discussed in the previous sections. Thus, we consider a set Ω and the groupoid of pairs $\mathbf{G}(\Omega) \rightrightarrows \Omega$ whose objects are the elements x of Ω and the morphisms α are pairs $(y,x) \in \Omega \times \Omega$. The isotropy groups G_x consist only of the unit $1_x = (x,x)$. Since we can define a morphism between any pair of objects in the category, the groupoid is connected.

Consider now an equivalence relation $\mathcal{R}$ in Ω, that is, a subset of $\Omega \times \Omega$ satisfying the properties in 1.8. Then we get:

Proposition 3.4
Any equivalence relation $\mathcal{R}$ in Ω defines a subgroupoid $\mathbf{G}_{\mathcal{R}} = \{(y,x) \in \Omega \times \Omega \mid x\mathcal{R}y\}$ of the groupoid of pairs $\mathbf{G}(\Omega)$. Conversely, any subgroupoid $\mathbf{H} \subset \mathbf{G}(\Omega)$ defines an equivalence relation in Ω which is just the subset $\widetilde{\mathbf{H}}$ defined by the natural extension of $\mathbf{H}$.

Proof 3.5 Given an equivalence relation, the subset $\mathbf{G}_{\mathcal{R}} \subset \Omega \times \Omega$, defined by the equivalence relation, is a subgroupoid. We use Prop. 3.2: given $\alpha = (y,x) \colon x \to y$ and $\beta = (y,z) \colon z \to y$ in $\mathbf{G}_{\mathcal{R}}$, that is, $x\mathcal{R}y$ and $z\mathcal{R}y$, then $\beta^{-1} \circ \alpha = (z,y) \circ (y,x) = (z,x)$ but $x\mathcal{R}z$, hence $\beta^{-1} \circ \alpha \in \mathbf{G}_{\mathcal{R}}$.

Conversely, if $\mathbf{H} \subset \mathbf{G}(\Omega)$ is a subgroupoid, consider the equivalence relation in Ω defined by $\widetilde{\mathbf{H}}$, that is, $y \sim x$ if $(y,x) \in \widetilde{\mathbf{H}}$. Then (reflexive) $x \sim x$ because $1_x \in \widetilde{\mathbf{H}}$ for all $x \in \Omega$ (this is precisely the reason we have to use the extension $\widetilde{\mathbf{H}}$ and not just $\mathbf{H}$ whose space of objects could be strictly contained in Ω).

The relation is symmetric because if $\alpha = (y,x) \in \widetilde{\mathbf{H}}$, then $\alpha^{-1} = (x,y) \in \widetilde{\mathbf{H}}$ too and, finally, the relation is transitive because $x \sim y$ and $y \sim z$, implies that $(z,x) = (z,y) \circ (y,x) \in \widetilde{\mathbf{H}}$.

See Fig. 3.4 for the diagram of the example of an equivalence relation in the set $\{x,y,z,t,u\}$ with classes $\{x,y,z,t\}$ and $\{u\}$.

Given an equivalence relation $\mathcal{R}$ in Ω, each equivalence class is an orbit $\mathcal{O}$ of the groupoid $\mathbf{G}_{\mathcal{R}}$ in Ω, but, because each orbit is connected, the restriction of the groupoid $\mathbf{G}_{\mathcal{R}}$ to each equivalence class $\mathcal{O} \subset \Omega$ is the groupoid of pairs of $\mathcal{O}$. Hence, the first structure theorem for groupoids, Thm. 3.1, tells us that the subgroupoid $\mathbf{G}_{\mathcal{R}}$ is the direct union of the groupoids obtained by restricting it to the orbits, that is, the disjoint union of the groupoids of pairs of each equivalence class. Thus, we have proved the following particular form of the connected component decomposition of a subgroupoid of a groupoid of pairs:

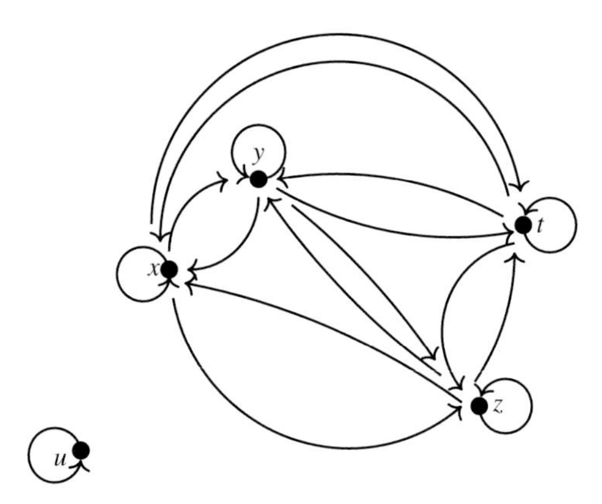

Figure 3.4: The graph of a groupoid based on an equivalence relation with two classes, $\{x,y,z,t\}$ and $\{u\}$.

Theorem 3.2
Let $\mathbf{G}(\Omega)$ *be the groupoid of pairs of the set* Ω*, then a subgroupoid* $\mathbf{H} \subset \mathbf{G}(\Omega)$ *is the disjoint union of groupoids of pairs* $\mathbf{G}(\mathcal{O})$ *where the disjoint sets* $\mathcal{O}$ *are the equivalence classes of an equivalence relation* $\sim$ *in* Ω*. Thus, we have:*

$$\widetilde{\mathbf{H}} = \bigsqcup_{\mathcal{O}\in\Omega/\sim} \mathbf{G}(\mathcal{O}) .$$

Then, each equivalence class in Ω defines the connected component of a subgroupoid. The coarsest example is obtained when $\mathcal{R} = \Omega \times \Omega$, which is the groupoid of pairs of Ω, while the finest one corresponds to $\mathcal{R} = \{(x,x) : x \in \Omega\}$. In this last case, the groupoid is completely disconnected and equivalent to the disjoint unions of trivial groups. If $|\Omega| = n$, the first case provides a finite groupoid, with n objects and n^2 morphisms, n of them units. Notice that the number of equivalence relations is finite, hence, the number of groupoids.

3.1.4 *Product of groupoids*

As in the case of groups, there is a canonical product of groupoids that uses the Cartesian product instead of the disjoint union.

Definition 3.13 Given two groupoids $\mathbf{G}_1 \rightrightarrows \Omega_1$ and $\mathbf{G}_2 \rightrightarrows \Omega_2$, we define their direct product (or just product) as the groupoid $\mathbf{G}$ whose morphisms are pairs of morphisms $(\alpha_1,\alpha_2)\colon (x_1,x_2) \to (y_1,y_2)$, $\alpha_1 \in \mathbf{G}_1$ and $\alpha_2 \in \mathbf{G}_2$, i.e., $\mathbf{G} = \mathbf{G}_1 \times \mathbf{G}_2$; whose space of objects Ω the space of pairs (x_1,x_2), $x_1 \in \Omega_1$, $x_2 \in \Omega_2$, i.e., $\Omega = \Omega_1 \times \Omega_2$; and whose composition law is given by:

$$(\alpha_1,\alpha_2)\circ(\alpha_1',\alpha_2') = (\alpha_1\circ\alpha_1',\alpha_2\circ\alpha_2'), \qquad \alpha_1,\alpha_1' \in \mathbf{G}_1\,,\alpha_2,\alpha_2' \in \mathbf{G}_2\,,$$

provided that $s(\alpha_1) = t(\alpha_1')$ and $s(\alpha_2) = t(\alpha_2')$. The source and target maps of the direct product are given by $s(a_1, \alpha_2) = (s(\alpha_1), s(\alpha_2))$ and $t(a_1, \alpha_2) = (t(\alpha_1), t(\alpha_2))$, respectively. The units are given by $(1_{x_1}, 1_{x_2})$ and the product will be denoted by $\mathbf{G}_1 \times \mathbf{G}_2 \rightrightarrows \Omega_1 \times \Omega_2$.

The definition can be extended without issue to a family of groupoids $\mathbf{G} \rightrightarrows \Omega_i$, $\mathbf{G} = \prod_i \mathbf{G}_i \rightrightarrows \prod_i \Omega_i$.

Example 3.19 *Consider the following two fundamental examples of products of groupoids: The product of groups and the product of the groupoid of pairs with a group.*

1. *If the groupoids $\mathbf{G}_i$ are groups, the direct product $\prod_i \mathbf{G}_i$ coincides with the direct product of groups.*

2. *Let $\mathbf{G}(\Omega)$ be the groupoid of pairs of Ω and G a group. The direct product $\mathbf{G}(\Omega) \times G$ is a groupoid whose space of objects is Ω (note that $\Omega \times \{e\} = \Omega$) and whose isotropy groups are G.*

Example 3.20 *Perhaps the most important example after the previous two is the groupoid obtained as the product of a set and a group:*

1. *Consider a set Ω and the totally disconnected groupoid of objects Ω. let G be a group, then the direct product of both groupoids $\Omega \times G$ is a totally disconnected groupoid whose space of objects is Ω, the composition law $(x, g) \circ (x, g') = (x, gg')$, $x \in \Omega$, $g, g' \in G$, and whose isotropy group is G (for all x). Note that this product of groupoids can also be seen, recall Eq. (3.6), as the direct union of copies of the group G along the set Ω, that is consider $G_x = G$ for all $x \in \Omega$, then:*

$$\Omega \times G = \sqcup_{x \in \Omega} G_x \,.$$

2. *Consider the product groupoid $\mathbf{G}(\Omega_1) \times \mathbf{G}(\Omega_2)$ with object space $\Omega_1 \times \Omega_2$. Its elements are pairs $(x_1, y_1; x_2, y_2)$, $x_1, y_1 \in \Omega_1$, and $x_2, y_2 \in \Omega_2$.*

Exercise 3.21 *Prove that the groupoids $\mathbf{G}(\Omega_1) \times \mathbf{G}(\Omega_2)$ and $\mathbf{G}(\Omega_1 \times \Omega_2)$ are isomorphic. In general $\prod_i \mathbf{G}(\Omega_i) = \mathbf{G}(\prod_i \Omega_i)$.*

Exercise 3.22 *Prove that given the direct product $\mathbf{G}_1 \times \mathbf{G}_2 \rightrightarrows \Omega_1 \times \Omega_2$ of groupoids, the subset $\widetilde{\mathbf{G}}_1 = \{(\alpha_1, 1_{x_2}) \mid \alpha_1 \in \mathbf{G}_1, x_2 \in \Omega_2\}$ is a (normal[4]) subgroupoid of $\mathbf{G}_1 \times \mathbf{G}_2$ (similarly with $\widetilde{\mathbf{G}}_2 = \{(1_{x_1}, \alpha_2) \mid \alpha_2 \in \mathbf{G}_2, x_1 \in \Omega_1\}$.*

Example 3.23 *Consider the direct product of Example 3.19.2, $\mathbf{G}(\Omega) \times G$, then using the notation of Exercise 3.22, it is obvious that $\widetilde{\mathbf{G}(\Omega)} = \mathbf{G}(\Omega)$ and $G \times \Omega$ is the fundamental groupoid of $\mathbf{G}(\Omega) \times G$. In particular, $\mathbf{G}(\Omega)$ is a connected (normal) subgroupoid of the product.*

[4]See Exercise 6.5

3.2 Puzzles and groupoids

So far we have found only relatively "few" groupoids which are not groups: Groupoids of pairs and disjoint unions of groups and groupoids. We will now discuss a family of different examples of groupoids that will show the richness and interest of the algebraic notions that we are unveiling. These groupoids are obtained considering new classes of games, related but different, to the games of permutations used to discuss the symmetric group. They will be the Loyd's and Rubik's groupoids.

3.2.1 The "15 puzzle"

A number of interesting groupoids arise in combinatorics. For the moment, and in order to keep extending our collection of examples, we will discuss another instance of board games. Certainly now the games we are going to describe are more sophisticated, and much more entertaining, we hope, than the connecting dots games described at the very beginning of this work (Sect. 1.1.1) or the "permutations" board game discussed at the beginning of this chapter (Sect. 2.2). We will use as a model the famous "15 puzzle" (see, for instance, [38]), that seems to have been erroneously adjudicated to Samuel Loyd [28], one of the greatest creators of puzzles.

The game consists of a 4×4 chessboard (see Fig. 3.5 (a)) with positions numbered from 1 to 16, and with 15 square pieces that can be moved at each step of the game to a nearest position, provided that is empty. Thus, a 'move' is reversible and it can be undone in the next step.

The objective of the puzzle is to move the pieces around so that we reach the final configuration depicted in Fig. 3.5 (b) starting, for instance, from the configuration in Fig. 3.5 (a).

Before attempting to solve it, we can 'groupoidify' the game. Actually, there is a beautiful groupoid describing the game. If all the pieces were identical (that is, if they were not numbered), the configurations of the puzzle would differ just because of the position of the empty square, so there would be 16 'states' labelled, respectively, $s_1,\ldots,s_{16}$ indicating the position of the empty square. Thus, the objects of the groupoid will be the 'states' of the chessboard $\Omega=\{s_1,s_2,\ldots,s_{15},s_{16}\}$. Each 'move' will define a morphism of the groupoid. Thus, from the starting state s_{16}, there are morphisms (moves) to the states s_{15} and s_{12}. We will denote each move as $m_{(k,j)}$, where k is the initial state and j is the final one (hence, if k is in the interior of the board, j can be any one of the four squares next to it, only three however if it is on the edges but not at the corners, in which case there will be just two). Then, the sequence of moves $m_{(16,12)},m_{(12,11)},m_{(11,15)},m_{(15,16)}$ takes the game into the configuration shown in Fig. 3.6.

There are also the trivial moves that do nothing to the board, and that will be denoted, respectively, by $1_1,\ldots,1_{16}$. The elements of the groupoid will be obtained by composing the previous moves, for instance, suppose that we perform the sequence of moves shown before: $m_{(16,12)}\circ m_{(12,11)}\circ m_{(11,15)}\circ m_{(15,16)}$ (the circle $\circ$ here is just the composition rule of the groupoid that consists of just performing one move after the other). The final state is s_{16} again, but the pieces have been moved around. Notice

1	2	3	4
5	6	7	8
9	10	11	12
13	14	15	

1	2	3	4
5	6	7	8
9	10	11	12
13	15	14	

Figure 3.5: (a) Starting position for the "15 puzzle" with the pieces numbered. (b) Can this position be reached?

1	2	3	4
5	6	7	8
9	10	15	11
13	14	12	

Figure 3.6: Final position of the pieces after the sequence of moves $m_{(16,12)}, m_{(12,11)}, m_{(11,15)}, m_{(15,16)}$ (the path followed by the empty square has been drawn on top of the board) leading to the 3-cycle $(11,12,15)$.

the explicity consistency condition on the composition of moves and that the source and target of the composition is the same state. In what follows, we will just denote the states s_i by the numerals i (unless risk of confusion).

To understand what has happened, we number the pieces from 1 to 15 to check how they have ended after the sequence of moves. Then, we see immediately (see Fig. 3.6) that the configuration of the pieces is different after performing the path that returns us to the original state s_{16}. The best way to take this into account is by introducing new morphisms that take care of this change in the 'inner' structure of the states. Thus, we may label the morphism corresponding to the path above by the corresponding permutation of the pieces in the permutation group S_{15} that, in this case, is just the 3-cycle $(11,12,15)$ (see Fig. 3.7). In other words, for each state, we remove the empty square and arrange the rows one after the other to get a board like that of the symmetric group Fig. 2.1 (a) with 15 indices. Thus the configuration shown in Fig. 3.6 is displayed in Fig. 3.7.

It is clear the isotropy group G_{s_k} of each state s_k will be obtained as the subgroup of the group of permutations generated by the permutations S_{15} obtained by wandering around the board and returning to the original state.

Then, we may define the 'Loyd's groupoid'[5] $\mathfrak{L}_4 = \{(j,\sigma,i) = \sigma_{ji} \mid i,j = 1,\ldots,16, \sigma \in S_{15}$ is the permutation corresponding to a sequence of allowed moves$\}$. The

[5]There is an obvious generalization of the theory to Loyd's groupoids $\mathfrak{L}_{(m,n)}$ over $m \times n$ boards.

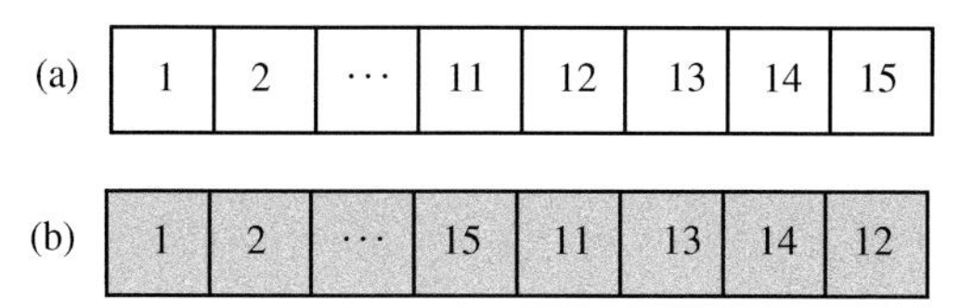

Figure 3.7: (a) A numbered board for the inner configurations of the game $\mathfrak{L}_4$. (b) Configuration reached after the sequence of moves $m_{(16,12)} \circ m_{(12,11)} \circ m_{(11,15)} \circ m_{(15,16)}$ corresponding to the 3-cycle $(11,12,15)$.

composition law of the groupoid is given as:

$$\sigma'_{kj} \circ \sigma_{ji} = (\sigma' \circ \sigma)_{ki} ,$$

for two composable morphisms. With this notation the units 1_k will be denoted also as e_{kk}.

We may, therefore, call the groupoid $\mathfrak{L}_4$, 'Loyd's groupoid of size 4×4'. It is easy to convince oneself that the groupoid $\mathfrak{L}_4$ is connected, however, it is not completely obvious that it has $|\mathfrak{L}_4| = 167382319104000$ elements (we will obtain an efficient way of computing the number of elements of a groupoid in Chap. 6).

After this elaboration, we can think back to the objective of the puzzle: Can we return to the original state after moving around the board but with the element in the isotropy group $G_{s_{16}}$ given by the transposition $(14,15)$ as shown in Fig. 3.5(b)?

A moment of reflection will convince us that closed paths around the board will always generate even permutations. Notice that each move determines transposition in the permutation group S_{15}. Thus, for instance, the first move in the previous sequence determines the transposition $(15,12)$. However, any closed path has an even number of steps (notice that any path can be 'deformed' to a union of rectangle paths), thus, the final permutation we get when we return to the original state is the composition of an even number of transpositions, therefore it must be even (recall Thm. 2.8). Then $G_{s_k} \subset A_{15}$.

Proposition 3.5
Loyd's "15 puzzle" has no solution.

Proof 3.6 The solution of the game consists of finding a closed path, such that the permutation obtained at the end will be the transposition $(14,15)$ which is odd. However because of the argument in the previous paragraph, this cannot happen, as any permutation obtained after a closed path is even.

Thus, not only the configuration proposed in the original game cannot be reached. Any odd permutation cannot be reached by following closed paths. Notice that the argument does not depend on the size of the game. The same will happen in a chessboard of size $n \times m$.

Exercise 3.24 *Can the "15 puzzle" be solved if we leave two empty squares?*

Exercise 3.25 *Prove that* $G_{s_k} = A_{15}$.

3.2.2 *The four squares puzzle: The groupoid* $\mathfrak{L}_2$

For the ease of the presentation, we will concentrate in what follows on the simplest non trivial Loyd's puzzle, that is, the square with four boxes, or, if you wish, on the groupoid $\mathfrak{L}_2$. We will number the positions of the empty square, that is, the states of the game, as the matrix $\begin{pmatrix} 1 & 2 \\ 3 & 4 \end{pmatrix}$, and mark the boxes with letters $\{a, b, c\}$ (or numbers 1,2,3 as before). Note that there are non allowed configurations, because of the given rules for the moves.

As explained before, a state is defined by the position of the empty box. There are four states s_1, s_2, s_3, s_4. Each state has different configurations, depending on the positions of the letters a, b, c. For instance, the state s_1 has three configurations, $\{s_1^1, s_1^2, s_2^3\}$ (see the first row in Fig. 3.8). The isotropy group is isomorphic to the alternating group in three elements, A_3, which, in this case, is Abelian and can be written as:

$$G_1 = \{e_{11}, (234)_{11}, (243)_{11}\}.$$

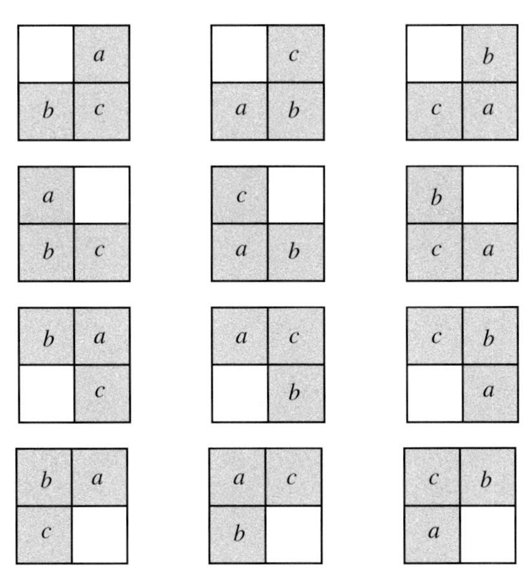

Figure 3.8: The four states in $\mathfrak{L}_2$ and their configurations. Each state has three configurations (the order of the alternating group $\mathcal{A}_3$). There are 12 configurations, (the order of the alternating group $\mathcal{A}_4$). Thus, for instance, the first row depicts the configurations corresponding to the state s_1.

Note that $(234) = (13)(34)(24)(12)$ and $(243) = (12)(24)(34)(13)$, are products of an even number of transpositions. Then, for instance, $(234)_{11} s_1^1 =$

s_1^2, $(243)_{11}s_1^1 = s_1^3$. The whole family of states and configurations is depicted in Fig. 3.8, and the isotropy groups are the following ($G_{s_i} \equiv G_i$):

$$\begin{aligned} G_1 &= \{e_{11}, (234)_{11}, (243)_{11}\}, \quad G_2 = \{e_{22}, (134)_{22}, (143)_{22}\}, \\ G_3 &= \{e_{33}, (124)_{33}, (142)_{33}\}, \quad G_4 = \{e_{44}, (123)_{44}, (132)_{44}\}. \end{aligned} \tag{3.9}$$

Notice that the order of the groupoid $\mathfrak{L}_2$ is 48.

The fundamental groupoid is obtained as the disjoint union of isotropy groups:

$$\begin{aligned} \mathbf{G}_0 =& \{e_{11}, (234)_{11}, (243)_{11}, e_{22}, (134)_{22}, (143)_{22}, \\ & e_{33}, (124)_{33}, (142)_{33}, e_{44}, (123)_{44}, (132)_{44}\}, \end{aligned}$$

and it is the maximal normal subgroupoid of $\mathfrak{L}_2$.

As we said above, apart from these morphisms, which leave the states invariant, we have other morphisms connecting the states (or, in other words, transforming a state into another). In fact, given the state s_1, the other states are reached by a sequence of moves from s_1 to the final state.

Written in terms of the symmetric group S_4, using the general convention above, the following transformations are morphisms between states (obtained by composing these morphisms with those in the isotropy groups)

$$\begin{aligned} (12)_{21}, (1342)_{21}, (1432)_{21} &: s_1 \to s_2, \\ (13)_{31}, (1243)_{31}, (1423)_{31} &: s_1 \to s_3, \\ (124)_{41}, (134)_{41}, (14)(23)_{41} &: s_1 \to s_4, \\ (132)_{32}, (243)_{32}, (14)(23)_{32} &: s_2 \to s_3, \\ (24)_{42}, (1342)_{42}, (1423)_{42} &: s_2 \to s_4, \\ (34)_{43}, (1243)_{43}, (1432)_{43} &: s_3 \to s_4, \end{aligned}$$

and their inverses:

$$\begin{aligned} (12)_{12}, (1234)_{12}, (1243)_{12} &: s_2 \to s_1, \\ (13)_{13}, (1342)_{13}, (1324)_{13} &: s_3 \to s_1, \\ (142)_{14}, (143)_{14}, (14)(23)_{14} &: s_4 \to s_1, \\ (123)_{23}, (234)_{23}, (14)(23)_{23} &: s_3 \to s_2, \\ (24)_{24}, (1243)_{24}, (1324)_{24} &: s_4 \to s_2, \\ (34)_{34}, (1234)_{34}, (1342)_{34} &: s_4 \to s_3, \end{aligned}$$

that correspond to the partition of the groupoid in the subsets $\mathbf{G}(j,i)$, for instance: $\mathbf{G}(1,2) = \{(12)_{12}, (1234)_{12}, (1243)_{12}\}$, $\mathbf{G}(2,1) = \{(12)_{21}, (1342)_{21}, (1432)_{21}\}$, $\mathbf{G}(3,1) = \{(13)_{13}, (1342)_{13}, (1324)_{13}\}$, etc., together with the isotropy groups $G_i = \mathbf{G}(i,i)$ listed above (3.9). Hence, the total number of morphisms, as computed before, is $12 + 36 = 48 = \frac{1}{2}(4!) \times 4$. In what follows, we will use the generic notation $\mathbf{G}$ for Loyd's groupoid $\mathfrak{L}_2$.

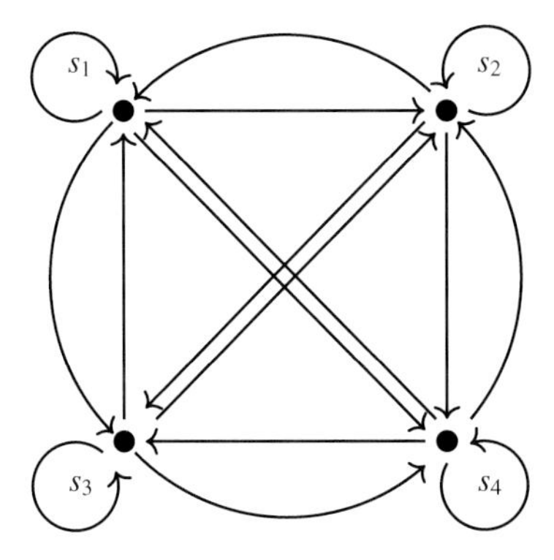

Figure 3.9: A graphical representation of the quotient groupoid $\mathfrak{L}_2/\mathbf{G}_0$ of the Loyd's groupoid $\mathfrak{L}_2$ over the normal subgroupoid $\mathbf{G}_0$ (i.e., the groupoid of pairs of 4 elements).

If we construct the quotient groupoid $\mathbf{G}/\mathbf{G}_0$, the classes are the sets of morphisms described above: $\mathbf{G}(i,j)$, $i,j = 1,\ldots,4$. They are in one-to-one correspondence with the set of moves (i,j). The canonical projection is $\pi : \mathbf{G} \to \mathbf{G}/\mathbf{G}_0$, and the quotient groupoid is isomorphic to the groupoid of pairs, $\mathbf{G}(\Omega)$: $(i,j) \to \mathbf{G}(i,j)$, $\pi(\alpha) = (s(\alpha),t(\alpha))$, which in this case corresponds to the complete graph K_4 (see Fig. 3.9 for a graphical representation).

3.2.3 Cyclic puzzles and cyclic groupoids

In the previous sections we have discussed some examples of groupoids under the name of Loyd's puzzles (certainly, jointly with the Rubik's cube which we will study in Section 3.2.4, simple gadgets that illustrates the notion of groupoids). We will discuss now a family of examples of groupoids that emerge from a family of puzzles closely related but simpler than Loyd's.

We will consider a game consisting of a board with n squares arranged periodically and a pointer signalling one of the squares of the board and such that we can move it from one square to the adjacent one (see Fig. 3.10 a). Because of the cyclic periodicity of the board, we can label the positions using elements $\mathbf{k}$ in $\mathbb{Z}_n$ and moving the pointer will consist in adding 1 to the actual position, that is passing from $\mathbf{k}$ to $\mathbf{k}+\mathbf{1}$.

In addition there is an auxiliary register R consisting of a finite number r of tags $a_1, a_2, \ldots, a_r$, that are transformed into themselves according with fixed rules each time that the pointer changes the position in the board, for instance each time we move in the board there is a permutation taking place in the register. The simplest situation is when the permutation σ is fixed and always the same. We will model the game as a groupoid with n objects (the positions in the board) and generated by a set of 'moves'. Each move will consist on moving the pointer from position $\mathbf{k}$ to $\mathbf{k}+\mathbf{1}$, $\mathbf{k} \in \mathbb{Z}_n$ and performing at the same time the permutation $\sigma \in S_r$ on the register R.

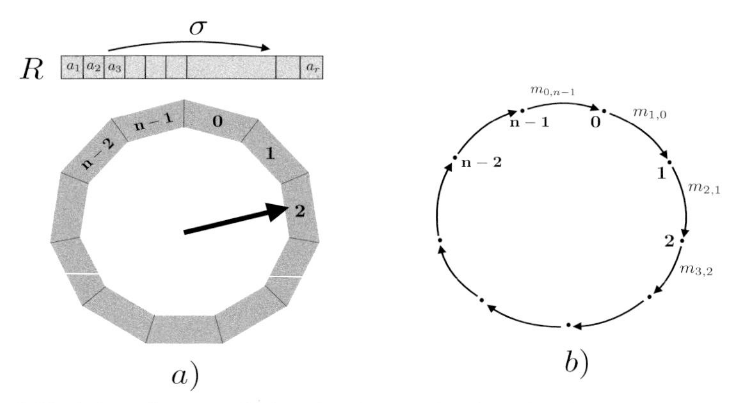

Figure 3.10: a) A cyclic game: the board, pointer and register. b) The cyclic quiver associated to the moves generating the cyclic groupoid.

The moves generating the groupoid will be denoted as $m_{k+1,k}$ and they will be composed from right to left, that is, moving the pointer from position 1 to 2 and later on, from 2 to 3, will be denoted by $m_{3,2}m_{2,1}$. Notice that the moves $m_{k+1,k}$, $k = 0,\ldots,n-1$ draw the circle quiver shown in Fig. 3.10 (b). The order of the permutation σ will be a finite number (a divisor of r) and will be denoted by $m = |\sigma|$, then if m is not a divisor of n, after we travel around the board we will not be back to the original configuration of the register R that will be reordered according to the permutation $\sigma^n \neq \mathrm{id}_R$.

Thus if we start with the pointer at position **0**, and we start travelling around the board, we will obtain the elements of the groupoid $m_{1,0}$, $m_{2,1}m_{1,0}$, etc. Notice that the source of all these elements is **0** while the target increases one unit at a time until we return to position **0**. We can keep traveling around the board and, eventually, after a finite number of turns, we will come back to the original configuration. If l is the least common multiple of n and m, then $l = sn = tm$ and, after s turns we will come back to the original configuration of the register R. We will have generated $ns = tm$ elements of the groupoid, all of them with source **0**.

We may proceed now by repeating the procedure starting at position **1**, and 2, and so on. Eventually we will construct a total number of $n^2 s$ elements in the groupoid (notice that the units and the inverses are already included!). The isotropy group will be the cyclic group C_s (that will be the subgroup generated by σ^n inside the cyclic group C_m generated by the permutation σ).

For instance if $n = 4$, the register R has three positions $1,2,3$, and $\sigma = (123)$, that is $m = 3$, we will generate a groupoid with 48 elements. We will denote this groupoid by $\mathfrak{C}_4(3)$.

Exercise 3.26 *Compare the previous game, i.e., the groupoid* $\mathfrak{C}_4(3)$ *with Loyd's groupoid* $\mathfrak{L}_2$.

Notice that if $n = 1$, i.e., there is only one position, we can play the game considering that we 'move' from position **0** to position **0** while the permutation σ takes place in the register. Thus after m moves we will be back to the original configuration and the groupoid generated by this game will be just the cyclic group of order m. Because of this it makes sense to define:

Definition 3.14 The groupoid generated by the cyclic quiver defined by the n moves $m_{1,0}, m_{2,1}, \ldots, m_{0,n-1}$, acting with a permutation σ of order m in a register R, will be called the cyclic groupoid with n positions and order of recurrence m.

This groupoid will be denoted by $\mathfrak{C}_n(m)$ and it has order $n^2 s = ntm$ where $l = sn = tm$ is the least common multiple of n and m. If n and m are relatively prime $l = nm$, the isotropy group would be the cyclic group of order m and the order of the groupoid is $|\mathfrak{C}_n(m)| = n^2 m$.

We will also say that the groupoid $\mathfrak{C}_n(m)$ has shift of order m over n positions.

Proposition 3.6
Let n, m be two positive integer numbers. Then the cyclic groupoid $\mathfrak{C}_n(m)$ is the direct product groupoid $\mathbf{G}(n) \times \mathbb{Z}_s$ (Example 3.19.2), with $\mathbf{G}(n)$ denoting the groupoid of pairs of a space with n elements (Example 1.11), and $l = sn = tm$ the least common multiple of n and m.

Proof 3.7 The proof is immediate by using the notation (k, σ, l) for elements in the groupoid, then the table of the groupoid is exactly the same as the table of the groupoid $\mathbf{G}(n) \times \mathbb{Z}_s$.

Example 3.27 *Some particular instances of cyclic groupoids are provided by $\mathfrak{C}_2(2) = \mathbf{G}(2)$ (since $s = 1$ as the least common multiple of 2 and 2 is 2). $\mathfrak{C}_2(3) = \mathbf{G}(2) \times \mathbb{Z}_3$, $\mathfrak{C}_2(4) = \mathbf{G}(2) \times \mathbb{Z}_2$, etc.*

It is interesting to note that $\mathfrak{C}_2(8) = \mathbf{G}(2) \times \mathbb{Z}_4$ and that the groupoid $\mathbf{G}(2) \times \mathbb{Z}_2 \times \mathbb{Z}_2$ is not a cyclic groupoid (it is obtained by using two commuting transpositions acting on the register R associated to each move).

3.2.4 Rubik's 'pocket cube'

Another family of amusing puzzles are provided by Rubik's cubes. We will discuss here the $2 \times 2 \times 2$ Rubik's cube (also called the "pocket cube", see Fig. 3.11) and leave the reader to explore the original one, i.e., the $3 \times 3 \times 3$ Rubik's cube and other variations. The states of the game will correspond to the coloring of the faces of the cube and there are $7! \times 3^7 = 11022480$.

We will assume that the cube is not oriented in space, thus if we label the vertex of the cube with numbers $1, 2, \ldots, 8$, there will be $8!/8 = 7!$ ways of rearranging the vertices because one can be always held fixed in space at will. Moreover each block of the cube (labelled by the corresponding vertex number) can have three positions (as can be rotated around the axis connecting it with its opposite block) so we must multiply the previous number by 3^8 except that, as in the previous considerations, we

Figure 3.11: Rubik's $2 \times 2 \times 2$ cube ('pocket cube'). Each face is colored with a different color.

can rotate the cube freely in space, thus we can select a particular orientation with respect to one vertex, and the total numbers of states (or colourings) of the cube will be $7! \times 3^7$ as indicated before.

Notations

We name each of the eight pieces with a capital letter, $\{\mathbf{A},\mathbf{B},\mathbf{C},\mathbf{D},\mathbf{E},\mathbf{F},\mathbf{G},\mathbf{H}\}$. In the fundamental configuration (see Fig. 3.12, left) they correspond to the vertices (we wrote $\bar{1} \equiv -1$):

$$
\begin{aligned}
\mathbf{A} &\to v_1 = (\bar{1}\bar{1}\bar{1}), \quad & \mathbf{E} &\to v_5 = (\bar{1}\bar{1}1) \\
\mathbf{B} &\to v_2 = (1\bar{1}\bar{1}), \quad & \mathbf{F} &\to v_6 = (1\bar{1}1) \\
\mathbf{C} &\to v_3 = (\bar{1}1\bar{1}), \quad & \mathbf{G} &\to v_7 = (\bar{1}11) \\
\mathbf{D} &\to v_4 = (11\bar{1}), \quad & \mathbf{H} &\to v_8 = (111)
\end{aligned}
$$

Each color is also named with a letter: $\{r,g,w,b,o,y\}$. Then, the eight pieces are (in the fundamental configuration)

$$
\begin{aligned}
&\mathbf{A}\,(-o,-b,-w) \equiv \mathbf{A}^1_{\bar{o}\bar{b}\bar{w}}, \qquad && \mathbf{E}\,(-o,-b,y) \equiv \mathbf{E}^5_{\bar{o}\bar{b}y} \\
&\mathbf{B}\,(r,-b,-w) \equiv \mathbf{B}^2_{r\bar{b}\bar{w}}, && \mathbf{F}\,(r,-b,y) \equiv \mathbf{F}^6_{r\bar{b}y} \\
&\mathbf{C}\,(-o,g,-w) \equiv \mathbf{C}^3_{\bar{o}g\bar{w}}, && \mathbf{G}\,(-o,g,y) \equiv \mathbf{G}^7_{\bar{o}gy} \\
&\mathbf{D}\,(r,g,-w) \equiv \mathbf{D}^4_{rg\bar{w}}, && \mathbf{H}\,(r,g,y) \equiv \mathbf{H}^8_{rgy}
\end{aligned}
$$

where the order of colors is the usual one (xyz) with the usual orientation. The vector $\bar{o} = -o$ is opposed to o.

States and configurations

A state of the cube is given by the relative position of the eight pieces (disregarding the color). We do not consider as different, states which can be related by a rotation in $\mathbb{R}^3$. Then, we can fix one of the pieces and permute the other seven. The number of states is 7!

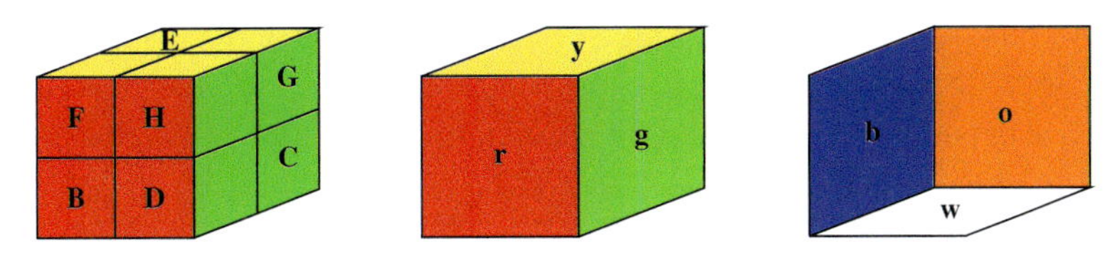

Figure 3.12: The elementary Rubik's cube in its fundamental configuration. The hidden faces are white (w), blue (b) and orange (o).

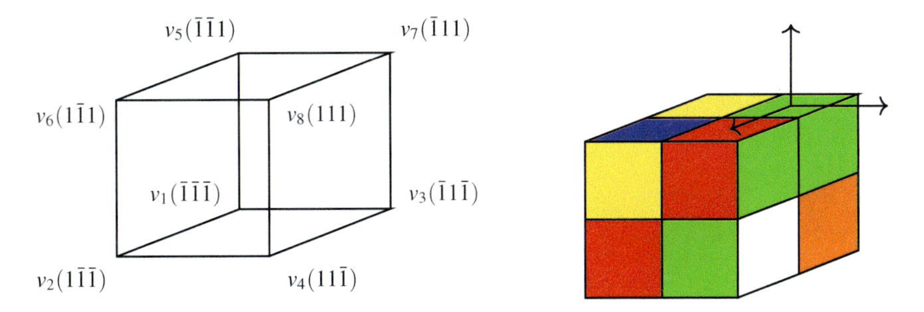

Figure 3.13: Vertex coordinates and the local reference systems for colors.

Now we consider the color of the faces of the pieces. Each piece has three different colors in its faces (which are given a priori). Then we could place each piece in three different ways.

Each configuration (that is, taking into account the color) can be represented by a permutation of the seven pieces and since each piece has three colors, we should also consider the orientation of the cube and get $7! \times 3^7$ possible configurations.

A state S is given by a permutation of the letters (we fix $\mathbf{A}$, but we will write it anyway) $\mathbf{A}, \mathbf{B}, \mathbf{C}, \mathbf{D}, \mathbf{E}, \mathbf{F}, \mathbf{G}, \mathbf{H}$. However, to easily recall which is the state, instead of permuting the letters, we add an index, taking care of the vertex v_i where the piece is. The state of the fundamental configuration is:

$$S_0 = |\mathbf{A}_1\mathbf{B}_2\mathbf{C}_3\mathbf{D}_4\mathbf{E}_5\mathbf{F}_6\mathbf{G}_7\mathbf{H}_8\rangle$$

A configuration in this state specifies the color position. For instance, the fundamental configuration C_0^0 is

$$C_0^0 = |\mathbf{A}^1_{\bar{o}\bar{b}\bar{w}}\mathbf{B}^2_{r\bar{b}\bar{w}}\mathbf{C}^3_{\bar{o}g\bar{w}}\mathbf{D}^4_{rg\bar{w}}\mathbf{E}^5_{\bar{o}\bar{b}y}\mathbf{F}^6_{r\bar{b}y}\mathbf{G}^7_{\bar{o}gy}\mathbf{H}^8_{rgy}\rangle$$

Moves

The moves are products of rotations with respect to the three orthogonal axes, fixing one slice (the four pieces on each face) and rotating the other one. The angles are $\pi/2$, π and $3\pi/2$, that is, the groups C_4^x, C_4^y and C_4^z, which, together with their products, form the symmetry group of the cube (excluding reflections), the octahedral group $\mathcal{O}$ (order 24). Note that a sequence of rotations could leave one of the pieces in its original site, but with a different color position (that is, the same state in a different configuration).

Objects and morphisms

An object in the structure we are constructing is a state, that is, a set formed by the corresponding configurations. A morphism is a sequence of moves, as described in the previous section, that is a map from a state to another one (or to the same state) carrying a configuration into another one (or leaving it invariant).

There are states with no morphism among them, that is, the action is not transitive. There is a clear distinction between the two spatial orientations, not related by proper rotations.

The morphisms fixing a state (but changing the configuration) form a group, which is not transitive.

The rotation matrices

The elementary moves are rotations with respect to the coordinate axes. This is the list of rotation matrices corresponding to these moves.

Identity:

$$\begin{pmatrix} 1 & 0 & 0 \\ 0 & 1 & 0 \\ 0 & 0 & 1 \end{pmatrix}$$

Secondary axes:

$$\begin{pmatrix} \bar{1} & 0 & 0 \\ 0 & 0 & 1 \\ 0 & 1 & 0 \end{pmatrix}, \quad \begin{pmatrix} \bar{1} & 0 & 0 \\ 0 & 0 & \bar{1} \\ 0 & \bar{1} & 0 \end{pmatrix}, \quad \begin{pmatrix} 0 & 0 & 1 \\ 0 & \bar{1} & 0 \\ 1 & 0 & 0 \end{pmatrix},$$

$$\begin{pmatrix} 0 & 0 & \bar{1} \\ 0 & \bar{1} & 0 \\ \bar{1} & 0 & 0 \end{pmatrix}, \quad \begin{pmatrix} 0 & 1 & 0 \\ 1 & 0 & 0 \\ 0 & 0 & \bar{1} \end{pmatrix}, \quad \begin{pmatrix} 0 & \bar{1} & 0 \\ \bar{1} & 0 & 0 \\ 0 & 0 & \bar{1} \end{pmatrix}.$$

Ternary axes:

$$\begin{pmatrix} 0 & 1 & 0 \\ 0 & 0 & 1 \\ 1 & 0 & 0 \end{pmatrix}, \quad \begin{pmatrix} 0 & 0 & 1 \\ 1 & 0 & 0 \\ 0 & 1 & 0 \end{pmatrix}, \quad \begin{pmatrix} 0 & \bar{1} & 0 \\ 0 & 0 & 1 \\ \bar{1} & 0 & 0 \end{pmatrix}, \quad \begin{pmatrix} 0 & 0 & \bar{1} \\ \bar{1} & 0 & 0 \\ 0 & 1 & 0 \end{pmatrix},$$

$$\begin{pmatrix} 0 & 1 & 0 \\ 0 & 0 & \bar{1} \\ \bar{1} & 0 & 0 \end{pmatrix}, \quad \begin{pmatrix} 0 & 0 & \bar{1} \\ 1 & 0 & 0 \\ 0 & \bar{1} & 0 \end{pmatrix}, \quad \begin{pmatrix} 0 & 0 & 1 \\ \bar{1} & 0 & 0 \\ 0 & \bar{1} & 0 \end{pmatrix}, \quad \begin{pmatrix} 0 & \bar{1} & 0 \\ 0 & 0 & \bar{1} \\ 1 & 0 & 0 \end{pmatrix}.$$

Cuaternary axes:

$$\begin{pmatrix} 1 & 0 & 0 \\ 0 & 0 & \bar{1} \\ 0 & 1 & 0 \end{pmatrix}, \quad \begin{pmatrix} 1 & 0 & 0 \\ 0 & \bar{1} & 0 \\ 0 & 0 & \bar{1} \end{pmatrix}, \quad \begin{pmatrix} 1 & 0 & 0 \\ 0 & 0 & 1 \\ 0 & \bar{1} & 0 \end{pmatrix},$$

$$\begin{pmatrix} 0 & 0 & 1 \\ 0 & 1 & 0 \\ \bar{1} & 0 & 0 \end{pmatrix}, \quad \begin{pmatrix} \bar{1} & 0 & 0 \\ 0 & 1 & 0 \\ 0 & 0 & \bar{1} \end{pmatrix}, \quad \begin{pmatrix} 0 & 0 & \bar{1} \\ 0 & 1 & 0 \\ 1 & 0 & 0 \end{pmatrix},$$

$$\begin{pmatrix} 0 & \bar{1} & 0 \\ 1 & 0 & 0 \\ 0 & 0 & 1 \end{pmatrix}, \quad \begin{pmatrix} \bar{1} & 0 & 0 \\ 0 & \bar{1} & 0 \\ 0 & 0 & 1 \end{pmatrix}, \quad \begin{pmatrix} 0 & 1 & 0 \\ \bar{1} & 0 & 0 \\ 0 & 0 & 1 \end{pmatrix}.$$

A move could be written as:

$$(R_1 \otimes \cdots \otimes R_8)(v_1 \otimes \cdots \otimes v_8) = R_1 v_1 \otimes \cdots \otimes R_8 v_8, \tag{3.10}$$

where four matrices R_i are equal to the identity and the other four equal to a matrix of the set above. The set $\mathfrak{V} = \{v_1, \ldots, v_8\}$ are the eight vertices (in cartesian coordinates). We need to know the initial state (that is, where the pieces $\mathfrak{P} = \{\mathbf{A}_{\alpha_1}^{v_{i_1}}, \ldots, \mathbf{H}_{\alpha_8}^{v_{i_8}}\}$ are placed (the vertex v_{i_j}) and their colors (α_j)). Note that the orientation is not changed under the rotations (then, if the orientation of one piece is different from that corresponding to the fundamental configuration there is no allowed transformation between both of them).

For instance, take as initial configuration the fundamental one and make a rotation of angle $\pi/2$ around the x axis:

$$R = \begin{pmatrix} 1 & 0 & 0 \\ 0 & 0 & \bar{1} \\ 0 & 1 & 0 \end{pmatrix}$$

Then,

$$R_1 = R_3 = R_5 = R_7 = I, \quad R_2 = R_4 = R_6 = R_8 = R$$

and

$$R_1 \otimes \cdots \otimes R_8 (v_1 \otimes \cdots \otimes v_8) = v_1 \otimes R v_2 \otimes v_3 \otimes R v_4 \otimes v_5 \otimes R v_6 \otimes v_7 \otimes R v_8$$

$$\begin{aligned}
v_1 &= (\bar{1}\bar{1}\bar{1}) \to (\bar{1}\bar{1}\bar{1}), & \mathbf{A}^1_{\bar{o}\bar{b}\bar{w}} &\to \mathbf{A}^1_{\bar{o}\bar{b}\bar{w}} \\
v_2 &= (1\bar{1}\bar{1}) \to (11\bar{1}), & \mathbf{B}^2_{r\bar{b}\bar{w}} &\to \mathbf{B}^4_{rw\bar{b}} \\
v_3 &= (\bar{1}1\bar{1}) \to (\bar{1}1\bar{1}), & \mathbf{C}^3_{\bar{o}g\bar{w}} &\to \mathbf{C}^3_{\bar{o}g\bar{w}} \\
v_4 &= (11\bar{1}) \to (1\bar{1}1), & \mathbf{D}^4_{rg\bar{w}} &\to \mathbf{D}^6_{rwg} \\
v_5 &= (\bar{1}\bar{1}1) \to (\bar{1}\bar{1}1), & \mathbf{E}^5_{\bar{o}\bar{b}y} &\to \mathbf{E}^5_{\bar{o}\bar{b}y} \\
v_6 &= (1\bar{1}1) \to (111), & \mathbf{F}^6_{r\bar{b}y} &\to \mathbf{F}^8_{r\bar{y}\bar{b}} \\
v_7 &= (\bar{1}11) \to (\bar{1}11), & \mathbf{G}^7_{\bar{o}gy} &\to \mathbf{G}^7_{\bar{o}gy} \\
v_8 &= (111) \to (1\bar{1}\bar{1}), & \mathbf{H}^8_{rgy} &\to \mathbf{H}^2_{r\bar{y}g}
\end{aligned}$$

Then, the piece, disregarding the color, transforms under a rotation. The colors also transform under the same rotation: The changes in the rotated pieces, **B**, **D**, **F**, **H**, are:

$$\begin{pmatrix} r \\ -b \\ -w \end{pmatrix} \to \begin{pmatrix} r \\ w \\ -b \end{pmatrix}, \quad \begin{pmatrix} r \\ g \\ -w \end{pmatrix} \to \begin{pmatrix} r \\ w \\ g \end{pmatrix},$$

$$\begin{pmatrix} r \\ -b \\ y \end{pmatrix} \to \begin{pmatrix} r \\ -y \\ -b \end{pmatrix}, \quad \begin{pmatrix} r \\ g \\ y \end{pmatrix} \to \begin{pmatrix} r \\ -y \\ g \end{pmatrix}.$$

Then we can write:

$$\mathcal{R}|\mathbf{A}_{\alpha_1}^{v_{i_1}} \cdots \mathbf{H}_{\alpha_8}^{v_{i_8}}\rangle = |\mathbf{A}_{R_1\alpha_1}^{R_1 v_{i_1}} \cdots \mathbf{H}_{R_8\alpha_8}^{R_8 v_{i_8}}\rangle$$

where $R_1, \ldots R_8$ are rotations, four of them equal to the identity and the other four equal to a three dimensional rotation R of the set described above.

Composing moves

Given two arbitrary states (two objects in the category language) we cannot assure that there exists a morphism (a sequence of moves) passing from one configuration of the first state to a configuration of the second one. In fact, given two configurations of the same state, we cannot assure the existence of a morphism passing from one to the other.

The inverse problem is easy to solve. Given a state and a move, we can pass from this state to another one and the correspondence (among configurations) is bijective (any morphism has an inverse). But, how can we construct a morphism (apart from the identity) from a state into the same state?

A move (a rotation, not equal to the identity) changes the state (rotations leave only an axis invariant). Then a morphism changing the configuration inside the same state has to be composed of a sequence of moves. In Fig. 3.14 we show a morphism, from a configuration of one state to another configuration of the same state, that is an element of the isotropy group. This morphism is composed of elementary moves.

An example of morphism

Consider a configuration of the cube as in Fig. 3.14, left. We will describe a morphism, T, transforming this configuration into the fundamental one. In fact, T is composed by a series of 32 moves, which we will not write here explicitly, just its consequences over the pieces of the cube.

The matrices providing the action over the eight pieces are the following. The pieces A, B, C, and D do not move:

$$T_{\mathbf{A}} = T_{\mathbf{B}} = T_{\mathbf{C}} = T_{\mathbf{D}} = I$$

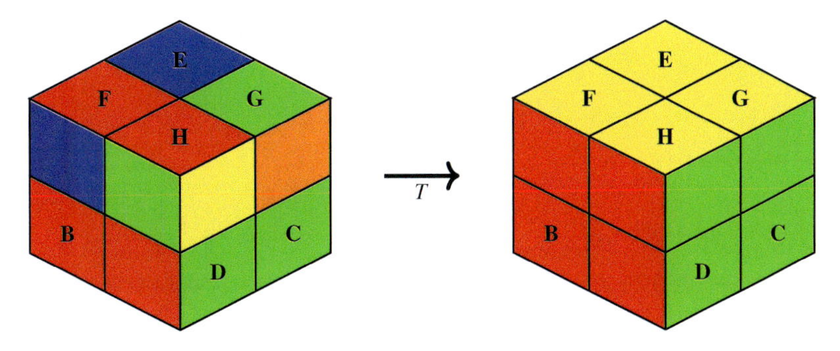

Figure 3.14: Changing the configuration in the same state.

The other four pieces do not change their position but change the orientation.

$$T_{\mathbf{E}} = \begin{pmatrix} 0 & 1 & 0 \\ 0 & 0 & -1 \\ -1 & 0 & 0 \end{pmatrix}, \quad T_{\mathbf{F}} = \begin{pmatrix} 0 & 0 & 1 \\ -1 & 0 & 0 \\ 0 & -1 & 0 \end{pmatrix},$$

$$T_{\mathbf{G}} = \begin{pmatrix} 0 & -1 & 0 \\ 0 & 0 & 1 \\ -1 & 0 & 0 \end{pmatrix}, \quad T_{\mathbf{H}} = \begin{pmatrix} 0 & 0 & 1 \\ 1 & 0 & 0 \\ 0 & 1 & 0 \end{pmatrix}.$$

Then

$$T_{\mathbf{E}}|\mathbf{E}^5_{\bar{y}\bar{o}b}\rangle = |\mathbf{E}^{T_{\mathbf{E}}v_5}_{T_{\mathbf{E}}(\bar{y}\bar{o}b)}\rangle = |\mathbf{E}^5_{\bar{o}\bar{b}y}\rangle$$

$$T_{\mathbf{F}}|\mathbf{F}^6_{b\bar{y}r}\rangle = |\mathbf{F}^{T_{\mathbf{F}}v_6}_{T_{\mathbf{F}}(b\bar{y}r)}\rangle = |\mathbf{F}^6_{r\bar{b}y}\rangle$$

$$T_{\mathbf{G}}|\mathbf{G}^7_{\bar{y}og}\rangle = |\mathbf{G}^{T_{\mathbf{G}}v_7}_{T_{\mathbf{G}}(\bar{y}og)}\rangle = |\mathbf{G}^7_{\bar{o}gy}\rangle$$

$$T_{\mathbf{H}}|\mathbf{H}^8_{gyr}\rangle = |\mathbf{H}^{T_{\mathbf{H}}v_8}_{T_{\mathbf{H}}(gyr)}\rangle = |\mathbf{H}^8_{rgy}\rangle.$$

Note that these transformations are rotations of angle $\pm 2\pi/3$ with axes through the vertices of the cube. Find the graphical description of the proposed solution in Appendix D.

Chapter 4

Actions of Groups and Groupoids

In this chapter we will continue the exploration of the main features of groupoids. For that we will start by recalling one of the main concepts behind the abstract notion of groups, that of symmetry, and we will work our way through it to understand in what sense groupoids constitute the main natural abstract structure embracing all the subtle issues revolving around the notion of symmetry and actions of groups.

First the notion of actions of groups will be introduced. Then we will extend it and the corresponding generalization of actions of groupoids will be analyzed.

Some significant examples will be presented such as the groupoid of permutations and Weinstein's tilings.

4.1 Symmetries, groups and groupoids

4.1.1 Groups and symmetries

Historically the mathematical notion of group has been closely related to the notion of symmetry. In fact in many instances 'group' is used for 'group of symmetries'. It also has been related to harmony, beauty, etc. Quoting H. Weyl [70]:

Symmetry, as wide or as narrow as you may define its meaning, is one idea by which man through the ages has tried to comprehend and create order, beauty, and perfection.

Unfortunately, the notion of group, even if widely identified with the notion of symmetry as it will be discussed below, is too coarse to capture some obvious 'symmetry' structures present on many problems under study (see later on the section devoted to Weinstein's tilings in this very Chapter). We feel that such situations deserve a proper treatment and it is convenient to discuss them in the context of this book.

Let us begin by discussing the problem of describing the symmetry of an structure from the perspective of this book.

When we think about the 'group of symmetry' of a given structure, very often we are considering a subgroup of a given group of transformations[1] of the space where this structure is defined and that preserves it.

For instance, we may think on the Euclidean plane E^2 as a carrier space and its group of isometries, that is, the group formed by all 'rigid' motions in E^2, in other words: translations, rotations and reflections. The 'structure' mentioned before is given by a figure on E^2 defined by a certain subset $S \subset E^2$, and the group of symmetry of the figure S will consist of all those isometries φ that map S onto itself, that is such that $\varphi(S) \subset S$.

In the particular instance that our figure is an equilateral triangle T, the group of symmetries of T will consists of $2\pi/3$ rotations around the center of the triangle and reflexions around the altitudes of the triangle. This group, that we have already found, is the dihedral group D_3. In general, the group of symmetries of a regular polygon with n edges is the dihedral group D_n (see Example 2.2.3).

As we consider other figures it could happen that their symmetry groups coincide, for instance the symmetry group of a regular star (equal size arms) is the same as the symmetry group of a square, the dihedral group D_4 (see Fig. 4.1). However if we deform the triangle or the square, the group of symmetries of the corresponding figures reduces drastically. For instance if we squeeze the triangle along one altitude to get an isosceles triangle, its group of symmetries reduces to the group $\mathbb{Z}_2$ consisting just on the identity and reflexion along that altitude. Furthermore, if we tilt the triangle, its group of symmetries consists of just the identity.

Notice that in the previous examples we have restricted ourselves to consider isometries, that is, transformations already preserving another structure: the Euclidean scalar product (or the Euclidean distance, if you prefer). In this sense, the group of isometries of E^2, also called the Euclidean group in two dimensions $E(2)$, is the symmetry group of the Euclidean plane and the group D_4 is the subgroup of the Euclidean group describing the symmetry of a square.

The structures that will be used to define symmetry groups can be a combination of different classes: Geometrical structures like metrics or any given tensor (a skewsymmetric bilinear form for instance); subspaces of a given set like families of points, shapes, submanifolds, etc.; or other properties of the carrier space of the theory, a given equivalence relation, a partial order, a causality relation, etc. We will have the opportunity to discuss some examples of each situation in this chapter.

Thus, in all situations considered, the main ingredients of the theory will consist of a group of transformations of a carrier space Ω and a given structure $\mathcal{S}$ on Ω that will determine the subgroup G, leaving it invariant. In what follows, we will assume

[1] In this context 'transformations' always refer to a class of bijective maps defined on the space supporting the structure under study, called the 'carrier' space, let it be an abstract structure or a particular model.

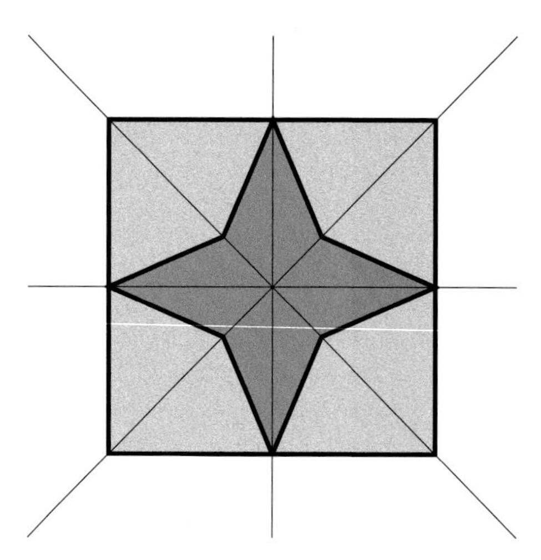

Figure 4.1: A square and a regular four-pointed star sharing the same symmetry group D_4.

that our structure is given by a tensorial object, a metric or a tensor for instance, or a specific subset in Ω (we will be more specific about that later).

One of the tasks that will be undertaken here is to 'groupoidify' this setting. For that purpose, we will follow the inspiration and insight offered by A. Weinstein's paper whose title, *Groupoids: Unifying internal and external symmetry* [66], already gives a hint on how to proceed.

The basic idea that, will be developed will consist of the reduction of a group symmetry to a groupoid. That is, we may consider, for instance, a tiling possessing a certain symmetry group (see for instance Fig. 4.2 displaying a beautiful tiling with a nice symmetry group with hexagonal symmetry[2]).

Exercise 4.1 *Describe the symmetry group of the Alhambra tiling in Fig. 4.2.*

It is clear that, by cutting the tiling (as shown in the picture against the ideal 'infinite' one) or introducing modifications (for instance replacing some of the damaged tiles by ones of different colors), the original symmetry group will be lost (or severely reduced), however, a clearly visible pattern will remain that is best described by a groupoid.

We will proceed as follows. After succinctly reviewing the notion of action of groups, we proceed to construct the action groupoid of the given group action and, following this inspiration, we will identify groupoids as the appropriate notion of symmetry for structures that lack a suitable group theoretical description. Other patterns that do not have an obvious group theoretical description, as the non-periodic Penrose tilings (see, for instance, the example shown in Fig. 4.3 of a non-periodic

[2]Gruban - https://www.flickr.com/photos/gruban/11341048/. Created: 1 January 2007. Tessellation in the Alhambra. This was one of the tilings sketched by M. C. Escher in 1936. License CC BY-SA 2.0.

Figure 4.2: A beautiful tiling found at the Alhambra in Spain.

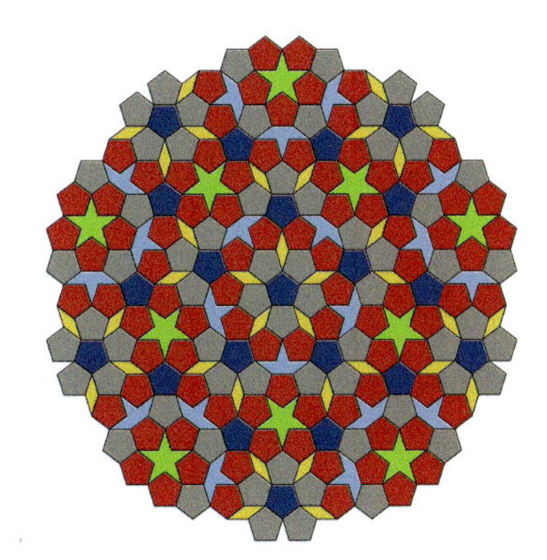

Figure 4.3: The non-periodic Penrose Tiling P1 with pentagonal symmetry.

tiling[3]) could be also considered. Along the way, we will discuss some interesting questions about 'triviality' and the structure of groupoids that will be laying the road to more elaborate discussions to follows. At the same time, the present discussions and examples will introduce us to a fundamental categorical notion to be discussed in full in the coming chapter, that of 'functor' among categories.

4.1.2 Actions of groups

In the previous section, we argued that the symmetry of a given structure is obtained as a subgroup of a group of transformations of the carrier space where the structure whose symmetry we want to describe is defined. We are now ready to formalize the notion of 'group of transformations' under the name of the action of a group on a set.

[3] `https://commons.wikimedia.org/wiki/File:PenroseTiling(P1).svg`

Definition 4.1 Given a group G and a set Ω, we define an action of G on Ω as a map:

$$\Psi\colon G\times\Omega\to\Omega\,,$$

such that the maps $\Psi_g\colon \Omega\to\Omega$, $\Psi_g(x)=\Psi(g,x)$ are bijective and the following properties hold:

$$\text{(a)}\,\Psi_e=\mathrm{id}_\Omega\,,\qquad \text{(b)}\,\Psi_{gg'}=\Psi_g\circ\Psi_{g'},\quad \forall g,g'\in G\,.$$

We will usually write $\Psi(g,x)=\Psi_g(x)\equiv gx$. Then, properties (a) and (b) above can just be written in a more condensed and appealing way, such as:

$$ex=x\,,\qquad (gg')x=g(g'x)\,,\quad \forall x\in\Omega,\quad g,g'\in G\,. \tag{4.1}$$

We could understand the maps Ψ_g associated to each group element g as the way that such an element 'acts' on the set, that is, how the element of the group transforms the points $x\in\Omega$. This interpretation leads to the name of 'group of transformations' to the family of maps Ψ_g (they reproduce the algebraic structure of the group G). The action of G is essentially a family of 'permutations' of the elements of Ω that reproduces the structure of the group, making it a generalization of Cayley's Theorem, Thm. 2.7. Note that it is not asked that the group G is isomorphic to the set of maps $\{\Psi_g\mid g\in G\}$. For instance, it could happen that $\Psi_g=\mathrm{id}_\Omega$ for some $g\neq e$. We will discuss these aspects in what follows.

If the set Ω possesses any additional structure of interest, we will often impose Ψ_g to preserve such structure. For instance, if Ω is a linear space, we may ask the maps Ψ_g to be linear.

We can choose Ω to be the group G itself. Then, there are a number of canonical actions of G on itself. One of them is simply the operation on the group on the left: $\Psi_g(g')=gg'$.

Exercise 4.2 *Prove that the family of maps $\Psi_g(g')=gg'$ define an action of G on itself. Show that the family of maps $\Psi'_g(g')=g'g^{-1}$ define an action of the group G on itself.*

Definition 4.2 Let G be a group, the multiplication map of the group, $\Psi\colon G\times G\to G$, $\Psi(g,g')=gg'$, $g,g'\in G$, defines an action of G on itself called the left action and denoted as L_g, that is, $L_g(g')=gg'$. Similarly, the right action of G on itself will be defined as $R_g(g')=g'g^{-1}$ for all $g,g'\in G$.

Note that the maps L_g and R_g are not group automorphisms, that is, $L_g(g'g'')\neq L_g(g')L_g(g'')$. However, the conjugation maps, Def. 2.24, define an action of the group G on itself by automorphisms, in fact:

$$\Psi_g(g')=\mathrm{Int}\,(g)(g')=gg'g^{-1}=L_gR_g(g')\,.$$

We will now introduce some background terminology for actions that will be tested against the examples provided below (see Examples 4.10).

Definition 4.3 An action is faithful if, for $g \neq g'$, there exists $x \in \Omega$ such that $gx \neq g'x$. An action is free if, for $g \neq g'$, $gx \neq g'x$ for all $x \in \Omega$. An action is transitive if given $x, x' \in \Omega$ there exists $g \in G$ such that $x' = gx$.

Definition 4.4 Given a group G and an action of G over a set Ω, the orbit through $x \in \Omega$ of the group G is the subset of Ω: $Gx = \{gx \in \Omega \colon g \in G\}$.

Exercise 4.3 *Prove that if* $Gx \cap Gx' \neq \emptyset$*, then* $Gx = Gx'$.

Exercise 4.4 *Let the group* G *act on the set* Ω*. Prove that the relation* $x \sim y$ *iff* $y = gx$ *for some* $g \in G$ *is an equivalence relation.*

The set Ω is the union of orbits which can have different sizes and structures. Different orbits do not intersect and, if $x, x' \in \Omega$ belong to different orbits, they are not connected by the group action, that is, there is no $g \in G$ such that $x' = gx$. As we will see in the coming section, there is a close relation between the notion of orbit of a point under the action of a group and the notion of orbit of a groupoid, Def. 3.9. Not only that, there is also a similar notion of isotropy group.

Definition 4.5 A point $x \in \Omega$ is fixed for $g \in G$ if $gx = x$. A point $x \in \Omega$ is fixed under an action of a group G if it is fixed for all $g \in G$.

Exercise 4.5 *Prove that if the action of* G *on* Ω *is faithful, then the set of fixed points of* $g \in G$*,* $g \neq e$*, is different from* Ω.

Prove that if the action is free, the set of fixed points of any $g \in G$*,* $g \neq e$*, is empty.*

Exercise 4.6 *Prove that given a point* $x \in \Omega$*, the set of elements* $g \in G$*, such that* x *is fixed by* g*, is a subgroup of* G.

Definition 4.6 Given an action of a group G on a set Ω and a point x in Ω, the subgroup $G_x = \{g \in G \colon gx = x\}$, of elements in the group fixing x, is called the isotropy group[4] of x.

Then, the isotropy group of a point x is the set of elements of G which have x as a fixed point. If the action is free, the isotropy group of any point is $\{e\}$. In addition, if the action is not faithful, this implies that there are $g \neq g'$ such that $gx = g'x$ for all $x \in \Omega$, that is $g^{-1}g' \in G_x$ for all x. In fact, the action will be faithful if $\bigcap_{x \in \Omega} G_x$ is equal to $\{e\}$ and unfaithful otherwise.

Exercise 4.7 *Prove that* $G_u = \bigcap_{x \in \Omega} G_x$ *is a normal subgroup of* G*. We will call such normal subgroup the unfaithful subgroup of* G.

[4]The reader may object that we should call G_x the 'isotopy' group, from 'iso' *same* – 'topos' *place*, rather than 'isotropy' group, however the terminology is already well rooted and as old habits die hard, we will not insist on it.

Proposition 4.1
Given an action of the group G on Ω, there exists a canonical faithful action induced from the given action of the quotient group G/G_u on Ω, where G_u is the unfaithful subgroup of G.

Proof 4.1 The induced action will be defined as $(gG_u)x = gx$. Notice that it is well defined because if $g' = gg_u$, with $g_u \in G_u$, then $(g'G_u)x = (gg_uG_u)x = (gG_u)x$, and this is certainly a faithful action.

Because of the previous proposition, all actions will be considered to be faithful in what follows because, if they were not, it suffices to consider the group obtained by taking the quotient with respect the unfaithful group of the action and the canonical induced action which is faithful.

Proposition 4.2
If two points $x, x' \in \Omega$ are related through a group element $x' = gx$, that is, they are in the same orbit, their isotropy groups are conjugate: $G_{x'} = gG_xg^{-1} = \mathrm{Int}(g)G_x$. If the action of the group G on Ω is transitive, all isotropy groups G_x are conjugate.

Exercise 4.8 *Prove the previous statement.*

Proposition 4.3
Let G be a group acting on the set Ω. Then, for any $x \in \Omega$, there is a canonical bijection between the space of cosets G/G_x and the orbit Gx, given by $gG_x \mapsto gx$. If G is finite, we have $|G| = |G_x||Gx|$ with $|Gx|$ the number of elements in the orbit Gx.

Exercise 4.9 *Prove the previous proposition.*

Example 4.10

1. *Left and right actions. The left and right actions of a group on itself are free and transitive, thus, $G_g = e$ for any $g \in G$ and $Gg = G$.*

2. *The symmetric group S_n has a natural action on the space of indexes $\Omega = \{1, 2, \ldots, n\}$, $(\sigma, k) \mapsto i = \sigma(k)$, sometimes called the defining action of S_n. The action is clearly transitive and the isotropy group of the index k is the subgroup of all permutations that leave invariant k, that is, the group of permutation of the remaining $n-1$ indexes, S_{n-1}. It is clear that the action is faithful as the intersection of all isotropy groups, that is, the set of permutations that fix all indexes is just the identity permutation.*

3. *Let Ω be a set. We denote by S_Ω the group of all bijective maps $p\colon \Omega \to \Omega$ (we may call them "permutations" of Ω). Then S_Ω is a group whose cardinal is denoted by $|\Omega|!$ (where $|\Omega|$ denotes again the cardinal of Ω). If Ω is a finite set with n elements, then S_Ω is the symmetric group S_n and its cardinal is just $n!$ The group S_Ω of permutations of Ω acts on Ω in the obvious way:*

$$\Phi\colon S_\Omega \times \Omega \to \Omega; \quad \Phi(p, x) = p(x), \quad \forall x \in \Omega, \quad \forall p \in S_\Omega. \tag{4.2}$$

The orbits $S_\Omega \cdot x = \Omega$, $\forall x \in \Omega$ *and the action is transitive. The action is also faithful and* $(S_\Omega)_x \simeq S_{\Omega_x}$, *where* $\Omega_x = \Omega \setminus \{x\}$, *i.e., the isotropy subgroup of any element* x *is (isomorphic to) the group of permutations of the remaining elements in the group.*

4. *Group of rotations in the Euclidean plane* $SO(2)$. *Consider the group of rotations of the Euclidean plane* E^2 *determined by the matrices* $R(\theta) = \begin{bmatrix} \cos\theta & -\sin\theta \\ \sin\theta & \cos\theta \end{bmatrix}$, *that transform a point* P *with coordinates* (x,y) *into the point* $P' = R(\theta)(P)$ *with coordinates* (x',y') *given by:* $x' = x\cos\theta - y\sin\theta$, $y' = x\sin\theta + y\cos\theta$. *The action is clearly faithful. It is not free, since the origin is fixed by any rotation, and it is not transitive.The orbits are the origin and the circles with center at the origin.*

5. *Group of Euclidean transformations. Consider the group* $E(2)$ *of Euclidean transformations in the plane, that is* $\varphi\colon E^2 \to E^2$, $P \in E^2 \mapsto \varphi(P) \in E^2$, *is either a rotation, a translation or a reflection on the plane. Then, given a point* $P \in E^2$, *the isotropy group of* P *consists of all rotations and reflections around* P. *Rotations are parametrized by an angle and the corresponding group is denoted by* $SO(2)$ *and reflections are defined by the set of lines passing through* P *that, as a set, is the real projective space* $\mathbb{RP}^1$. *The action of the Euclidean group is faithful and transitive.*

6. *General Linear group. Consider a linear space* V *and the corresponding group of isomorphisms* $L\colon V \to V$, *also called the general linear group of* V *(see Apendix A.2.3). The group* $GL(V)$ *acts naturally on* V *as* $\Psi(L,v) = L(v)$. *Such action is transitive but not free. The orbit of a given vector* v *is the set of vectors* $L(v)$, L *an isomorphism. There are two orbits: The zero vector* $\mathbf{0}$ *and* $V \backslash \mathbf{0}$. *Notice that, if* $v \neq 0$, $GL(V)/GL(V)_v \simeq V \setminus \mathbf{0}$.

An immediate application of the previous ideas is found when considering a groupoid $\mathbf{G} \rightrightarrows \Omega$ and the set $\mathbf{G}_-(x)$ of morphisms with target x, then the isotropy group G_x acts on the left on $\mathbf{G}_-(x)$ by composition, that is $\Psi_\gamma(\alpha) = \gamma \circ \alpha$ for any $\gamma \in G_x$ and $\alpha\colon y \to x$. Then we can prove:

Proposition 4.4
Let $\mathbf{G} \rightrightarrows \Omega$ *be a groupoid. For any* $x \in \Omega$ *there is a canonical indentification between the space of orbits* $\mathbf{G}_-(x)/G_x$ *(similarly for* $\mathbf{G}_+(x)/G_x$*) and the orbit* $\mathcal{O}_x = \mathbf{G}x$ *of* $\mathbf{G}$ *through* x.

Proof 4.2 We should prove that there is a natural bijection between the orbits of G_x on $\mathbf{G}_-(x)$ and $\mathcal{O}_x$. An orbit $G_x\alpha$ of G_x on $\mathbf{G}_-(x)$ has the form $\gamma \circ \alpha$, with $\alpha\colon y \to x$, $\gamma \in G_x$, then we associate to the orbit $G_x\alpha$ the point y, that is, we define a map $\psi\colon \mathbf{G}_-(x)/G_x \to \mathcal{O}_x$ as $\psi(G_x\alpha) = s(\alpha) = y$. This map is well defined (all morphisms $\gamma \circ \alpha$ in the orbit $G_x\alpha$ have the same source y), it is injective because if

$G_x\alpha$ and $G_x\alpha'$ are such that α and α' have the same source, say y, then both α and α' are in $\mathbf{G}(x,y)$ and there will exist γ such that $\alpha' = \gamma \circ \alpha$, but then $G_x\alpha = G_x\alpha'$. Finally, the map ψ is surjective, because if $y \in \mathcal{O}_x$, then there exists $\alpha : x \to y$, clearly $\psi(G_x\alpha^{-1}) = y$.

Exercise 4.11 *Show that the action of G_x on $\mathbf{G}_x(x)$ is free.*

Because of the previous exercise, Exe. 4.11, and Prop. 4.3, we conclude that there is bijection between G_x and its orbit on $\mathbf{G}_-(x)$. Then, if the groupoid $\mathbf{G}$ is finite, the orbits $\mathcal{O}_x$ will also be finite, as well as $\mathbf{G}_-(x)$. Hence, because of Prop. 4.4, we have: $|\mathbf{G}_-(x)| = |G_x||\mathcal{O}_x|$. Moreover, because of Eq. (3.3), we get: $|\mathbf{G}| = \sum_{x\in\Omega} |\mathbf{G}_-(x)| = \sum_{x\in\Omega} |G_x||\mathcal{O}_x|$. If $\mathbf{G}$ is connected, all its isotropy groups are isomorphic and $\mathcal{O}_x = \Omega$ for all $x \in \Omega$. We have proved:

Corollary 4.1
Let $\mathbf{G} \rightrightarrows \Omega$ be a finite connected groupoid. Then $|\mathbf{G}| = |\Omega|^2|G_x|$, x an arbitrary point in Ω.

Given a group G acting on a set Ω and a subset $S \subset \Omega$, we will call the symmetry group of S (with respect to G) the subgroup G_S of elements in G leaving S invariant, that is $g \in G_S$ if $gS \subset S$. We will also say that G_S is the group of symmetries of S (with respect to G).

Example 4.12

1. *Semidirect product of groups. Let K be a group and H another group. Consider an action Ψ of K on H. Suppose that we require that the bijective maps $\Psi_k : H \to H$ defined by the action of any element $k \in K$ preserve the group structure of H, that is, they are group isomorphisms, then the action Ψ defines a map $\omega : K \to \mathrm{Aut}(H)$, $\omega(k) = \Psi_k$. In such case, we may define the semidirect product of the groups K and H, usually denoted as $K \ltimes_\omega H$ (K acts on H) as the group structure on $H \times K$ given by $(h,k) \star (h',k') = (h\omega_k(h'), kk')$ (recall Sect. 2.5.3).*

2. *Linear representations of groups. From the perspective of group actions, a linear representation of a group G, the main theme of the second part of this book, is an action of the group G on a linear space V by linear maps, that is, if V is a linear space, we may consider an action Ψ of G on V such that the bijections $\Psi_g : V \to V$ are linear maps. This constitutes the definition of a linear representation of the group G with support on V.*

4.2 The action groupoid

As it happens, it is easy to 'groupoidify' the notion of a group action. Consider a group[5] Γ acting on the set Ω and define a groupoid associated to this action as follows.

Definition 4.7 Given a group Γ acting on a set Ω, we will denote by $\mathbf{G}(\Gamma,\Omega)$ the groupoid whose objects are the points of Ω and whose morphisms are triples $(y,g,x) \in \Omega \times \Gamma \times \Omega$, such that $y = gx$. The source and target maps $s,t\colon \mathbf{G}(\Gamma,\Omega) \to \Omega$ are defined in the obvious way: $s(y,g,x) = x$ and $t(y,g,x) = y$, as well as the composition law:

$$(z,g',y) \circ (y,g,x) = (z,g'g,x)\,. \tag{4.3}$$

The groupoid $\mathbf{G}(\Gamma,\Omega) \rightrightarrows \Omega$ will be called the action groupoid[6] associated to the action of the group Γ on Ω.

Depending on the context, alternative notations will be used. In particular, and in addition to the standard notation $\alpha\colon x \to y$ for the morphism (y,g,x) (also written as $\alpha_g\colon x \to y$), we will often use the notation g_{yx} for this morphism, which is reminiscent of the notation used for moves (Sects. 2.2 and 3.2.1). Notice that, with this notation, the composition law (4.3) becomes:

$$g'_{zy} g_{yx} = (g'g)_{zx}\,.$$

First, we will notice that the isotropy groups $\mathbf{G}(\Gamma,\Omega)_x$ of the action groupoid $\mathbf{G}(\Gamma,\Omega)$ coincide with the isotropy groups Γ_x of the group Γ acting on Ω. Actually, if $\alpha = (y,g,x) \in \mathbf{G}(\Gamma,\Omega)_x$, then $y = x = gx$, hence, $g \in \Gamma_x$, and conversely. We will avoid the cumbersome notation $\mathbf{G}(\Gamma,\Omega)_x$ and we will simply denote the isotropy groups of the action groupoid as the isotropy groups Γ_x of the original group action.

In the same vein, we observe that the orbit $\mathcal{O}_x$ through x of the action groupoid $\mathbf{G}(\Gamma,\Omega)_x$ consists of those points $y = t(\alpha)$ with $\alpha = (y,g,x)$, that is $y = gx$. Hence, the orbit through x of the action groupoid coincides with the orbit Γx of the group action. We will use both notations in what follows without further comments.

Recall that the action of a group Γ on a set Ω is transitive if there is just one orbit, i.e., $\Omega = \Gamma x$, but this is the definition of a transitive (or connected) groupoid, thus, the action groupoid is connected if and only if the group action is transitive.

Finally, we recall that, according to Cor. 3.1, isotropy groups of the action groupoid corresponding to points in the same orbit are isomorphic, Prop. 4.2. However, they are not canonically isomorphic. Remember, Prop. 4.2 that if $x' = gx$, then $G_{x'}$ is conjugate of G_x, $G_{x'} = \mathrm{Int}\,(g)(G_x)$, but any g such that $x' = gx$, provides a way of idnetifying $G_{x'}$ and G_x by means of the inner automorphism $\mathrm{Int}\,(g)$.

Thus, the action groupoid of a group action provides an easy way to construct a variety of examples of groupoids. Let us look at some examples.

[5] In the following paragraphs we will change the notation slightly in order to make them more readable. Groups will be denoted using capital greek letters: $\Gamma, \Sigma, \ldots$

[6] A. Weinstein used the terminology "transformations groupoid".

Example 4.13 *1. Consider the canonical action of the symmetric group S_n on the set $\Omega = \{1,2,\ldots,n\}$. The corresponding action groupoid is given by $S_n \times \Omega$, and its arrows $\alpha\colon k \to j$, are given by triples (j,σ,k) where $\sigma \in S_n$, $\sigma(k) = j$. Clearly, the isotropy group of k is the set of permutations σ, such that $\sigma(k) = k$, and it is isomorphic to S_{n-1}.*

2. *Let G be a group, Ω a set and consider the trivial action of G on Ω, that is $gx = x$ for all $x \in \Omega$, $g \in G$. Then, the action groupoid is the totally disconnected groupoid $\Omega \times G$ discussed in Example 3.14.2.*

3. *Let G be a group and $H \subset G$ a subgroup. Consider the action of G on the space of cosets G/H by left multiplication, that is $\Psi_g(g'H) = gg'H$, $g,g' \in G$. Then, the action groupoid is the groupoid defined on the set $G \times G/H$ over the space of objects $\Omega = G/H$. The source and target maps $s(y,g,x) = x$, $t(y,g,x) = y = gx$, where x and y denote the cosets $g'H$ and $gg'H$ respectively. Notice that the isotropy group of the group at the coset $x = gH$ is given by the subgroup gHg^{-1} of G, which is the conjugate subgroup of H by the element g. The groupoid $G \times G/H$ is transitive.*

4. *Consider again Example 4.12.1 of a semidirect product, that is, the action of a group K on another group H defined by an homomorphism $\omega\colon K \to \mathrm{Aut}(H)$. The action groupoid is defined on $H \times K$, its arrows are triples $(\omega_k(h),k,h)$ and the composition law is given by $(\omega_{k'k}(h),k',\omega_k(h))(\omega_k(h),k,h) = (\omega_{k'k}(h),k'k,h)$, which is quite different to the semidirect product structure on $H \times K$ defined by the previous action.*

The previous observations state clearly that the notion of group action is encoded perfectly in the structure of the action groupoid, this being the clue to consider groupoids as 'generalized group actions'. We will discuss this idea in the sections to follow and use it to construct useful examples of generalized symmetries by means of groupoids.

4.3 Symmetries and groupoids

4.3.1 Groupoids and generalised actions

In this section, we will review the notion of groupoid from the perspective of symmetries following the lead laid down by the example of the action groupoid of a group action. In this sense, a groupoid can be considered right away as an abstraction of the notion of an action of a group on a set, with the corresponding notions of orbits, isotropy groups, etc., as in the particular instance of the action groupoid of a group action. However, we can even push this idea a bit further and try to think anew what could be the notion of an 'action of a groupoid on a set'.

Mimicking the notion of a group acting on a set, we could define the action of a groupoid $\mathbf{G} \rightrightarrows \Omega$ on a set M by assigning to any pair (α,m) in $\mathbf{G} \times M$ another point

$\alpha m \in M$. For instance, we may try to reproduce the canonical left and right action of a group on itself, Def. 4.2, by chosing $M = \mathbf{G}$ and defining the action map as $(\alpha, \beta) \mapsto \alpha \circ \beta$. However, we immediately see the problem: It is not always possible to compose α and β! So, we need a 'reference' map $\mu\colon M \to \Omega$, such that only the elements $m \in M$ over $x = s(\alpha)$ could be acted upon by α, that is indicating which elements on M can be acted upon by a given element α in the groupoid, in particular those over the source of α.

The associativity-like property of actions, Def. 4.1 (2), would require that $\alpha(\beta m) = (\alpha \circ \beta)m$ whenever α and β are composable, that is, whenever $s(\alpha) = t(\beta)$, for instance $\beta\colon x \to y$ and $\alpha\colon y \to z$. Then $\mu(\beta m) = y = s(\alpha) = t(\beta)$.

Finally, the same condition applied to β and 1_x will imply that $(\beta \circ 1_x)m = \beta(1_x m)$ for all $\beta\colon x \to y$, $x = \mu(m)$. Then composing with β^{-1} on the left, we get $1_x m = m$.

Summarizing, to define an action of the groupoid $\mathbf{G} \rightrightarrows \Omega$ on the set M we need a map $\mu\colon M \to \Omega$ on the set M, that serves to indicate which elements can be acted upon by morphisms on the groupoid, satisfying a few natural conditions. Then, we can mimic the notion of a group action, Def. 4.1 by means of the following definition.

Definition 4.8 Let $\mathbf{G} \rightrightarrows \Omega$ be a groupoid and $\mu\colon M \to \Omega$ a map. Define the composable set $\mathbf{G} \circ M = \{(\alpha, m) \in \mathbf{G} \times M \mid s(\alpha) = \mu(m)\}$ of pairs of morphisms α in the groupoid and elements m in the set that lie on top of the same object. An action of $\mathbf{G}$ on M is a map $\Psi\colon \mathbf{G} \circ M \to M$, such that the family of maps $\Psi_\alpha\colon \mu^{-1}(x) \to M$, $\alpha\colon x \to y$, satisfy:

1. $\mu(\Psi_\alpha(m)) = y = t(\alpha)$,
2. $\Psi_{1_x} m = m$ for any $m \in \mu^{-1}(x)$, and
3. $\Psi_\alpha \Psi_\beta = \Psi_{\alpha \circ \beta}$, for any $\beta\colon z \to x$.

The map $\mu\colon M \to \Omega$ will be called the moment map of the action[7]. The sets $\mu^{-1}(x) \subset M$ will be called the fibres of the moment map and denoted by M_x.

As in the case of groups acting on sets, we will prefer to use the notation $\alpha m = \Psi_\alpha(m) = \Psi(\alpha, m)$ unless there is risk of confusion. Thus, with this notation, the properties defining a groupoid action becomes (compare with (4.1)):

$$\mu(\alpha m) = t(\alpha), \qquad 1_{\mu(m)} m = m, \qquad \beta(\alpha m) = (\beta \circ \alpha) m, \tag{4.4}$$

for all $m \in M$, $\mu(m) = s(\alpha)$ and $t(\alpha) = s(\beta)$.

[7]This terminology was introduced by A. Weinstein and it is reminiscent of the momentum map in mechanics and symplectic geometry. In fact, it has its origin in the English translations of Souriau's French term "moment" [61]: "moment map" and "momentum map". Both expressions were used interchangeably in the literature, however Weinstein chooses to use 'momentum map' as in its original sense and restrict 'moment map' for the reference map introduced to define a groupoid action [65].

Example 4.14 *1. Any groupoid $\mathbf{G} \rightrightarrows \Omega$ acts on Ω with moment map the identity map. In particular the action groupoid $\mathbf{G}(\Gamma,\Omega)$ of a group Γ acting on Ω, acts on Ω with moment map the identity.*

2. *The left action of a groupoid. The groupoid $\mathbf{G} \rightrightarrows \Omega$ acts on itself on the left with moment map μ the target map $t\colon \mathbf{G} \to \Omega$ as $L_\alpha \beta = \alpha \circ \beta$, whenever $\mu(\beta) = t(\beta) = s(\alpha)$, hence, whenever α and β are composable (note that with the notation Definition 3.1 we have consistently that $\mathbf{G} \circ \mathbf{G} = \mathbf{G}_2$).*

3. *Similarly, we define the right action of a groupoid on itself as $R_\alpha \beta = \beta \circ \alpha^{-1}$. Note that the moment map μ of the right action is now the source map because $\beta \circ \alpha^{-1}$ is defined, provided that $\mu(\beta) = s(\alpha) = t(\alpha^{-1}) = s(\beta)$.*

4. *Consider the Loyd's groupoid $\mathfrak{L}_2$. We consider the set $M = \Omega \times R$, where $\Omega = \{1,2,3,4\}$ are the states of the puzzle and $R = \{1,2,3\}$ is a register. Consider the moment map $\mu\colon M \to \Omega$ given by the projection on the first factor, that is $\mu(x,l) = x$. Then, we define a natural action of $\mathfrak{L}_2$ on M with moment map μ as $\sigma_{ji}(i,l) = (j,\sigma(l))$, where σ_{ji} is an element of the groupoid, $i,j = 1,\ldots,4$ and $l = 1,2,3$ (we are following the conventions laid down in Sect. 3.2.1).*

In the following section, we will describe a rich class of examples of groupoid actions beyond those described here but, before that, we will discuss a particular instance of the previous definition that will be instrumental in the remaining of the book.

Notice that, because of the definition, given an action Ψ of the groupoid $\mathbf{G}$, the map Ψ_α maps $\mu^{-1}(x) = M_x$ into $\mu^{-1}(y) = M_y$. Moreover, because $\alpha^{-1} \circ \alpha = 1_x$ and (4.4), we get: $\Psi_{1_x} = \mathrm{id}_{M_x}$ and $\Psi_{\alpha^{-1}}\Psi_\alpha = \Psi_{\alpha^{-1}\circ\alpha} = \Psi_{1_x} = \mathrm{id}_{M_x}$ and, similarly $\Psi_\alpha \Psi_{\alpha^{-1}} = \Psi_{\alpha\circ\alpha^{-1}} = \Psi_{1_y} = \mathrm{id}_{M_y}$. Hence, a groupoid action defines a family of bijective maps $\Psi_\alpha\colon M_x \to M_y$ that implement the groupoid algebraic structure (see Fig. 4.4 for a pictorical description of a groupoid action).

The moment map $\mu\colon M \to \Omega$ defines an equivalence relation in M, $m \sim m'$ iff $\mu(m) = \mu(m')$, whose equivalence classes are the fibres of μ. Each fibre of μ, denoted as usual by $M_x = \mu^{-1}(x)$, can be labelled by the point $x \in \Omega$ common to all the elements on the fibre, i.e., such that $\mu(\alpha) = x$. The set M is partitioned in equivalence classes, or fibres of μ, and we can write $M = \bigsqcup_{x\in\Omega} M_x$.

A particular instance of this situation is provided when the fibres of the moment map are linear spaces. Let E be a set and $\mu\colon E \to \Omega$ a surjective map such that $\mu^{-1}(x)$ is a linear space for any $x \in \Omega$. We denote by E_x the fibre of μ over x. We will assume that Ω is finite (or discrete) and we will say that $\mu\colon E \to \Omega$ is a vector bundle over Ω with projection μ. Consider an action of the groupoid $\mathbf{G} \rightrightarrows \Omega$ on the vector bundle E with moment map the projection map $\mu\colon E \to \Omega$. This will imply that for any $\alpha\colon x \to y$, the map $\Psi_\alpha\colon E_x \to E_y$, defined by the action is bijective. However because the fibres E_x are linear spaces, we would like that in addition the maps Ψ_α would be linear maps. In such case, we will say that the action Ψ is a linear representation of the groupoid $\mathbf{G}$.

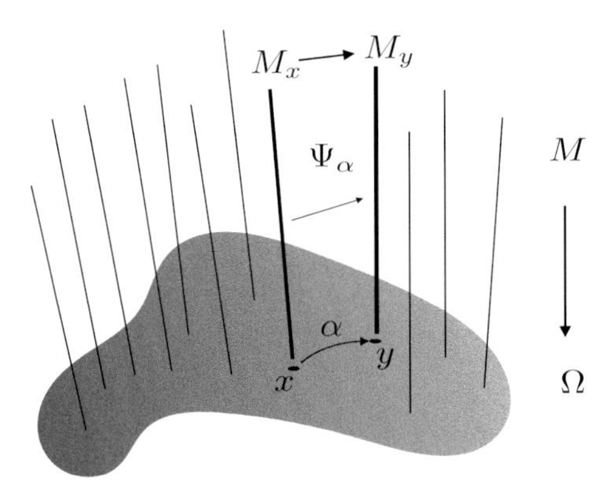

Figure 4.4: We may visualize a groupoid action as a family of bijective maps $\Psi_\alpha : M_x \to M_y$ that reproduce the algebraic relations of the groupoid.

Thus, we have encountered for the first time the notion we want to analyze further in this book, that of a linear representation of a groupoid. A linear representation of a groupoid is, on one hand a particular instance of an action of a groupoid, this time in a 'bundle' of linear spaces, hence, the obvious name of 'vector bundle' for such structure. On the other hand, such notion, extends in a natural way the notion of action of a group, hence the notion of linear representation of a group. Then, we expect that some deep relation among them will emerge.

The notion of action of groupoids, in particular that of linear representation of groupoids, even if natural, takes more time to explain and justify that what one would expect for such a fundamental concept. In fact, we will see in the coming chapter, Chap. 5, that there is a much more concise way of expressing it by using the categorical approach. However, before we move to that, we would like to conclude this chapter by analyzing some non-trivial examples and constructions relating the notion of action of groupoids and symmetries and offering a first taste of Cayley's theorem for groupoids.

4.3.2 Groupoids and symmetries: The restriction of an action groupoid

As it was mentioned above, and we are going to repeat now, group actions on sets provide a very good source of groupoid examples. We will complete this idea with a restriction procedure that will provide, not only a larger source of examples, but of quite interesting ones.

Remember that a symmetry group of a given structure is the subgroup of a given group Γ acting on the carrier space Ω leaving invariant that structure. This notion is best understood in the setting of groupoids, that is, the action of the group Γ is described by the action groupoid $\mathbf{G}(\Gamma,\Omega)$ and, given a subset B of Ω, there is a natural notion of the restriction of the action groupoid to it (recall Def. 3.7).

Definition 4.9 Let Γ be a group acting on a set Ω and $\mathbf{G}(\Gamma,\Omega)$ the corresponding action groupoid. Then, given any subset $B \subset \Omega$, there is a canonical subgroupoid of the action groupoid $\mathbf{G}(\Gamma,\Omega)$ consisting of all morphisms (y,g,x) such that $x,y \in B$ that will be called the restriction of the action groupoid $\mathbf{G}(\Gamma,\Omega)$ to B and will be denoted by $\mathbf{G}_B(\Gamma,\Omega) = \{(y,g,x) \in \mathbf{G}(\Gamma,\Omega) \mid x,y \in B\}$. The source and target maps are just the restriction of the source and target maps to the subset and the composition is the composition law of the action groupoid.

The isotropy groups of the elements $x \in B$ in the restriction of the action groupoid $\mathbf{G}(\Gamma,\Omega)$ coincide with those of the original groupoid, thus they will be denoted again with the same symbol Γ_x.

It is clear that, if B is an orbit of the action of the group Γ, then the restriction of the action groupoid to the orbit, recall Def. 3.7 and Exercise 3.16, is just the subgroupoid $\mathbf{G}(\Gamma,B)$ and coincides with the restriction of the action groupoid to B, that is $\mathbf{G}_B(\Gamma,\Omega) = \mathbf{G}(\Gamma,B)$, but, in general, $\mathbf{G}_B(\Gamma,\Omega)$ is not the action groupoid of a group action on a subset of Ω.

Lemma 4.1

If $\mathcal{O}'$ is an orbit of the restriction groupoid $\mathbf{G}_B(\Gamma,\Omega)$, then $\mathcal{O}' \subset \mathcal{O}$ where $\mathcal{O}$ is an orbit of Γ. In addition, given the orbit $\mathcal{O}'$ of the restriction groupoid, there is a subgroup $\Gamma' \subset \Gamma$, such that $\Gamma'/\Gamma'_x \cong \mathcal{O}'$.

Proof 4.3 That $\mathcal{O}' \subset \mathcal{O}$ is a subset of an orbit $\mathcal{O}$ of Γ is obvious because $y \in \mathcal{O}'$ if there exists $\alpha = (y,g,x) \in \mathbf{G}_B(\Gamma,\Omega)$, then $y = gx$ for some $g \in \Gamma$ and y is in the Γ-orbit of x.

Given the orbit $\mathcal{O}'$ of $\mathbf{G}_B(\Gamma,\Omega)$ passing through $x \in B$, consider now the set Γ' of elements of Γ such that $gx \in B$. Clearly $\Gamma' \subset \Gamma$ and Γ' is a subgroup.

However, contrary with what happens with the action groupoid, the subgroup Γ' may change from orbit to orbit and the restriction groupoid of an action groupoid is not, in general, the action groupoid of a subgroup of the original action, that is, in general $\mathbf{G}_B(\Gamma,\Omega) \neq \mathbf{G}(\Gamma',B)$. Thus, if a group Γ is the symmetry group of a given structure, it is possible that restricting the setting of the structure will make the symmetry group trivial. However, there will be a groupoid obtained by restriction that will provide a description of the symmetry. These facts will be clarified by the examples exhibited below.

We will close this section by emphasizing again that not every groupoid can be obtained as the restriction of an action groupoid. Notice that, because of the discussion above, the orbits of the groupoids obtained by restriction of action groupoids are homogeneous spaces for standard groups Γ', hence, if we consider, for instance, the groupoid of pairs of a finite set Ω, its orbit is the space Ω, and there is no group Γ' whose quotient is Ω.

4.4 Weinstein's tilings

In this section, we will discuss in detail a simple but illustrative example of the use of groupoids in order to analyze the symmetry of structures for which the standard group theoretical interpretation fails or is too poor to render an appropriate account of it. The discussion below is a slightly adapted version of an example analyzed by A. Weinstein, emphasizing the power of groupoids to capture local symmetries [66].

4.4.1 *Tilings and groupoids*

A particularly interesting source of group actions are provided by tilings. We will concentrate here on a particularly simple tiling of the plane (even though the arguments can be extended easily to any tiling of the plane or of higher dimensional spaces).

Consider a tiling of the plane by 2×1 rectangles. The tiling is specified by the set X defining the grout between the tiles (idealised as 1-dimensional). Then, X is the union of horizontal and vertical lines $X = H \cup V$, with $H = \mathbb{R} \times \mathbb{Z}$ (horizontal lines) and $V = 2\mathbb{Z} \times \mathbb{R}$ (vertical lines). Each connected component in $\mathbb{R}^2 \backslash X$ is a tile (see Fig. 4.5).

As discussed at the beginning of this chapter, the symmetry group of the tiling is traditionally defined as the set of all rigid motions Γ that leave invariant the tiling, that is, Γ consists of horizontal translations by even integers, vertical translations by integers, reflections with respect to horizontal or vertical lines in H and V, respectively, i.e., translations with respect to elements in the lattice $\Lambda = 2\mathbb{Z} \times \mathbb{Z}$; and reflections with respect to horizontal and vertical lines accross the points in the lattice $\frac{1}{2}\Lambda = \mathbb{Z} \times \frac{1}{2}\mathbb{Z}$ (see Fig. 4.6).

Now, consider what we have lost if we pass from the actual tiling determined by X to the lattice of corner points Λ. The same symmetry group Γ would arise from the lattice Λ, that is $\Gamma \cong \Lambda \ltimes R$ where R denotes the reflections across the points in Λ and $\frac{1}{2}\Lambda$ as commented above. The group Γ has no information on the local structure of the tiled plane, such as the fact that neighbourhoods of points inside the tiles will look identical if the tiles are uniform, while they may look different if the tiles are painted with a design (that could still be invariant under Γ).

Hence, the action groupoid corresponding to the action of Γ on $\mathbb{R}^2$ will be:

$$\mathbf{G}(\Gamma, \mathbb{R}^2) = \{(y, g, x) \in \mathbb{R}^2 \times \Gamma \times \mathbb{R}^2 \mid y = gx\}.$$

In addition to this, there is yet the action groupoid corresponding to the restriction of the action of Γ to the lattice of corner points. This groupoid is the restriction of the action groupoid $\mathbf{G}(\Gamma, \mathbb{R}^2)$ to Λ, however, in this case, $\mathbf{G}_\Lambda(\Gamma, \mathbb{R}^2)$ to $\Lambda = \mathbf{G}(\Gamma, \Lambda)$ and the restriction groupoid is indeed the action groupoid of a group (the same group Γ) to a subset (the lattice Λ).

If, as is the case with real floors, the tiling is restricted to a finite part of the plane, such as $B = [0, 2m] \times [0, n]$, the symmetry group shrinks dramatically. The subgroup of Γ leaving $X \bigcap B$ invariant contains just 4 elements, even though a repetitive pattern on the floor is clearly visible.

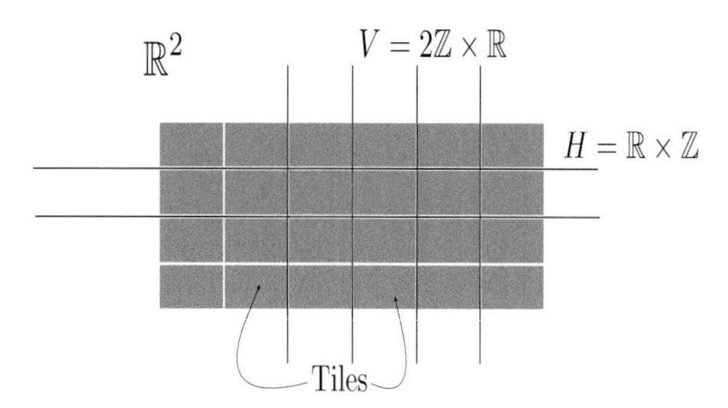

Figure 4.5: Weinstein's tiling of the plane.

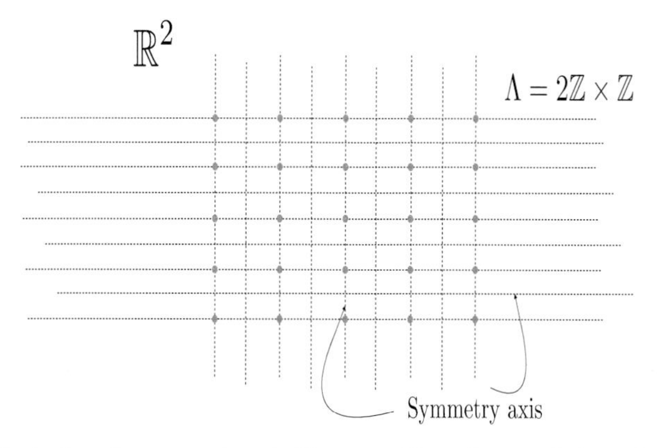

Figure 4.6: The lattice of corner points Λ. The group Γ acts by translations along the axis and reflections along the indicated dashed lines.

The restriction of the action groupoid will, however, allow us to describe the symmetry of the finite tiled structure. Thus, the restriction of the action groupoid $\mathbf{G}(\Gamma, \mathbb{R}^2)$ to B will be:

$$\mathbf{G}_B(\Gamma, \mathbb{R}^2) = \{(y, g, x) \in B \times \Gamma \times B \mid y = gx\}.$$

Notice that, contrary to the situation with the restriction to the lattice Λ, the restriction of the action groupoid to B is not the action groupoid of a group acting on B. This fact will become evident when analysing the orbit structure of $\mathbf{G}_B(\Gamma, \mathbb{R}^2)$. Consider, for instance, the case $m = 5$, $n = 4$ depicted in Fig. 4.7. Two points are in the same orbit of $\mathbf{G}_B(\Gamma, \mathbb{R}^2)$ if they are related by an allowed transformation: Translations, rotations or reflections.

The isotropy group of $x \in B$, denoted by $G_x = \{(x, g, x) \mid gx = x\}$, is trivial except for the points in $\frac{1}{2}\Lambda$, for which it is $\mathbb{Z}_2 \times \mathbb{Z}_2$.

The space of orbits $\mathcal{M}$ can be parametrised by the 'quarter tile', $\mathcal{M} = [0,1] \times [0,1/2]$. Notice that each orbit of a point $x \in B$ lying inside a tile consists of 80 points. Thus, if there were a group Γ', such that the restriction groupoid $\mathbf{G}_B(\Gamma, \mathbb{R}^2)$

would be the action groupoid $\mathbf{G}(\Gamma', B)$, then $|\Gamma'| = 80$ (because the isotropy group of x is trivial). However, if x lies in the corners, then the orbit contains 30 points (all corners of the floor) and because the isotropy group now has four elements, then the group Γ' should have 120 elements.

4.4.2 Local symmetries

We can describe the local symmetry of our tiling by introducing yet another groupoid closely related to the symmetry groupoids described in the previous paragraphs. For this purpose, we consider the plane $\mathbb{R}^2$ as decomposed in the disjoint union of $P_1 = B \bigcap X$ (the grout), $P_2 = B \backslash P_1$ (the tiles), and $P_3 = \mathbb{R}^2 \backslash B$ (the exterior). Let E be the group of all Euclidean motions of the plane and define the local symmetry groupoid $\mathbf{G}_{loc}(B)$ as the set of triples: $(y, g, x) \in B \times E \times B$ with $y = gx$, g is an Euclidean motion, and for each x there is a neighborhood $U \subset \mathbb{R}^2$ such that $g(U \cap P_i) \subset P_i$, $i = 1, 2, 3$. The composition is given by the same law. The idea behind this notion of "local groupoid transformations" is that the structure of the tiling around a point is mapped into the structure of the tiling around the transformed point. Hence, all interior points to the tiles are in the same orbit of the local groupoid as we may take a small ball around each point contained in the tile that is mapped by a translation into a ball around any other selected point inside the tile.

The local groupoid $\mathbf{G}_{loc}(B)$ has only 6 different orbits (see Fig. 4.7):

1. $\mathcal{O}_1$, consisting of the interior points of the tiles.
2. $\mathcal{O}_2$, consisting of the interior edge points.
3. $\mathcal{O}_3$, consisting of the interior corner points.
4. $\mathcal{O}_4$, consisting of the boundary edge points.
5. $\mathcal{O}_5$, consisting of the boundary "T" points.
6. $\mathcal{O}_6$, consisting of the boundary corner points.

The isotropy group of a point in $\mathcal{O}_1$ is $O(2)$. The isotropy group of a point in $\mathcal{O}_2$ is $\mathbb{Z}_2 \times \mathbb{Z}_2$. The isotropy group of a point in $\mathcal{O}_3$ is D_4 and $\mathbb{Z}_2$ for points in $\mathcal{O}_4$, $\mathcal{O}_5$ and $\mathcal{O}_6$.

If we consider now the restriction of the local groupoid to $\Lambda \bigcap B$ (see Fig. 4.8), then this groupoid has just three orbits $\mathcal{O}_3$, $\mathcal{O}_5$ and $\mathcal{O}_6$ with isotropy groups D_4, $\mathbb{Z}_2$ and $\mathbb{Z}_2$, respectively. If we denote this groupoid just by $\mathbf{G}$ and the restriction to each orbit by $\mathbf{G}_3$, $\mathbf{G}_5$ and $\mathbf{G}_6$ respectively, then the order of each connected component is given by the standard formula $|\Omega|^2 |G_x|$ (see Fig. 4.8):

$$|\mathbf{G}_3| = 12^2 \times |D_4| = 1152, \quad |\mathbf{G}_5| = 14^2 \times |\mathbb{Z}_2| = 392, \quad |\mathbf{G}_6| = 4^2 \times |\mathbb{Z}_2| = 32.$$

Hence, $|\mathbf{G}| = 1576$. The set of points $\Lambda \bigcap B$, corresponding to the vertices of the tiling in the finite floor, have no structure at all except for that provided by the groupoid $\mathbf{G}$.

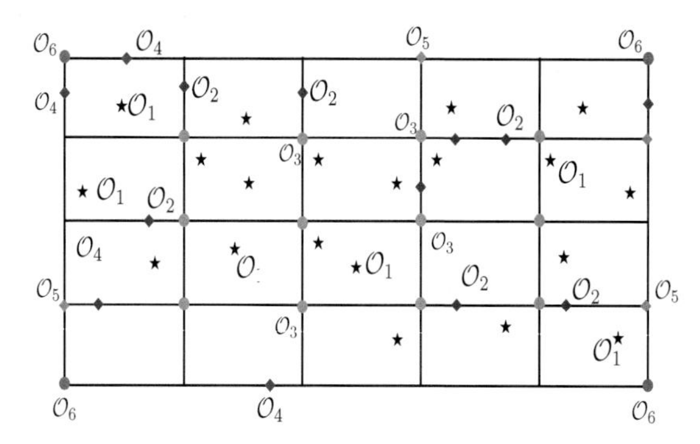

Figure 4.7: The orbits of the local symmetries groupoid $\mathbf{G}_{loc}(B)$.

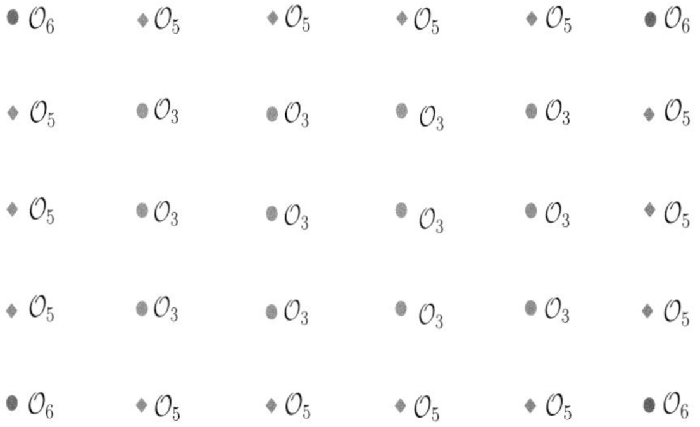

$\mathcal{O}_6$ $\mathcal{O}_5$ $\mathcal{O}_5$ $\mathcal{O}_5$ $\mathcal{O}_5$ $\mathcal{O}_6$

$\mathcal{O}_5$ $\mathcal{O}_3$ $\mathcal{O}_3$ $\mathcal{O}_3$ $\mathcal{O}_3$ $\mathcal{O}_5$

$\mathcal{O}_5$ $\mathcal{O}_3$ $\mathcal{O}_3$ $\mathcal{O}_3$ $\mathcal{O}_3$ $\mathcal{O}_5$

$\mathcal{O}_5$ $\mathcal{O}_3$ $\mathcal{O}_3$ $\mathcal{O}_3$ $\mathcal{O}_3$ $\mathcal{O}_5$

$\mathcal{O}_6$ $\mathcal{O}_5$ $\mathcal{O}_5$ $\mathcal{O}_5$ $\mathcal{O}_5$ $\mathcal{O}_6$

Figure 4.8: The lattice of points $\Lambda \cap B$ and the orbits of its local symmetries groupoid.

4.5 Cayley's theorem for groupoids

Is there a Cayley's theorem for groupoids? That is, given a groupoid $\mathbf{G} \rightrightarrows \Omega$ can a realisation (representation) of this groupoid in terms of maps on sets be found as it happens with groups? (Remember Cayley's Theorem for groups, Thm. 2.7).

The answer to this question is positive again. We can provide an explicit construction, as in the case of Cayley's theorem for groups, for groupoids. Let us go back to the problem at hand and recall that, if we are given a set Ω, the family of bijective maps $\varphi : \Omega \to \Omega$ form a group, called the symmetric group of Ω. Now, if we are given a map $\mu : M \to \Omega$, we may consider the family of bijective maps $\varphi_{xy} : M_x \to M_y$, where M_x, M_y are the fibres $\mu^{-1}(x)$ and $\mu^{-1}(y)$, respectively, and we will denote them by $\mathbf{G}(\mu)$ or, even better, $\mathrm{Aut}(\mu)$. It is easy to check that $\mathbf{G}(\mu)$ is a groupoid over Ω, the source map defined as $s(\varphi_{xy}) = x$, the target map defined as $t(\varphi_{xy}) = y$ and the composition law the standard composition of maps. Units will be the identity maps $\mathrm{id}_x : M_x \to M_x$ and the inverse of φ_{xy} the inverse map φ_{xy}^{-1}. The groupoid $\mathrm{Aut}(\mu) = \mathbf{G}(\mu)$ will be called the groupoid associated to the map $\mu : M \to \Omega$ and also the groupoid of automorphisms of the map $\mu : M \to \Omega$.

In Sect. 5.3.4 the groupoid of automorphisms of various maps will be considered in the broader context of homomorphisms of groupoids and formal definitions provided, Def. 5.8. In the present context, we are going to use it just in a very specific situation.

Finally, let us recall that Cayley's theorem for groups was just the statement that the left (or right) action of the group G on itself allows us to identify the elements of the group with bijective maps of G on itself, hence, with a subgroup of the symmetric group of G. Then, we may state:

Theorem 4.1
(Cayley's theorem for groupoids) Let $\mathbf{G} \rightrightarrows \Omega$ be a groupoid. The right action of the groupoid $\mathbf{G}$ on itself with moment map the source map of the groupoid, $\mu = s\colon \mathbf{G} \to \Omega$, allows us to identify the groupoid $\mathbf{G}$ with a subgroupoid of the groupoid of automorphisms $\mathrm{Aut}\,(s) = \mathbf{G}(s)$ of the map s. Similarly, the left action of the groupoid $\mathbf{G}$ on itself with moment map the target map of the groupoid, allows us to identify the groupoid $\mathbf{G}$ with a subgroupoid of the groupoid associated to the target map $t\colon \mathbf{G} \to \Omega$, that is with a subgroupoid of $\mathrm{Aut}\,(t)$.

Proof 4.4 Let us consider, for instance, the second assertion. The proof lies in the observation that the groupoid associated with the target map of $\mathbf{G}$ consists of all bijective maps from $t^{-1}(x)$ to $t^{-1}(y)$, but the right action R_α of $\mathbf{G}$ on itself with moment map $t\colon \mathbf{G} \to \Omega$ (recall Example 4.14.2), will map $t^{-1}(x)$ into $t^{-1}(y)$ if $\alpha\colon y \to x$.

Chapter 5

Functors and Transformations

'Functor' is the keyword for group homomorphisms, group actions, groupoid actions, linear representations of groups, etc. In this chapter, the notion of functor and its closely-related notion of natural transformations will be discussed. As a natural application of the notion of functor an abstract setting for databases will be described.

The basic properties of functors as homomorphisms of groupoids will be discussed too, in particular, the language of exact sequences so convenient in the realm of group theory. In this context, the notion of linear representations of groups and groupoids, and its associated notion of equivalence of representations, will be anticipated.

The significative notion of the groupoid of automorphisms of a groupoid will be discussed and the role played by the fundamental subgroupoid will be emphasized.

5.1 Functors

5.1.1 Functors: Definitions and first examples

We may say that the notion of functor extends the notion of homomorphism of algebraic structures to the realm of categories, but it is better to think of a functor as an abstraction of the notion of transformation that preserves the corresponding composition laws, thus, functors transform categories into categories bringing the composition of two arrows into the composition of the new ones.

In the previous chapters we were pointing out that various disparate notions could be described better by taking a more abstract approach. This is exactly what the notion of functor does for us. It provides the conceptual framework to deal simultaneously with the notion of actions, homomorphism and representations in a single and natural way. Thus we have:

Definition 5.1 Let $\mathbf{C}$ and $\mathbf{C}'$ be two categories. A functor $F : \mathbf{C} \to \mathbf{C}'$ from the category $\mathbf{C}$ to the category $\mathbf{C}'$ is an assignment of a object $F(x)$ in the category $\mathbf{C}'$ to any object x in the category $\mathbf{C}$, and of a morphism $F(\alpha)\colon F(x) \to F(y)$ to any morphism $\alpha\colon x \to y$, such that:

1. $F(\beta \circ \alpha) = F(\beta) \circ F(\alpha)$,
2. $F(1_x) = 1_{F(x)}$,

for all objects x and composable arrows α, β in the category $\mathbf{C}$.

We will also use the notation $F\colon x \to x'$ to indicate that $x' = F(x)$[1].

Let us consider first some examples that will illustrate why the notion of functor is central.

Example 5.1 1. *Group homomorphisms. If we consider a group G as a groupoid $\mathbf{G} \rightrightarrows \{*\}$ with a single object $\Omega = \{*\}$ (note that the unit corresponding to $*$ is $1_* = e$, the neutral element of the group), then a group homomorphism $f\colon G \to G'$ (recall Def. 2.16) is a functor from the category $\mathbf{G}$ to the category $\mathbf{G}'$. Let us denote, for the purpose of this discussion, the functor defined by the homomorphism f by F, thus, F maps the object $*$ of the first group into the object $*'$ of the second. Then, $F(e) = F(1_*) = 1_{F(*)} = 1_{*'} = e'$ and condition (2) in Def. 5.1 is satisfied. Moreover the functor F assigns the element $g \in G$, considered as a morphism, into the element $f(g) \in G'$, thus, $F(gg') = f(gg') = f(g)f(g') = F(g)F(g')$ and we get Def. 5.1.1. In what follows, we will not distinguish between the homomorphism of groups f and the corresponding functor F.*

2. *Group actions. Given a group action Ψ of the group G on the set Ω, we may conceive it as a functor from the group category $\mathbf{G}$ (i.e., a groupoid with a single object) into the category of sets.*

   *We will denote, as usual, by **Sets** the category of all sets, i.e., its objets are sets (recall Example 1.7). Morphisms in this category are just maps between sets, $\mathbf{Sets}(\Omega, \Omega') = \{f\colon \Omega \to \Omega' \mid f \text{ is a map}\}$ and the composition law is the composition of maps.*

   Then a functor $F\colon \mathbf{G} \to \mathbf{Sets}$ assigns a set Ω to the object $*$ of the group, i.e., $\Omega = F(*)$, and it assigns a map $F(g)\colon F(*) = \Omega \to F(*) = \Omega$ to any $g \in G$. Moreover $F(1_*) = \mathrm{id}_{F(*)} = \mathrm{id}_\Omega$. Finally, Def. 5.1.1, implies that if we denote the maps $F(g)$ by Ψ_g, they satisfy Def. 4.1(a) and (b), that is F defines an action of G on Ω.

[1]Notice that even if we use the notation $F\colon x \to x'$, a functor F is not in general a map because $\mathbf{C}$ and $\mathbf{C}'$ do not have to be sets.

In Chapt. 1 we have encountered various examples of categories: **FinSet**, **FinVect**, $\mathbb{N}$, $\mathbf{G}(S)$, etc. We will consider various other categories, some of them very abstract, and some natural functors among them.

Example 5.2 *Let* $\mathbf{C}$ *be a category whose objects are sets and whose morphisms are maps (not necessarily small), then, we define the functor* $\text{Forget}\colon \mathbf{C} \to \mathbf{Sets}$ *that assigns to any object* $x \in \mathbf{C}$*, itself just considered as a set:* $\text{Forget}(x) = \{x\}$*. In the same way, we assign to each morphism, the map between sets defined by it. This functor (rather trivial indeed) is called the "forgetful functor".*

For instance, we have the forgetful functor from **FinVect** $\to$ **Sets** *that assigns to any finite-dimensional linear space its own set.*

Example 5.3 *Another example is provided by the category* $\mathbf{C}$*. This (rather amazing) category is the category of all categories, i.e., the objects in the category* $\mathbf{C}$ *are categories. Given two categories* $\mathbf{C}$ *and* $\mathbf{C}'$*, then the morphisms between* $\mathbf{C}$ *and* $\mathbf{C}'$ *are all functors from the category* $\mathbf{C}$ *in the category* $\mathbf{C}'$*.*

Notice that given any category $\mathbf{C}$*, there is a functor, denoted by* $1_{\mathbf{C}}\colon \mathbf{C} \to \mathbf{C}$*, that assigns to any object x itself:* $1_{\mathbf{C}}(x) = x$*, and to any morphism* α*, itself:* $1_{\mathbf{C}}(\alpha) = \alpha$*.*

The composition of functors $F\colon \mathbf{C} \to \mathbf{C}'$*,* $G\colon \mathbf{C}' \to \mathbf{C}''$ *is given as follows:* $G \circ F\colon \mathbf{C} \to \mathbf{C}''$ *is a functor from the category* $\mathbf{C}$ *to the category* $\mathbf{C}''$*, it assigns to any object* $x \in \mathbf{C}$ *the object* $G(F(x))$ *in* $\mathbf{C}''$ *(we use the notation "*$x \in \mathbf{C}$*" meaning that x is an object belonging to the category* $\mathbf{C}$*, not that* $\mathbf{Ob}(\mathbf{C})$ *is a set and x an element in it). Clearly, the morphism* $(G \circ F)(\alpha) = G(F(\alpha))$ *satisfies that,*

$$\begin{aligned}(G \circ F)(\beta \circ \alpha) &= G(F(\beta \circ \alpha)) = G(F(\beta) \circ F(\alpha)) \\ &= G(F(\beta)) \circ G(F(\alpha)) = (G \circ F)(\beta) \circ (G \circ F)(\alpha),\end{aligned}$$

and

$$(G \circ F)(1_x) = G(F(1_x)) = G(1_{F(x)}) = 1_{G(F(x))}. \tag{5.1}$$

The composition law $G \circ F$ *is clearly associative*

$$\begin{aligned}&(H \circ (G \circ F))(x) = H((G \circ F)(x)) = H(G(F(x))), \\ &((H \circ G) \circ F)(x) = (H \circ G)(F(x)) = H(G(F(x))).\end{aligned} \tag{5.2}$$

the same easy checks apply to morphisms.

The category $\mathbf{C}$ *is already a "wild thing", but it is not the most bizarre thing we could think up. You may imagine that we might construct categories of categories of categories, etc., ad nauseam. This is, however, the "easy trip". The structures that we actually want to consider are those coming from "equivalences" between functors, that is, when two functors between categories are "equivalent".*

We will try to exploit this fact in order to get a proper understanding of the notion of equivalent functors, but, before that, let us introduce another example of functor that contains many of the insights leading us into this book.

5.1.2 Functors and realisations of categories

The example showing that a group action is a particular instance of functor, Example 5.1, shows that we may think of functors as realizations of a given category in terms of objects and arrows of another one. Thus, a functor $F\colon \mathbf{C} \to \mathbf{Sets}$ from the category $\mathbf{C}$ to the category of sets will be called a (set-theoretical) realisation of the category $\mathbf{C}$. We may discuss now different properties of such realisations, similar to the properties of actions of groups on sets (faithful, free, etc.). We will not do that, however, as we will concentrate on a few specific situations that are going to be central to the purposes of this book.

Linear representations of groups

Let G be a group (finite or infinite as a set). Let us recall (see Example 4.10.6) that a linear representation of G is a map $\rho\colon G \to \mathrm{End}(V)$, associating to each element g of G a linear map $\rho(g)\colon V \to V$, where V is a linear space (finite dimensional or not), satisfying:

$$\rho(e) = \mathrm{id}_V\,, \qquad \rho(gh) = \rho(g)\rho(h)\,, \qquad \forall g,h \in G\,. \tag{5.3}$$

Two representations, ρ, ρ' of G in the linear spaces V and V' respectively, are said to be equivalent if there exists a linear isomorphism $\phi\colon V \to V'$ such that $\phi \circ \rho(g) = \rho'(g) \circ \phi$, for all $g \in G$ or, more pictorically, if the following diagram is commutative:

$$\begin{array}{ccc} V & \xrightarrow{\rho(g)} & V \\ \phi\downarrow & & \downarrow\phi \\ V' & \xrightarrow{\rho'(g)} & V' \end{array} \tag{5.4}$$

The space V is called the support of the representation μ. Later on, we will provide a full discussion of the theory of linear representations of finite groups (see Chaps. 7 and 8). The theory of representations of groups is a key element in the use of groups in mathematics as groups are often 'represented' in other problems (most of the times linearly, or even more 'unitarily').

Consider, as we did before, the category of all linear spaces **Vect**, finite or infinite dimensional. We will not be worried about finer requisites on the linear spaces we are working with. Now the main observation is that a representation μ of a group G is nothing but a functor between categories:

Proposition 5.1
A linear representation $\rho\colon G \to \mathrm{End}(V)$ is a functor from the category $\mathbf{G}$ (the group is considered as a groupoid with only one object) to the category $\mathbf{Vect}$ of all linear spaces.

Proof 5.1 A functor $\rho\colon \mathbf{G} \to \mathbf{Vect}$, associates a linear space V to the object $*$: $\rho(*) = V$, and a morphism $\rho(g)\colon \rho(*) \to \rho(*)$ to any morphism $g \in G$. Hence, $\rho(g)$

is a linear map from $V \to V$, and the functor property establishes that:

$$\rho(gh) = \rho(g)\rho(h), \qquad \rho(1_*) = 1_{\rho(*)} = \mathrm{id}_V, \tag{5.5}$$

for any $g, h \in G$, and the functor ρ defines a linear representation of G with support V.

A first taste of linear representations of groupoids

Thus, linear representations of groups are just a particular instance of functors. This fact allows us to discuss the notion of linear representation of groupoids.

Definition 5.2 Let $\mathbf{G} \rightrightarrows \Omega$ be a groupoid. A linear representation R of the groupoid $\mathbf{G}$ is a functor $R\colon \mathbf{G} \to \mathbf{Vect}$, that is, an assignment of a linear space V_x to any object $x \in \Omega$; $R(x) = V_x$; and a linear isomorphism $R(\alpha)\colon V_x \to V_y$ to any morphism $\alpha\colon x \to y$, such that

$$R(\beta \circ \alpha) = R(\beta)R(\alpha), \quad R(1_x) = \mathrm{id}_{V_x}, \tag{5.6}$$

$$\begin{array}{ccccc} x & \xrightarrow{\alpha} & y & \xrightarrow{\beta} & z \\ \downarrow R & & \downarrow R & & \downarrow R \\ V_x & \xrightarrow{R(\alpha)} & V_y & \xrightarrow{R(\beta)} & V_z \end{array} \tag{5.7}$$

Example 5.4 *For instance, consider the groupoid of pairs $\mathbf{G}(\Omega)$ of the set Ω. Consider the linear space $\mathbb{C}$ and assign to any $x \in \Omega$ the linear space $\mathbb{C}_x \equiv \mathbb{C}$. To any pair (y,x) we assign the isomorphism $\mathrm{id}\colon \mathbb{C}_x \to \mathbb{C}_y$. Then, this functor R is clearly a linear representation of $\mathbf{G}(\Omega)$.*

This point of view about representations of groupoids can be nicely discussed from the point of view of actions of groupoids.

Functors and actions of groupoids

We will offer now the natural definition of an action of a groupoid introduced in Sect. 4.3.1.

Consider a groupoid $\mathbf{G} \rightrightarrows \Omega$ (we will assume that Ω is a set) and a map $\mu\colon M \to \Omega$. We will consider now the category $\mu\mathbf{Sets}$ whose objects are the fibres $M_x = \mu^{-1}(x)$ of the map μ (that will be called accordingly μ-sets) and whose morphisms are maps $\varphi_{yx}\colon \mu^{-1}(x) \to \mu^{-1}(y)$. The composition law is the standard composition of maps.

Proposition 5.2
An action of the groupoid $\mathbf{G} \rightrightarrows \Omega$ on a set M is a functor $F\colon \mathbf{G} \to \mu\mathbf{Sets}$.

Proof 5.2 In fact, given a functor $F\colon \mathbf{G} \to \mu\mathbf{Sets}$, to any morphism $\alpha\colon x \to y$ on the groupoid $\mathbf{G}$ we associate a map $F(\alpha)\colon \mu^{-1}(x) \to \mu^{-1}(y)$ or, using the notation from Sect. 4.3.1, $F(\alpha)\colon M_x \to M_y$. Then, the defining properties of the functor F coincide with properties 1, 2 and 3 in Def. 4.8. That is, denoting by Ψ_α the map $F(\alpha)$, we get: $\mu(\Psi_\alpha(m)) = \mu(F(\alpha)(m)) = y$ provided that $\alpha\colon x \to y$, because $F(\alpha)\colon \mu^{-1}(x) \to \mu^{-1}(y)$. Moreover,

$$\Psi_{1_x} = F(1_x) = 1_{F(x)} = \mathrm{id}_{M_x},$$

then $\Psi_{1_x}(m) = m$ and, finally, condition (3) is the immediate consequence of $F(\alpha \circ \beta) = F(\alpha) \circ F(\beta)$.

We will end up this section by observing that a linear representation R of a groupoid $\mathbf{G} \rightrightarrows \Omega$ can be thought as an action of $\mathbf{G}$ on a μ-set $: E \to \Omega$, whose fibres $E_x = \mu^{-1}(x)$ are linear spaces[2], each fiber being the linear space $R(x)$. This bundle as a whole is best depicted as in Fig. 4.4, with fibres being linear spaces. A morphism $\alpha : x \to y$, becomes a linear map, $R(\alpha)\colon R(x) \to R(y)$.

For instance, if we consider the groupoid of pairs $\mathbf{G}(\Omega)$ and Ω is finite or countable, then we may consider the bundle of linear spaces $E = \sqcup_{x \in \Omega} V_x$, where each linear space V_x is the 1-dimensional complex linear space $\mathbb{C}$. The moment map $\mu\colon E \to \Omega$, is the map $\mu(v_x) = x$, where $v_x \in V_x$.

There is a canonical linear representation of $\mathbf{G}(\Omega)$ on $\mu\colon E \to \Omega$, given by the functor $R(y,x)\colon V_x \to V_y$, $R(y,x) = 1$ (note that both V_x and V_y are just $\mathbb{C}$). This linear representation is called the fundamental representation of $\mathbf{G}(\Omega)$ and will be denoted by π_0 in Chaps. 9 and 10.

5.2 An interlude: Categories and databases

We have found a number of examples and applications of the notions introduced so far: From the 'connecting dots' categories, to the groups of permutations passing through graphs and symmetries. There is, however, a way of looking at categories, consequently to everything we are discussing here, that goes beyond these examples. Such perspective comes from information theory.

Usually, we organize information in databases. Loosely speaking, a (relational) database (DB) is a bunch of tables and relationships between them. A table is an array of data (strings of symbols from some alphabets) whose rows are called 'records' or 'instances'. The columns are called 'attributes' and they could contain 'pure data' or they may be 'keys'. We will take the convention that every table has a distinguished key, the primary key of the table, and it will be its first column. A table could have foreign key columns that link it to other tables.

Let us consider the following example:

[2]We prefer to call such sets a "bundle" of linear spaces over the set Ω. The notion of vector bundle is exactly this, even if in the presence of topology, a local triviality property is usually required.

5.2.1 A simple database: Classes and courses

We are asked to help to organise the teaching activity of our institution, so we prepare a database with three tables: A table containing the courses to be taught, a table containing information about the professor who are going to teach them and another one with the schedule.

Thus, the records of the table 'Courses' will contain a primary key (see Table 5.1, first column): 101, 102, etc., and several attributes like the name of the course, the responsible professor for the course, the teaching assistants that will colaborate, the classroom where the course will be held, etc. The table 'Professors' will contain the information on the professors in the Department. There will be a primary key (see Table 5.2) with a code like: p01, p02, etc., and the usual personal data, name, surname, office, phone, etc. The table 'Schedule' will consists of a primary key (see Table 5.3) containing the day of the week (we will assume that the schedule is the same for all weeks) and the period, i.e., Monday1, Monday2, etc., and the attributes will consist of the course, classroom, professor, etc. Thus, our tables will look like Tables 5.1, 5.2, 5.3.

Table 5.1: Table of 'courses'.

Courses	Title	ProfC	TA1	TA2	Class1
101	Algebra	Evans	Parker	Diaz	H012
102	Calculus	Jin	Danilo	Dupont	H014
⋮	⋮	⋮	⋮	⋮	⋮

Courses table of the database 'Classes and courses'.

Table 5.2: Table of 'professors'.

Professors	Name	Surname	Office	Phone	e-mail
p01	Peter	Evans	D027	4489	evans@univ.com
p02	Jin	Hu	D002	4478	jin@univ.com
p03	Mario	Diaz	D191	4456	diaz@univ.com
⋮	⋮	⋮	⋮	⋮	⋮

Professors table of the database 'Classes and courses'.

Table 5.3: Table of 'schedules'.

Schedule	Course	ProfS	Class2
Monday1	Algebra	Evans	H012
Monday2	Calculus	Diaz	H014
⋮	⋮	⋮	⋮
Thursday4	Algebra	Danilo	HS11
⋮	⋮	⋮	⋮

Schedule table of the database 'Classes and courses'.

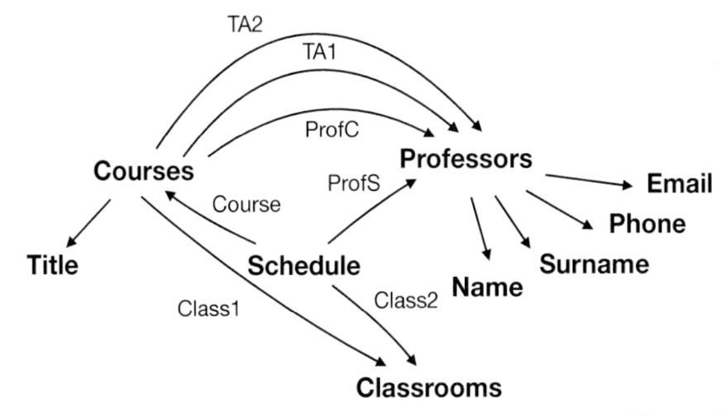

Figure 5.1: Scheme for DB 'Schedule'. Notice the consistency conditions: Course $\circ$ Class1 $=$ Class2, TA1 $\circ$ Course $=$ ProfS, TA2 $\circ$ Course $=$ ProfS, ProfC $\circ$ Course $=$ ProfS.

Notice that each table has a primary key, the first column, and various data attributes and foreign key columns. Some consistency constraints should be satisfied, for instance, in the table 'Schedule', the professor and the classroom should coincide with the information stored in the table 'Courses'.

Actually, the distinction between data columns and foreign keys is unnecessary as we may consider any data attribute as a 1-column table. In the previous examples, we may consider the column 'phone' in the table 'Professors' as a foreign key of the one column table 'Phones' with records the Department telephone numbers, or we may consider the columns 'Class1', 'Class2', in the tables 'Courses' and 'Schedule' as foreign keys of a table 'Classrooms' that will consist of the code of the classroom (and maybe other attributes like capacity, equipment, etc.). Thus, we will assume that each column in a table is a key (the first primary, the rest foreign).

Hence, we will visualise the structure of the database by drawing a scheme displaying all keys and with arrows linking the primary keys to the foreign keys on each table (see Fig. 5.1).

Now we see that we have converted the structure of the DB classes and courses in a category in the same way we proceed for the connecting dots games in Chapter 1. More formally, we may define a database schema as a diagram consisting of nodes and oriented links where each node represents a table (or, if you prefer, the primary key of a table). Each table (or primary key) is an object in a category and has some attributes (foreign keys) related to primary keys. Each directed link represents a foreign key and is a morphism in the category. The consistency of the database (data integrity) is guaranteed by constraints determined by path equalities that determine the composition rule of the category. In this sense, we may conclude that there is a perfect match between categories and database schemas.

5.2.2 Databases and functors

Once we have shown that databases schemas are categories: Tables (entities) are objects, columns (attributes) are arrows and consistency constraints determine the composition law, it remains to understand how to fill a given database with data.

Filling a database schema $\mathbf{S}$ with data is the same as defining a functor $F: \mathbf{S} \to$ **FinSet**, that is to each object (primary key) we assign a finite set and to each column (arrow) a map from the primary key to the set associated to the related key (a primary key in another table). In other words, a functor $F: \mathbf{S} \to$ **FinSet** fills the schema with data in a consistent way because the relations established in the database schema (the category structure) are preserved by the functor.

Thus, in our previous example of the database schema $\mathbf{S} =$ 'classes and courses', the functor F assigns to the primary key 'Courses' the set

$$F(\text{Courses}) = \{101, 102, 103, \ldots\},$$

to the key 'Professors' the set $F(\text{Professors}) = \{p01, p02, \ldots\}$, to the key 'Classrooms', the set $F(\text{Classrooms}) = \{H010, H011, H012, \ldots\}$, to the key 'Name' in table 'Professors', etc.

5.3 Homomorphisms of groupoids

Equipped with the background provided by the previous discussion, we are ready now to introduce the notion of homomorphism of groupoids as a further example of the notion of functor. This will provide us with concepts that were so fruitful when discussing the properties and structure of groups.

5.3.1 *Homomorphisms of groupoids: Basic notions*

It is clear that we should follow the lead offered by the first example of functors as group homomorphisms (Example 5.1.1):

Definition 5.3 Let $\mathbf{G} \rightrightarrows \Omega$, $\mathbf{G}' \rightrightarrows \Omega'$ be two groupoids, a homomorphism from the groupoid $\mathbf{G}$ to the groupoid $\mathbf{G}'$ is a functor $F: \mathbf{G} \to \mathbf{G}'$.

Note that, because of the properties of a functor, a homomorphism $F: \mathbf{G} \to \mathbf{G}'$ assigns to any object $x \in \Omega$ an object $x' = F(x) \in \Omega'$. Moreover, it assigns to any morphism $\alpha: x \to y$, a morphism $\alpha' = F(\alpha): F(x) \to F(y)$ such that $F(1_x) = 1_{F(x)}$ for any $x \in \Omega$, and $F(\alpha \circ \beta) = F(\alpha) \circ F(\beta)$ for any pair of composable α, β. In addition, we have that $F(\alpha^{-1}) = F(\alpha)^{-1}$ because any morphism α is invertible.

As is customary in this book, we will just consider groupoids that are sets, then a groupoid homomorphism F is a map from $\mathbf{G}$ to $\mathbf{G}'$ satisfying the conditions above.

Example 5.5 *1. Group homomorphisms are groupoid homomorphisms.*

2. *Let $f: \Omega \to \Omega'$ be any map, then the assignment $F(x) = f(x)$ and $F(y,x) = (f(y), f(x))$, defines a homomorphism of groupoids $F: \mathbf{G}(\Omega) \to \mathbf{G}(\Omega')$ between the groupoids of pairs of Ω and Ω'.*

3. *Consider a subgroupoid* $\mathbf{H} \subset \mathbf{G}$ *of the groupoid* $\mathbf{G} \rightrightarrows \Omega$*, then the natural inclusion map* $i \colon \mathbf{H} \to \mathbf{G}$ *mapping any object and any morphism of* $\mathbf{H}$ *into themselves is a homomorphism of groupoids.*

4. *Consider the groupoid action* $\mathbf{G}(\Gamma,\Omega)$ *of the action of the group* Γ *in* Ω*, and the restriction* $\mathbf{G}_S(\Gamma,\Omega)$ *of the groupoid action to a subset* S*, then the canonical inclusion map* $i \colon \mathbf{G}_S(\Gamma,\Omega) \to \mathbf{G}(\Gamma,\Omega)$ *is a homomorphism of groupoids.*

Exercise 5.6 *Let* $F \colon \mathbf{G} \to \mathbf{G}'$ *be a homomorphism of groupoids, show that* F *induces a group homomorphism* $F_x \colon G_x \to G'_{F(x)}$*, with* G_x*,* $G'_{F(x)}$ *the corresponding isotropy groups.*

Proposition 5.3
Let $F \colon \mathbf{G} \to \mathbf{G}'$ *be a homomorphism of groupoids, then* F *induces a homomorphism of groupoids* $F_0 \colon \mathbf{G}_0 \to \mathbf{G}'_0$*, with* $\mathbf{G}_0$*,* $\mathbf{G}'_0$ *the fundamental groupoids of* $\mathbf{G}$ *and* $\mathbf{G}'$ *respectively.*

Proof 5.3 It follows immediately from Exercise 5.6. Define $F_0(\gamma_x) = F(\gamma_x)$, $\gamma_x \in G_x$, then $F_0(\gamma_x) \in G_{F(x)}$.

Then, as in the case of groups, given a homomorphism of groupoids $F \colon \mathbf{G} \to \mathbf{G}'$, we have two distinguished subgroupoids:

Proposition 5.4
Let $F \colon \mathbf{G} \to \mathbf{G}'$ *be a groupoid homomorphism, then* $\ker F = \{\alpha \colon x \to y \in \mathbf{G} \mid F(\alpha) = 1_{F(x)}\}$ *and* $\operatorname{ran} F = \{F(\alpha) \in \mathbf{G}' \mid \alpha \in \mathbf{G}\}$ *are subgroupoids of* $\mathbf{G}$ *and* $\mathbf{G}'$ *respectively.*

Proof 5.4 To show that $\ker f$ is a subgroupoid, we consider $\alpha : z \to y$, $\beta : x \to y$, $\alpha, \beta \in \ker F$. Note that, because $t(\alpha) = y = t(\beta)$, $\alpha^{-1} \circ \beta$ is defined and $F(\alpha^{-1} \circ \beta) = F(\alpha^{-1}) \circ F(\beta)$. Since $\alpha, \beta \in \ker F$, we have $F(\beta) = 1_{F(x)}, F(\alpha) = 1_{F(z)}$ and $F(x) = F(y) = F(z)$. Since $F(\alpha^{-1} \circ \alpha) = F(1_z) = 1_{F(z)}$, we have $F(\alpha^{-1}) \circ F(\alpha) = 1_{F(z)}$, that is, $F(\alpha^{-1}) = F(\alpha)^{-1}$ and then $F(\alpha^{-1}) = 1_{F(z)} = 1_{F(x)}$. Then we conclude that $F(\alpha^{-1} \circ \beta) = 1_{F(x)}$, that is, $\alpha^{-1} \circ \beta \in \ker F$ and because of Prop. 3.2, $\ker F$ is a subgroupoid of $\mathbf{G}$.

Similarly we prove that $\operatorname{ran} F$ is a subgroupoid of $\mathbf{G}'$.

Exercise 5.7 *Prove that the groupoid of objects* Ω *of the object space of the groupoid* $\mathbf{G} \rightrightarrows \Omega$*, is always contained in the subgropoid* $\ker F$*, with* $F \colon \mathbf{G} \to \mathbf{G}'$ *a homomorphism of groupoids.*

Contrary to the situation with homomorphism of groups, we will define the analogue notions of monomorphisms, epimorphisms and isomorphisms by using the properties of the subgroupoids $\ker F$ and $\operatorname{ran} F$.

Definition 5.4 We will say that the homomorphism of groupoids $F\colon \mathbf{G} \to \mathbf{G}'$ is a monomorphism if $\ker F = \Omega$.

We will say that $F\colon \mathbf{G} \to \mathbf{G}'$ is an epimorphism if $\operatorname{ran} F = \mathbf{G}'$.

Finally we will say that $F\colon \mathbf{G} \to \mathbf{G}'$ is an isomorphism of groupoids if it is a monomorphism and an epimorphism.

That the notions introduced in the previous definition are consistent with the corresponding ones in the case of groups is shown in the following proposition.

Proposition 5.5
Let $F\colon \mathbf{G} \to \mathbf{G}'$ be a monomorphism of groupoids[3]*, then the corresponding map sending morphisms in $\mathbf{G}$ in morphisms in $\mathbf{G}'$ is injective.*

Similarly, if $F\colon \mathbf{G} \to \mathbf{G}'$ is an epimorphism of groupoids the corresponding map is surjective.

Finally, if $F\colon \mathbf{G} \to \mathbf{G}'$ is an isomorphism of groupoids, then there exists another homomorphism of groupoids $F'\colon \mathbf{G}' \to \mathbf{G}$ such that $F \circ F' = \mathrm{id}_{\mathbf{G}'}$ and $F' \circ F = \mathrm{id}_{\mathbf{G}}$. Moreover the map is bijective.

Proof 5.5 Suppose that $F\colon \mathbf{G} \to \mathbf{G}'$ is a monomorphism, then $\ker F = \Omega$. Consider the map defined by the functor, that will be denoted by F again. Suppose that $F(\alpha) = F(\beta) = 1_x$, then if $\alpha\colon x \to y$, $\beta\colon x \to z$, $\beta \circ \alpha^{-1}\colon y \to z$, and $F(\beta \circ \alpha^{-1}) = F(\beta) \circ F(\alpha^{-1}) = 1_{F(x)}$, with $F(x) = F(y) = F(z)$. But then $\beta \circ \alpha^{-1} \in \ker F = \Omega$, and $\beta \circ \alpha^{-1} = 1_y$, with $y = z$. Then, $\alpha = \beta$ and F is injective.

Similarly, we prove that if F is an epimorphism, the corresponding map is surjective.

To show that, if F is an isomorphism, there exists F' with the stated properties, we will observe first that the corresponding map is bijective (it is injective and surjective). Let us denote the corresponding inverse map by F^{-1}. Then, we define $F'\colon \mathbf{G}' \to \mathbf{G}$ as $F'(x') = F^{-1}(x')$ for any object $x' \in \Omega'$ and $F'(\alpha') = F^{-1}(\alpha')$ and a routine computation shows that F' is a functor with the desired properties.

Example 5.8

1. *Groups. The notions of monomorphism, epimorphism and isomorphism of groupoids, become the corresponding ones for groups.*

2. *The canonical inclusion $i\colon \mathbf{G}_0 \to \mathbf{G}$ defined by the fundamental subgroupoid of a given groupoid $\mathbf{G} \rightrightarrows \Omega$ is a groupoid monomorphism, i.e.,* $\ker i = \Omega$.

3. *The canonical projection maps $\pi_a\colon \mathbf{G}_1 \times \mathbf{G}_2 \to \mathbf{G}_a$, $a = 1,2$ are epimorphisms.*

4. *Consider the cyclic groupoid $\mathfrak{C}_n(m)$ and the functor $F\colon \mathbf{G}(n) \times \mathbb{Z}_s \to \mathfrak{C}_n(m)$, with $s = l/m$, $l = \mathrm{l.c.m.}(n,m)$, given by: $F(i) = i-1$, $i = 1,2,\ldots,n$, $F(i+$*

[3]We assume here that both $\mathbf{G}$ and $\mathbf{G}'$ are small categories, so that the space of morphisms and objects are sets.

$1,\mathbf{1},i) = m_{i,i-1}$, $i = 1,\dots,n-1$, $F(0,\mathbf{1},n) = m_{n,0}$, with $m_{i+1,i}$ the move of the cyclic groupoid that passes from state i to $i+1$ and acts with the permutation σ of order m in the auxiliary register R (recall Sect. 3.2.3, Prop. 3.6). Then F induces an isomomorphism of groupoids.

Two groupoids $\mathbf{G} \rightrightarrows \Omega$ and $\mathbf{G}' \rightrightarrows \Omega'$, such that there exists an isomorphism of groupoids $F\colon \mathbf{G} \to \mathbf{G}'$, could be identified: Their objects are in one-to-one correspondence as well as their morphisms and the corresponding composition laws are identical. Clearly the relation 'being isomorphic' is an equivalence relation in the class of groupoids. Stated in these terms, the problem of classification of groupoids consists of determining when two groupoids are isomorphic and describing the corresponding equivalence classes, that is exhibiting a representative for each one.

The problem of classification of groupoids includes the problem of classification of groups, so it is a hard problem. We will be able to prove a first theorem leading to the classification of groupoids in the next chapter. For that, it will be useful to extend the language of exact sequences to the class of groupoids.

5.3.2 Exact sequences of homomorphisms of groupoids

As in the case of homomorphisms of groups, we can define exact sequences of homomorphisms of groupoids (recall Def. 2.22).

Definition 5.5 Let $F_1\colon \mathbf{G}_1 \to \mathbf{G}_2$ and $F_2\colon \mathbf{G}_2 \to \mathbf{G}_3$ be two homomorphism of groupoids, we will say that they define an exact sequence if $\ker F_2 = \operatorname{ran} F_1$. In such case, we will denote it by $\mathbf{G}_1 \xrightarrow{F_1} \mathbf{G}_2 \xrightarrow{F_2} \mathbf{G}_3$.

The definition of exact sequence can be extended without difficulty to include various steps, that is $\cdots \to \mathbf{G}_i \xrightarrow{F_i} \mathbf{G}_{i+1} \xrightarrow{F_{i+1}} \mathbf{G}_{i+2} \to \cdots$, with $\ker F_{i+1} = \operatorname{ran} F_i$, $i = 1,2,\dots,n$.

As in the case of groups, there is a fancy way of expressing that a given homomorphism of groupoids is monomorphism, epimorphism or isomorphism (Prop. 2.12).

Proposition 5.6
Let $\mathbf{G} \rightrightarrows \Omega$ and $\mathbf{G}' \rightrightarrows \Omega'$ be two groupoids and $F\colon \mathbf{G} \to \mathbf{G}'$ a groupoid homomorphism, then:

1. *The sequence of homomorphisms of groupoids $\mathbf{1}_\Omega \to \mathbf{G} \xrightarrow{F} \mathbf{G}'$, where the first homomorphism is the canonical inclusion, is exact iff F is a monomorphism.*

2. *The sequence of homomorphisms of groupoids $\mathbf{G} \xrightarrow{F} \mathbf{G}' \to \mathbf{1}_*$, where $\mathbf{1}_*$ is the trivial groupoid with only one object $\{*\}$ and one unit 1_*, and the last homomorphism is the canonical functor sending each morphism $\alpha'\colon x' \to y' \in \mathbf{G}'$ into 1_*, is exact iff F is an epimorphism.*

3. *The sequence of homomorphisms of groupoids* $\mathbf{1}_\Omega \to \mathbf{G} \xrightarrow{F} \mathbf{G}' \to \mathbf{1}_*$ *is exact iff* F *is an isomorphism.*

Proof 5.6 Regarding the first statement, we see clearly that if $\mathbf{1}_\Omega \to \mathbf{G} \xrightarrow{F} \mathbf{G}'$ is exact, then $\ker F = \operatorname{ran} i$, with $i\colon \mathbf{1}_\Omega \to \mathbf{G}$ the canonical inclusion functor, but $\operatorname{ran} i = \mathbf{1}_\Omega$ and F is a monomorphism.

To prove that F is epimorphism, we notice that the canonical functor $\mathbf{G}' \to \mathbf{1}_*$, is such that its kernel is $\mathbf{G}'$, but if the sequence is exact, then $\operatorname{ran} F = \mathbf{G}'$ and F is epimorphism.

Finally, if $\mathbf{1}_\Omega \to \mathbf{G} \xrightarrow{F} \mathbf{G}' \to \mathbf{1}_*$ is exact F is both monomorphism and epimorphism, thus, it is an isomorphism.

The results in Prop. 5.6 show that the subgroupoid of units $\mathbf{1}_\Omega$ plays the same role that the trivial subgroup $\{e\}$ of a group regarding monomorphisms while the trivial subgroupoid $\mathbf{1}_*$ with only one object and arrow, plays the role of the trivial subgroup with respect to epimorphisms.

Definition 5.6 A short exact sequence of groupoids is an exact sequence of homomorphisms of groupoids of the form:

$$\mathbf{1} \to \mathbf{H} \xrightarrow{F} \mathbf{G} \xrightarrow{F'} \mathbf{K} \to \mathbf{1}.$$

Clearly F is a monomorphism, F' is an epimorphism and $\ker F' = \operatorname{ran} F$. In such case, we will say that the groupoid $\mathbf{G}$ is an extension of the groupoid $\mathbf{K}$ by the groupoid $\mathbf{H}$.

Example 5.9 1. *If* G, H, K *are groups and we have a short exact sequence* $1 \to H \to G \to K \to 1$, *of homomorphisms of groups, it is also a short exact sequence of groupoids.*

2. *Given a groupoid* $\mathbf{G} \rightrightarrows \Omega$, *we have the exact sequence* $\mathbf{1}_\Omega \to \mathbf{G}_0 \to \mathbf{G}$ *as the canonical inclusion map of the fundamental subgroupoid in* $\mathbf{G}$ *is a monomorphism.*

3. *Consider the groupoid* $\mathbf{G}(\Omega) \times G$, *with* G *a group. Then the fundamental groupoid* $\mathbf{G}_0$ *is* $\Omega \times G$ *and the sequence of homomorphisms of groupoids* $\mathbf{1} \to \Omega \times G \to \mathbf{G}(\Omega) \times G \to \mathbf{G}(\Omega) \to \mathbf{1}$ *is a short exact sequence.*

4. *Let* $\mathbf{G} \rightrightarrows \Omega$ *be a groupoid and the subgroupoid* $\mathbf{G}_S(\Omega)$ *obtained by restricting* $\mathbf{G}$ *to a subset* $S \subset \Omega$ *(see Sect. 4.3.2), and Def. 3.7), then the sequence* $\mathbf{1} \to \mathbf{G}_S(\Omega) \to \mathbf{G}(\Omega)$ *is exact.*

5. *Consider a subgroupoid* $\mathbf{H} \rightrightarrows \Omega'$ *of a groupoid* $\mathbf{G} \rightrightarrows \Omega$, *then the sequence* $\mathbf{1}_{\Omega'} \to \mathbf{H} \to \widetilde{\mathbf{H}}$ *is exact where* $\widetilde{\mathbf{H}}$ *is the natural extension of the subgroupoid* $\mathbf{H}$, *Def. 3.5.*

We do not know yet how to construct good examples of epimorphisms of groupoids. Remember that an epimorphism of groups is defined by a normal subgroup, Thm. 2.12, a notion that we have not discussed yet for groupoids (see Sect. 6.1 for a detailed description). However, there is a general instance for canonical epimorphisms of groupoids as shown in the following exercise.

Exercise 5.10 *Given a groupoid $\mathbf{G} \rightrightarrows \Omega$, consider the canonical homomorphism of groupoids $\pi\colon \mathbf{G} \to \mathbf{G}(\Omega)$, $\pi(\alpha) = (y,x)$ if $\alpha\colon x \to y \in \mathbf{G}$. Prove that π is an epimorphism of groupoids. Hence, the sequence $\mathbf{G} \xrightarrow{\pi} \mathbf{G}(\Omega) \to \mathbf{1}$, is exact.*

5.3.3 Homomorphisms of groupoids, direct unions and products of groupoids

As in the case of groups, we would like to understand the structure of products and disjoint unions of groupoids by using the powerful notion of homomorphisms.

Let us recall from Exercise 3.22 that given the product $\mathbf{G}_1 \times \mathbf{G}_2 \rightrightarrows \Omega_1 \times \Omega_2$ of groupoids, the subset $\widetilde{\mathbf{G}}_1 = \{(\alpha_1, 1_{x_2}) \mid \alpha_1 \in \mathbf{G}_1, x_2 \in \Omega_2\}$ is a subgroupoid of $\mathbf{G}_1 \times \mathbf{G}_2$, and similarly $\widetilde{\mathbf{G}}_2 = \{(1_{x_1}, \alpha_2) \mid \alpha_2 \in \mathbf{G}_2, x_1 \in \Omega_1\}$. Alternatively we may define a canonical monomorphism that describes the groupoid $\mathbf{G}_1 \rightrightarrows \Omega_1$ as a subgroupoid of $\mathbf{G}_1 \times \mathbf{G}_2$.

First, note that the object space of $\mathbf{G}_1 \times \mathbf{G}_2$ is $\Omega_1 \times \Omega_2$ while the object space of $\mathbf{G}_1$ is just Ω_1. The natural way to 'enlarge' the object space of $\mathbf{G}_1$ to $\Omega_1 \times \Omega_2$ without changing the groupoid structure is to take the product with the groupoid of objects Ω_2. In fact, doing that, the product $\mathbf{G}_1 \times \Omega_2$ has object space $\Omega_1 \times \Omega_2$ and its morphisms are pairs $(\alpha_1, 1_{x_2})$, which is clearly the subset $\widetilde{\mathbf{G}}_1$ above. Thus denoting $\mathbf{G}_1 \times \Omega_2$, as $\mathbf{G}_1 \times \mathbf{1}_{\Omega_2}$ (or just $\mathbf{G}_1$ if there is no risk of confusion) we define a natural monomorphism $i_1\colon \mathbf{G}_1 \to \mathbf{G}_1 \times \mathbf{G}_2$ as $i_1(\alpha_1, 1_{x_2}) = (\alpha_1, 1_{x_2})$. Clearly $\ker i = 1_{\Omega_1 \times \Omega_2}$ and $\mathrm{ran}\,(i) = \widetilde{\mathbf{G}}_1$.

It is however easier to define the canonical projection homomorphism of groupoids $\pi_2\colon \mathbf{G}_1 \times \mathbf{G}_2 \to \mathbf{G}_2$ (similarly with $\pi_1\colon \mathbf{G}_1 \times \mathbf{G}_2 \to \mathbf{G}_1$) as $\pi_2(\alpha_1, \alpha_2) = \alpha_2$ (recall Example 5.8.2).

Exercise 5.11 *Prove that the short sequence $\mathbf{1}_{\Omega_1 \times \Omega_2} \to \mathbf{G}_1 \times \mathbf{1}_{\Omega_2} \xrightarrow{i_1} \mathbf{G}_1 \times \mathbf{G}_2 \xrightarrow{\pi_2} \mathbf{G}_2 \to \mathbf{1}_*$ is exact, i.e.,* $\ker \pi_2 = \mathrm{ran}\,(i_1)$.

We will find the following notion very useful.

Definition 5.7 Let $\pi\colon \mathbf{G} \to \mathbf{G}'$ be an epimorphism of groupoids. A section (also called a cross-section) of π is a map $\sigma\colon \mathbf{G}' \to \mathbf{G}$, such that $\sigma(1_{x'}) = 1_x$ with $\pi(x) = x'$, and for any $\alpha'\colon x' \to y' \in \mathbf{G}'$, then $\pi(\sigma(\alpha')) = \alpha'$. In general, a section is not a homomorphism, that is, in general $\sigma(\beta' \circ \alpha') \neq \sigma(\beta') \circ \sigma(\alpha')$, but, if it is, we will say that the short exact sequence $\mathbf{1} \to \ker \pi \to \mathbf{G} \to \mathbf{G}' \to \mathbf{1}$, splits.

In the cases we are dealing with, that is, all groupoids are sets (or even small categories), sections always exist. All that is needed in order to define a section is to choose for any $\alpha' : x' \to y' \in \mathbf{G}'$ a morphism $\alpha : x \to y$, with $\pi(x) = x'$ and $\pi(y) = y'$.

Example 5.12 *1. Let $\mathbf{G} \rightrightarrows \Omega$ be a groupoid and $\mathbf{G}_0 \rightrightarrows \Omega$ its totally disconnected fundamental groupoid, i.e., $\mathbf{G}_0 = \bigsqcup_{x\in\Omega} G_x$. Consider the canonical epimorphism $\pi_0 : \mathbf{G}_0 \to \mathbf{1}_\Omega$ given by $\pi_0(\gamma_x) = 1_x$, $\gamma_x \in G_x$. Then π_0 has a unique section, $\sigma(1_x) = 1_x$.*

2. The map $\sigma : \mathbf{G}_2 \to \mathbf{G}_1 \times \mathbf{G}_2$, given by $\sigma(\alpha_2) = (1_{x_1}, \alpha_2)$ (x_1 is a fixed point in Ω_1) is a section of π_2. Notice that σ is also a homomorphism, but there is no canonical choice.

Let $\mathbf{G} \rightrightarrows \Omega$ be a groupoid and consider the canonical epimorphism $\pi : \mathbf{G} \to \mathbf{G}(\Omega)$ described in Exercise 5.10, that is, $\pi(\alpha) = (y,x)$, with $\alpha : x \to y$. Then a section of π is a map $\sigma : \mathbf{G}(\Omega) \to \mathbf{G}$, such that $\sigma(x,x) = 1_x$ and $\sigma(y,x) : x \to y$ (because $\pi(\sigma(y,x)) = (y,x)$). The following proposition shows one of the many interesting uses of sections.

Proposition 5.7
Let $\mathbf{G} \rightrightarrows \Omega$ be a connected groupoid and σ a section of the canonical epimorphism $\pi : \mathbf{G} \to \mathbf{G}(\Omega)$, then, fixing a point $x_0 \in \Omega$, the section σ induces an isomorphism of groupoids $\varphi_{\sigma,x_0} : \mathbf{G}_0 \to \Omega \times G_{x_0}$, defined as $\varphi_{\sigma,x_0}(\gamma_x) = (1_x, \sigma(x_0,x) \circ \gamma_x \circ \sigma(x_0,x)^{-1})$.

Proof 5.7 We check that φ_{σ,x_0} is an isomorphism of groupoids. First it maps units into units: $\varphi_{\sigma,x_0}(1_x) = (1_x, \sigma(x_0,x) \circ 1_x \circ \sigma(x_0,x)^{-1}) = (1_x, \sigma(x_0,x) \circ \sigma(x_0,x)^{-1}) = (1_x, 1_{x_0})$ (where 1_{x_0} is the neutral element of G_{x_0}).

Moreover, it is a simple exercise to check that $\varphi_{\sigma,x_0}(\gamma_x\gamma_x') = \varphi_{\sigma,x_0}(\gamma_x)\varphi_{\sigma,x_0}(\gamma_x')$.

Finally, if $\gamma_x \in \ker \varphi_{\sigma,x_0}$, then $\sigma(x_0,x) \circ \gamma_x \circ \sigma(x_0,x)^{-1} = 1_x$, which implies that $\gamma_x = 1_x$, hence, φ_{σ,x_0} is a monomorphism and, clearly, $\mathrm{ran}(\varphi_{\sigma,x_0}) = \Omega \times G_{x_0}$.

In plain words, what Prop. 5.7 asserts is that, if the groupoid $\mathbf{G} \rightrightarrows \Omega$ is connected, we can identify the fundamental subgroupoid $\mathbf{G}_0$ with the simple product of groupoids $\Omega \times G$, by choosing a section of π, that is, a morphism on each set $\mathbf{G}(x,y)$, and a reference point x_0; then G is the isotropy group G_{x_0}. It is important to appreciate that all isotropy groups G_x of connected groupoids are isomorphic, but not canonically isomorphic, that is, there is not a 'preferred' way of identifying them. Sections provide a way of doing that, which shows that, in fact, there is not a canonical identification.

In the next chapter, it will be shown that if a section can be chosen to be a homomorphism, this will have dramatic consequences for the structure of the groupoid.

5.3.4 Groupoids of automorphisms

We will use the notion of homomorphism of groupoids to offer another way of understanding the action of a groupoid on a fibered set $\mu : M \to \Omega$.

When discussing the extension of Cayley's theorem for groupoids, Sect. 4.5, we introduced a very interesting notion, the groupoid $\mathrm{Aut}(\mu)$ of automorphisms of the map μ as the family of bijective maps $\varphi_{y,x} : M_x \to M_y$. That is, the groupoid of automorphisms of the map μ is the groupoid of the category μ**Sets** defined in Sect. 5.1.2. Recall that the morphisms of the category μ**Sets** are all maps $\varphi_{y,x} : M_x \to M_y$, thus, the groupoid of automorphisms of μ consists of the invertible morphisms of the category μ**Sets**. We will restate all these notions more formally.

Definition 5.8 Given a map $\mu : M \to \Omega$, we call the groupoid of the category μ**Sets**, the groupoid of automorphisms of μ. It will be denoted by $\mathrm{Aut}(\mu)$ or $\mathrm{Aut}(M)$. The space of objects of $\mathrm{Aut}(\mu)$ is Ω and the source and target maps are given by $s(\varphi_{y,x}) = x$, $t(\varphi_{y,x}) = y$. The composition law is the standard composition of maps and $1_x = \mathrm{id}_{M_x}$.

If the fibres M_x of the map $\mu : M \to \Omega$ carry a distinguished structure we will consider just maps $\varphi_{y,x} : M_x \to M_y$, preserving it. For instance if $\mu : E \to \Omega$ is a vector bundle, that is, $E_x = \mu^{-1}(x)$ is a linear space, the space of automorphisms of μ will consists of all linear isomorphisms $\varphi_{y,x} : E_x \to E_y$.

Example 5.13 *1. Consider the Cartesian product $\Omega_1 \times \Omega_2$ and the projection maps π_a, $a = 1, 2$, on each one of the factors. The groupoid of automorphisms of the map π_1, consists of all triples (y_1, φ, x_1), where $x_1, y_1 \in \Omega_1$ and $\varphi : \Omega_2 \to \Omega_2$ is a bijection. The composition law is given by $(z_1, \varphi, y_1) \circ (y_1, \psi, x_1) = (z_1, \varphi \circ \psi, x_1)$.*

2. *Consider a set M and an equivalence relation $\sim$ on it. Consider the set $\Omega = M/\sim$ of equivalence classes and the canonical projection map $\pi : M \to \Omega$. The groupoid of automorphisms of π is the set of triples $([y], \varphi_{yx}, [x])$, where $[x]$, $[y]$ denote the equivalence classes containing x and y, respectively, and $\varphi_{y,x} : [x] \to [y]$ is a bijection mapping the class $[x]$ into the class $[y]$. Notice that, if the set M is finite, then, if the equivalence classes $[x]$ and $[y]$ have different number of elements, there are no bijective maps from $[x]$ to $[y]$ and the connected components of the groupoid* $\mathrm{Aut}(\pi)$ *correspond to the connected components of the equivalence relation (considered as a groupoid, recall Prop. 3.4).*

Exercise 5.14 *Let G be a group and $H \subset G$ a subgroup. Describe the groupoid of automorphisms of the canonical projection map $\pi : G \to G/H$.*

Now, we can rephrase the description of an action of a groupoid provided by Prop. 5.2, by simply stating that an action of a groupoid $\mathbf{G} \rightrightarrows \Omega$ with moment map $\mu : M \to \Omega$ is a homomorphism of groupoids $\Psi : \mathbf{G} \to \mathrm{Aut}(\mu)$.

Example 5.15 *The left and right action of a groupoid on itself. We may recall again the left and right action of a groupoid* $\mathbf{G} \rightrightarrows \Omega$ *on itself, Examples 4.14.2 and 3, as the natural homomorphisms of groupoids* $L\colon \mathbf{G} \to \mathrm{Aut}\,(t)$ *and* $R\colon \mathbf{G} \to \mathrm{Aut}\,(s)$*, given by* $L_\alpha\beta = \alpha \circ \beta$ *and* $R_\alpha\beta = \beta \circ \alpha^{-1}$*, respectively.*

We will end this section by considering a particularly interesting instance of groupoid of automorphisms. Consider a groupoid $\mathbf{G} \rightrightarrows \Omega$ and the corresponding fundamental subgroupoid $\mathbf{G}_0$. Since the source and target maps coincide (the groupoid is totally disconnected), we may consider them as a moment map for an action: $\pi_0\colon \mathbf{G}_0 \to \Omega$. From this perspective, the fundamental group is a bundle of groups, the isotropy groups G_x. Then we may consider its groupoid of automorphisms.

Definition 5.9 We will denote by $\mathrm{Aut}\,(\mathbf{G}_0)$ the groupoid of automorphisms of the fundamental subgroupoid of the groupoid $\mathbf{G}$. It consists of all group isomorphisms $\varphi_{yx}\colon G_x \to G_y$.

5.4 Equivalence: Natural transformations

5.4.1 Equivalence of categories

The notion of equivalence of structures finds its natural realm in categorical language. Thus, two groups are isomorphic, the fancy name that the notion of equivalence takes when dealing with groups, if there is an isomorphism between them, that is a functor among them with an inverse. Exactly the same works with groupoids: Two groupoids are 'equivalent', that is, isomorphic, if there is a functor between them with an inverse. So, in full generality, we will say that two categories are equivalent if there is a functor between them with an inverse.

For instance, we can try our new notion with quivers. Consider two quivers $Q_1(\Omega_1, L_1)$ and $Q_2(\Omega_2, L_2)$, each of them generate a category $\mathbf{C}(Q_1)$ and $\mathbf{C}(Q_2)$. Then, we may say that the two quivers are equivalent if the categories $\mathbf{C}(Q_1)$ and $\mathbf{C}(Q_2)$ are equivalent, that is if there exists a functor $F\colon \mathbf{C}(Q_1) \to \mathbf{C}(Q_2)$ and another functor $G\colon \mathbf{C}(Q_2) \to \mathbf{C}(Q_1)$ such that $G(F(w_1)) = w_1$ for all $w_1 \in \mathbf{C}(Q_1)$ and $F(G(w_2)) = w_2$ for all $w_1 \in \mathbf{C}(Q_2)$. Note that, in particular, $G(F(x_1)) = x_1$ for all vertex $x_1 \in \Omega_1$ and $F(G(x_2)) = x_2$ for all vertex $x_2 \in \Omega_2$, however, it is not necessary that $G(F(l_k)) = l_k$ because the links of the quiver Q_1, that generate the category $\mathbf{C}(Q_1)$ are not necessarily transformed in the links generating $\mathbf{C}(Q_2)$ and viceversa[4].

In a sense, the notion of equivalence of categories, is nothing but the fact that we can use arbitrary 'names' for the objects and morphisms, and that the only relevant thing we are interested in is the 'relations' among them. There is, however, another aspect of the notion of equivalence that we should discuss now. When introducing the notion of functor, one of the many ways of thinking them is considering that a functor

[4]Note that the standard set-theoretical notion of equivalence of graphs is more restrictive than the categorical notion of equivalence of quivers used here.

$F\colon \mathbf{C}_1 \to \mathbf{C}_2$ describes a way of 'realizing' the category $\mathbf{C}_1$ in terms of the objects and arrows of the category $\mathbf{C}_2$, can we compare how similar two such 'realizations' are? That is, is there a natural way of comparing functors? The answer is positive and it comes by the hand of the notion of natural transformation among functors.

5.4.2 The notion of natural transformation

Definition 5.10 Given two functors, $F, G\colon \mathbf{C}_1 \to \mathbf{C}_2$ from the category $\mathbf{C}_1$ to the category $\mathbf{C}_2$, a natural transformation, denoted as $\Phi\colon F \Rightarrow G$, from the functor F to the functor G is an assignment to any object x in the category $\mathbf{C}_1$, of a morphism $\Phi(x)\colon F(x) \to G(x)$ in the category $\mathbf{C}_2$ such that:

$$\begin{array}{ccc} F(x) & \xrightarrow{F(\alpha)} & F(y) \\ {\scriptstyle \Phi(x)}\downarrow & & \downarrow{\scriptstyle \Phi(y)} \\ G(x) & \xrightarrow{G(\alpha)} & G(y) \end{array} \tag{5.8}$$

that is,

$$\Phi(y) \circ F(\alpha) = G(\alpha) \circ \Phi(x), \quad \forall \alpha \in \mathbf{C}_1(x,y). \tag{5.9}$$

We will say that the natural transformation $\Phi : F \Rightarrow G$ is an equivalence if $\Phi(x) : F(x) \to G(x)$ is an isomorphism for all objects x of the category $\mathbf{C}_1$.

Natural transformations can be composed. If $\Phi\colon F \Rightarrow G$ and $\Psi\colon G \Rightarrow H$, then we can define another natural transformation, denoted by $\Psi \circ \Phi\colon F \Rightarrow H$, as: $\Psi \circ \Phi(x) = \Psi(\Phi(x))$ for any object x in the category $\mathbf{C}_1$.

Exercise 5.16 *Prove that the composition $\Phi \circ \Psi$ of two natural transformations $\Phi\colon F \Rightarrow G$ and $\Psi\colon G \Rightarrow H$ is a natural transformation.*

If the natural transformation $\Phi\colon F \Rightarrow G$ is an equivalence, it is obvious that there exists another natural transformation $\Psi\colon G \Rightarrow F$, defined by $\Psi(x) = (\Phi(x))^{-1}$ such that $\Psi \circ \Phi(x) = 1_{F(x)}$ and $\Phi \circ \Psi(x) = 1_{G(x)}$. We will denote such natural transformation Ψ as Φ^{-1}.

Example 5.17 *1. Consider two groups G, G' and two group homomorphisms $f, f'\colon G \to G'$. From the point of view of categories, the homomorphisms $f, f'\colon G \to G'$ are functors from the category G to the category G'. The groups G, G', considered as categories, have just one object, denoted for instance, by $*$, identified with the neutral element, i.e., $1_* = e$. Thus, a natural transformation $\Phi\colon f \Rightarrow f'$, will be an assignment of a morphism $\Phi(*)\colon f(*) \to f'(*)$, that is, of a group element $g' \in G'$ (which are the morphisms in G'), such that, according to Eq. (5.9), $g' f'(g) = f(g) g'$, or equivalently, $f = \mathrm{Int}(g') \circ f'$.*

Note that, because any morphism on a group has an inverse, any natural transformation between group homomorphisms is an equivalence.

In other words, two group homomorphisms f, f' are equivalent if they are related by an inner automorphism, Def. 2.24. Thus, the equivalence classes of automorphisms of groups with respect to the natural equivalence provided by natural transformations is just the quotient group $\mathrm{Out}(G) = \mathrm{Aut}(G)/\mathrm{Int}(G)$ *of outer automorphisms.*

2. *If $\rho, \rho' \colon G \to$ **Vect** are two representations of the group G, a natural transformation $\Phi\colon \rho \Rightarrow \rho'$ is an assignment of a morphism $\Phi(*)\colon \rho(*) \to \rho'(*)$ such that:*

$$\begin{array}{ccc} V & \xrightarrow{\rho(g)} & V \\ {\scriptstyle \Phi(*)}\downarrow & & \downarrow{\scriptstyle \Phi(*)} \\ V' & \xrightarrow{\rho'(g)} & V' \end{array} \tag{5.10}$$

that is (compare with Eq. (5.4)):

$$\Phi(*) \circ \rho(g) = \rho'(g) \circ \Phi(*), \quad \forall g \in G. \tag{5.11}$$

We will also say that Φ intertwines the representations ρ and ρ' or is an intertwiner between the two representations ρ and ρ'.

The set of intertwiners between ρ and ρ' is just the family of natural transformations between the functors ρ and ρ', sometimes denoted by $\mathrm{Nat}(\rho, \rho')$.

Thus, if Φ is an isomorphism, we get the notion of equivalence of representations, i.e., ρ and ρ' are equivalent if there is a natural transformation between the functors ρ, ρ'.

There are many consequences of the previous examples that we would like to explore in what follows. We will postpone many of them because they will form the backbone of the second part of this work, all that are related to the notion and equivalence of representations of groupoids. However, we will close this section by using the previous arguments to establish the notion of equivalence of group (and groupoid) actions.

Thus, if we consider a group action as a functor $\Psi\colon G \to$ **Sets** (recall Example 5.1.2), where the set Ω upon which the group acts is the set assigned to the object $*$ of G, $\Omega = \Psi(*)$ and $\Psi(g)\colon \Omega \to \Omega$ is the morphism assigned to any $g \in G$, then two actions $\Psi\colon G \to$ **Sets**, $\Psi'\colon G \to$ **Sets** will be equivalent if there exists a natural transformation $\Phi\colon \Psi \Rightarrow \Psi'$, such that the morphism $\Phi(*)\colon \Psi(*) = \Omega \to \Psi'(*) = \Omega'$ is invertible, i.e., a bijection, and such that: $\Psi'(g) \circ \Phi(*) = \Phi(*) \circ \Psi(g)$. Denoting the bijective map $\Phi(*)\colon \Omega \to \Omega'$ just as $\varphi\colon \Omega \to \Omega'$, the previous conditions becomes $\Psi'_g \circ \varphi = \varphi \circ \Psi_g$, and, using the standard notation for actions, we obtain the simple expression: $g(\varphi(x)) = \varphi(gx)$ for all $x \in \Omega$, $g \in G$.

Exercise 5.18 *Work out the definition of equivalence for groupoid actions.*

Chapter 6

The Structure of Groupoids

This chapter will start with the delayed discussion of the notion of normal subgroupoids and quotient groupoids, notions that play an instrumental role in the theory of groups as the keys to understanding their structure.

The notion of simple groupoids will be introduced too and some of its consequences will be analyzed. Then, a general structure theorem that shows that a connected groupoid is an extension of a groupoid of pairs by its totally disconnect fundamental groupoid will be discussed.

A theorem of classification of groupoids up to order 20 will be stated and along its proof, notions from the general structure of groupoids will be applied and extended. For that, a simple cohomology theory will be developed.

6.1 Normal subgroupoids

As in the case of groups (recall the discussion in Sect. 2.1.2), given a subgroupoid **H** of a groupoid **G**, there are various equivalence relations associated to it. We may define the right equivalence relation, $\alpha \sim_R \alpha'$, where α is right equivalent to α' if there exists $\beta \in \mathbf{H}$ such that $\alpha' = \alpha \circ \beta$.

The relation $\sim_R$ is an equivalence relation: In fact $\alpha \sim_R \alpha$ for any α because $\alpha \circ 1_x = \alpha$ (recall that we may assume that the subgroupoid is full); it is transitive because if $\alpha \sim_R \alpha'$ (that is $\alpha' = \alpha \circ \beta$, $\beta \in \mathbf{H}$), and $\alpha' \sim_R \alpha''$ (again $\alpha'' = \alpha' \circ \beta'$, $\beta' \in \mathbf{H}$), then we get $\alpha'' = \alpha' \circ \beta' = \alpha \circ (\beta \circ \beta')$ with $\beta \circ \beta' \in \mathbf{H}$ and $\alpha'' \sim_R \alpha$ and, finally, we check that it is symmetric because if $\alpha \sim_R \alpha'$, then there exists $\beta \in \mathbf{H}$ such that $\alpha' = \alpha \circ \beta$ and, in consequence, $\alpha = \alpha' \circ \beta^{-1}$ and $\alpha' \sim_R \alpha$.

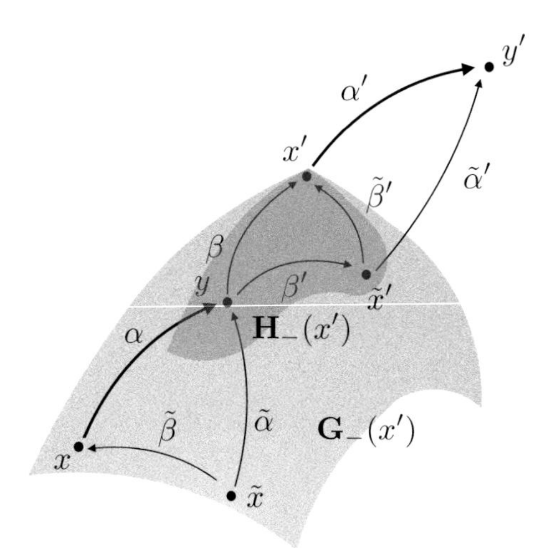

Figure 6.1: Pictorial description of the composability condition for right cosets. The coset $\alpha \circ \mathbf{H}$ will be composable with the coset $\alpha' \circ \mathbf{H}$ (whose representatives α and α' are represented by thick arrows) if there is $\beta \in \mathbf{H}$ 'connecting' α with α', that is, such that $s(\beta) = t(\alpha)$ and $t(\beta) = s(\alpha')$. In other words, the composition $(\alpha' \circ \mathbf{H}) \circ (\alpha \circ \mathbf{H})$ can be defined if $y = t(\alpha)$ is in the 'attraction basin' $s(\mathbf{H}_-(x'))$ of $x' = s(\alpha')$ with respect to $\mathbf{H}$ (dark shadow).

Definition 6.1 Let $\mathbf{H}$ be a subgroupoid of the groupoid $\mathbf{G} \rightrightarrows \Omega$, then we call the equivalence relation $\sim_R$, defined as $\alpha \sim_R \alpha'$ if there exists $\beta \in \mathbf{H}$ such that $\alpha' = \alpha \circ \beta$, the right equivalence relation associated to the subgroupoid $\mathbf{H}$ and 'right cosets' its corresponding equivalence classes $\alpha \circ \mathbf{H} = \{\alpha \circ \beta \mid \beta \in \mathbf{H}, t(\beta) = s(\alpha)\}$.

In a similar way we will define the left equivalence relation associated to the subgroupoid $\mathbf{H}$ as: $\alpha \sim_L \alpha'$ if there exists $\beta \in \mathbf{H}$ such that $\alpha' = \beta \circ \alpha$. The corresponding equivalence classes $\mathbf{H} \circ \alpha = \{\beta \circ \alpha \mid \beta \in \mathbf{H}, s(\beta) = t(\alpha)\}$ will be called left cosets.

The space of right cosets $\alpha \circ \mathbf{H}$ will be denoted by $\mathbf{G}/\mathbf{H}$. There is a canonical map $\pi_R \colon \mathbf{G} \to \mathbf{G}/\mathbf{H}$ given by $\alpha \mapsto \pi_R(\alpha) = \alpha \circ \mathbf{H}$. However, in general, π_R will not be a functor. For that, it would be necessary that $\mathbf{G}/\mathbf{H}$ would have the structure of a category, but, in general this will not be true. To understand why, notice that if $\mathbf{G}/\mathbf{H}$ were a category, there would be defined a composition law among right cosets $\alpha' \circ \mathbf{H}$ and $\alpha \circ \mathbf{H}$. For this to be possible it would be necessary for there to exist $\beta \in \mathbf{H}$ 'connecting' α with α', that is such that (see Fig. 6.1)

$$t(\alpha) = y = s(\beta) \qquad \text{and} \qquad t(\beta) = x' = s(\alpha'). \tag{6.1}$$

Using the terminology from actions of groupoids, we may also say that x' and y are in the same orbit of the groupoid $\mathbf{H}$. Then, we could define the composition

$$(\alpha' \circ \mathbf{H}) \circ (\alpha \circ \mathbf{H}) = (\alpha' \circ \beta \circ \alpha) \circ \mathbf{H}. \tag{6.2}$$

For this operation to be well defined we must check that if $\tilde{\alpha}' \circ \mathbf{H} = \alpha' \circ \mathbf{H}$ and $\tilde{\alpha} \circ \mathbf{H} = \alpha \circ \mathbf{H}$, that is $\tilde{\alpha}$ and $\tilde{\alpha}'$ are different elements in the cosets $\alpha \circ \mathbf{H}$ and $\alpha' \circ \mathbf{H}$ respectively, then $(\alpha' \circ \mathbf{H}) \circ (\alpha \circ \mathbf{H}) = (\tilde{\alpha}' \circ \mathbf{H}) \circ (\tilde{\alpha} \circ \mathbf{H})$. We realize first that, if $\tilde{\alpha}' \circ \mathbf{H} = \alpha' \circ \mathbf{H}$, then $\tilde{\alpha}' = \alpha' \circ \tilde{\beta}'$ for some $\tilde{\beta}' \in \mathbf{H}$ and, similarly, $\tilde{\alpha} \circ \mathbf{H} = \alpha \circ \mathbf{H}$, implies that $\tilde{\alpha} = \alpha \circ \tilde{\beta}$ (see Fig. 6.1). We notice immediately that $\tilde{\alpha} \circ \mathbf{H}$ and $\tilde{\alpha} \circ \mathbf{H}$ are composable as $\beta'' = (\tilde{\beta}')^{-1} \circ \beta \in \mathbf{H}$ such that $s(\beta'') = y = t(\tilde{\alpha})$ and $t(\beta'') = \tilde{x}' = s(\tilde{\alpha}')$. Moreover, let $\beta' \in \mathbf{H}$ such that $t(\beta') = \tilde{x}' = s(\tilde{\alpha}')$ and $s(\beta') = y = t(\tilde{\alpha})$, then:

$$\begin{aligned}
(\tilde{\alpha}' \circ \mathbf{H}) \circ (\tilde{\alpha} \circ \mathbf{H}) &= (\tilde{\alpha}' \circ \beta' \circ \tilde{\alpha}) \circ \mathbf{H} \\
&= (\alpha' \circ \tilde{\beta}' \circ \beta' \circ \alpha \circ \tilde{\beta}) \circ \mathbf{H} \\
&= (\alpha' \circ \beta \circ \beta^{-1} \circ \tilde{\beta}' \circ \beta' \circ \alpha) \circ \mathbf{H} \\
&= \alpha' \circ \beta \circ \alpha \circ (\alpha^{-1} \circ \beta^{-1} \circ \beta''' \circ \alpha) \circ \mathbf{H}, \qquad (6.3)
\end{aligned}$$

with $\beta''' = \tilde{\beta}' \circ \beta' \colon y \to x' \in \mathbf{H}$. Hence, because of (6.3), we get that $(\tilde{\alpha}' \circ \mathbf{H}) \circ (\tilde{\alpha} \circ \mathbf{H}) = \alpha' \circ \beta \circ \alpha \circ \mathbf{H} = (\alpha' \circ \mathbf{H}) \circ (\alpha \circ \mathbf{H})$ if and only if $\alpha^{-1} \circ \beta^{-1} \circ \beta''' \circ \alpha \in H_x$. Notice that $\beta^{-1} \circ \beta''' \colon y \to y \in H_y$.

A similar argument shows that if $\mathbf{H} \circ \alpha'$ and $\mathbf{H} \circ \alpha$ are two left cosets, then the composition $(\mathbf{H} \circ \alpha') \circ (\mathbf{H} \circ \alpha)$ could be defined if there exists $\beta \colon y \to x' \in \mathbf{H}$, in which case, we will define:

$$(\mathbf{H} \circ \alpha') \circ (\mathbf{H} \circ \alpha) = \mathbf{H} \circ (\alpha' \circ \beta \circ \alpha), \qquad (6.4)$$

and this composition law will be well defined if and only if $\alpha' \circ \beta''' \circ \beta^{-1} \circ \alpha'^{-1} \in H_{y'}$.

Thus, we have identified the condition that will allow us to define a composition law in the space of right (left) cosets of a connected groupoid: Given any two objects x, y of the subgroupoid $\mathbf{H}$, then it must happen that:

$$\alpha^{-1} \circ \beta \circ \alpha \in H_x \qquad \forall \alpha \colon x \to y, \quad \forall \beta \in H_y.$$

Notice that, if $\mathbf{H}$ and $\mathbf{G}$ are groups, the previous condition becomes $g^{-1}hg \in H$ for all $g \in G$, which is just the notion of normal subgroup, Def. 2.8. Thus, we will extend the terminology introduced for groups as follows.

Definition 6.2 We will say that the subgroupoid $\mathbf{H} \rightrightarrows \Omega'$ of the groupoid $\mathbf{G} \rightrightarrows \Omega$ is normal if the left and right equivalence relations defined by it coincide or, equivalently, if $\alpha^{-1} \circ \beta \circ \alpha \in H_x$ for all $\alpha \colon x \to y$, $\beta \in H_y$, and $x, y \in \Omega'$.

In other words, $\mathbf{H}$ is a normal subgroupoid if $\alpha^{-1} \circ \mathbf{H}_0 \circ \alpha = \mathbf{H}_0$ where $\mathbf{H}_0$ is the fundamental subgroupoid (recall Def. 3.12) of $\mathbf{H}$ or, written even in an alternative way, if $\alpha \circ \mathbf{H}_0 = \mathbf{H}_0 \circ \alpha$ for all $\alpha \in \mathbf{G}$.

Thus, we get the following characterization of quotient groupoids:

Theorem 6.1

Let $\mathbf{G} \rightrightarrows \Omega$ be a groupoid and $\mathbf{H} \rightrightarrows \Omega'$ a subgroupoid of $\mathbf{G}$. The family of right cosets $\mathbf{G}/\mathbf{H}$ (alternatively, the family of left cosets $\mathbf{H}\backslash\mathbf{G}$) inherits a groupoid structure such that the natural projection π becomes a homomorphism of groupoids if and only if the subgroupoid $\mathbf{H}$ is normal. In such case, the space of objects of $\mathbf{G}/\mathbf{H}$ (respec. of $\mathbf{H}\backslash\mathbf{G}$) is the space of $\mathbf{H}$-orbits $\Omega'/\mathbf{H}$ and the functor π_R (respec. π_L) maps each object $x \in \Omega$ into its orbit $\mathbf{H}x$.

Proof 6.1 First we will define a groupoid structure on $\mathbf{G}/\mathbf{H}$. The space of objects is the space of $\mathbf{H}$-orbits $\Omega/\mathbf{H}$, that is the space of equivalence classes of objects in Ω with respect to the action of the groupoid $\mathbf{H}$: x, y are in the same orbit if there exists $\beta : x \to y \in \mathbf{H}$. We will denote the $\mathbf{H}$-orbit passing through x as $[x]$ or $\mathbf{H}x$. Thus, the source and target maps $s, t \colon \mathbf{G}/\mathbf{H} \rightrightarrows \Omega/\mathbf{H}$ are defined as:

$$s(\alpha \circ \mathbf{H}) = \mathbf{H}x, \qquad t(\alpha \circ \mathbf{H}) = \mathbf{H}y, \qquad \alpha : x \to y.$$

Then, the cosets $\alpha \circ \mathbf{H}$ and $\alpha' \circ \mathbf{H}$ are composable if $t(\alpha \circ \mathbf{H}) = \mathbf{H}y = s(\alpha' \circ \mathbf{H}) = \mathbf{H}x'$, i.e., x' and y are in the same $\mathbf{H}$-orbit, or in other words, they are composable according to our previous definition, Def. 6.1. The composition of cosets is defined according to (6.2) and, as it was checked before, if $\mathbf{H}$ is normal, it is well defined (6.3). The unit elements are given by $1_{\mathbf{H}x} = 1_x \circ \mathbf{H} = \mathbf{H}_-(x)$, that is: $(\alpha \circ \mathbf{H}) \circ (1_x \circ \mathbf{H}) = \alpha \circ \mathbf{H}$. Finally, the inverse of $\alpha \circ \mathbf{H}$ is given by $\alpha^{-1} \circ \mathbf{H}$.

The canonical map $\pi_R \colon \mathbf{G} \to \mathbf{G}/\mathbf{H}$, $\pi_R(\alpha) = \alpha \circ \mathbf{H}$ is a homomorphism of groupoids (i.e., a functor). Notice that $\pi_R(1_x) = 1_x \circ \mathbf{H}$. and that if $\alpha : x \to y$, $\alpha' : y \to z$ are two composable elements, then $\pi_R(\alpha' \circ \alpha) = (\alpha' \circ \alpha) \circ \mathbf{H} = (\alpha' \circ \mathbf{H}) \circ (\alpha \circ \mathbf{H}) = \pi_R(\alpha') \circ \pi_R(\alpha)$.

We would proceed along similar lines to show that the space of left cosets $\mathbf{H}\backslash\mathbf{G}$ becomes a groupoid and the canonical map $\pi_L \colon \mathbf{G} \to \mathbf{H}\backslash\mathbf{G}$ is a groupoid homomorphism.

If the normal subgroupoid $\mathbf{H}$ is full, that is, its space of objects is Ω, the same as that of $\mathbf{G}$, then the space of objects of the quotient groupoid $\mathbf{G}/\mathbf{H}$ is just the space of orbits $\Omega/\mathbf{H}$.

The following examples are fundamental to understanding the notion of normal subgroupoid and its implications.

Example 6.1 *Normal subgroups. As it was indicated before, if $\mathbf{G}$ is a group, a subgroupoid $\mathbf{H}$ of G is just a subgroup $H \subset G$. Then, H is a normal subgroupoid if it is a normal subgroup, in which case, the space of right cosets G/H acquires a canonical group structure, that of the quotient group (Def. 2.9). It is a trivial exercise to check that such structure coincides with the groupoid structure defined above in the space of cosets $\mathbf{G}/\mathbf{H}$, i.e., $\mathbf{G}/\mathbf{H} = G/H$.*

Exercise 6.2 *Prove that if $\mathbf{H} \subset \mathbf{G}$ is a normal subgroupoid, then the isotropy groups H_x of $\mathbf{H}$ are normal subgroups of the isotropy groups G_x of $\mathbf{G}$.*

Proposition 6.1
Let $\mathbf{H} \subset \mathbf{G}$ be a normal subgroupoid, then the isotropy groups of the quotient groupoid $\mathbf{G}/\mathbf{H}$ are isomorphic to G_x/H_x.

Proof 6.2 The object space of $\mathbf{G}/\mathbf{H}$ is the space of $\mathbf{H}$-orbits, $\mathbf{H}x \in \Omega'/\mathbf{H}$. Let $\alpha \circ \mathbf{H} = \mathbf{H} \circ \alpha$ be a coset whose source and target is $\mathbf{H}x$, this means that $s(\alpha)$ and $t(\alpha)$ lie in $\mathbf{H}x$. Then there exists $\gamma \in H_x$ such that $\alpha \circ \gamma \colon x \to x$ or in other words, $\alpha \circ \gamma \in G_x$, that is $\alpha = g_x \gamma^{-1}$, with $g_x \in G_x$, and $\gamma \in H_x$. The coset $\alpha \circ \mathbf{H}$ can be identified with the coset $g_x H_x$.

As a consequence of the previous results, we get that, if $\mathbf{G}$ is finite and connected and the normal subgroup, $\mathbf{H}$ is full and all its isotropy subgroups have the same order, then the order of the quotient groupoid is $|\mathbf{G}/\mathbf{H}| = |\Omega/\mathbf{H}| \times |G_x/H_x|$. But $|\mathbf{G}| = |\Omega|^2|G_x|$, $\mathbf{H}| = |\Omega|^2|H_x|$ and $|\mathbf{G}/\mathbf{H}| = |\Omega/\mathbf{H}| \times |\mathbf{G}|/|\mathbf{H}|$ (compare with the situation for groups where, because of Lagrange's theorem, $[G:H] = |G/H| = |G|/|H|$).

However, normal subgroupoids of groupoids with non-trivial object space have new characteristics with respect to the situation with plain groups. Consider the following fundamental example.

Example 6.3 *Normal subgroups of the groupoid of pairs $\mathbf{G}(\Omega)$. Let us consider a subgroupoid $\mathbf{H}$ of a groupoid of pairs $\mathbf{G}(\Omega)$. The extension $\widetilde{\mathbf{H}}$ subgroupoid is the disjoint union of groupoids of pairs of subsets $\mathcal{O} \subset \Omega$ (where each subset $\mathcal{O}$ is an equivalence class of an equivalence relation on Ω, recall Thm. 3.2). Notice that the isotropy groups of the groupoid of pairs are trivial, and the same happens to the isotropy groups of any subgroupoid, hence the property defining normal subgroupoids, Def. 6.2, is trivially satisfied and we conclude that any subgroupoid of the groupoid of pairs is normal (even if the groupoid of pairs is far from being Abelian in any reasonable sense).*

Even more interesting is the description of the quotient groupoid $\mathbf{G}(\Omega)/\mathbf{H}$. For that, we will first consider the case that the equivalence relation defining the subgroupoid contains just one non-trivial equivalence class $\mathcal{O}$, all other equivalence classes consist of just a single element. Note that, in such case, if we denote by $\mathbf{H} = \mathbf{G}(\mathcal{O})$ the groupoid of pairs of $\mathcal{O}$, then the groupoid we are considering is just the extension of $\widetilde{\mathbf{H}} = \widetilde{\mathbf{G}(\mathcal{O})}$ of $\mathbf{H}$.

Then the space of cosets $\mathbf{G}(\Omega)/\widetilde{\mathbf{G}(\mathcal{O})}$ consist on the sets $\alpha \circ \mathbf{G}(\mathcal{O})$, but because of Thm. 6.1, its objects are the orbits of $\mathbf{G}(\mathcal{O})$, i.e., the equivalence classes themselves. Hence, we conclude that the quotient groupoid is the groupoid of pairs of the quotient set $\Omega/\sim$, i.e., the set obtained by identifying all elements in the subset $\mathcal{O}$ and collapsing all of them to a single point.

As in the case of groups, there are 'trivial' normal subgroupoids we are not interested in. In the case of groups, they were the total group G itself and the subgroup consisting of just the neutral element $\{e\}$. In the case of groupoids, apart from the total groupoid itself (which is always a normal subgroupoid), there is another 'trivial' normal subgroupoid.

Example 6.4 *The trivial subgroupoid of a groupoid. Given a groupoid* $\mathbf{G} \rightrightarrows \Omega$*, it always possesses a trivial subgroupoid which is the subgroupoid consisting of its units, that is the groupoid of objects* $\Omega = \mathbf{1}_\Omega$*. Clearly, this subgroupoid is normal and it plays the role of the trivial subgroup* $\{e\}$ *in the case of groups.*

Exercise 6.5

1. *Prove that given the direct product* $\mathbf{G}_1 \times \mathbf{G}_2 \rightrightarrows \Omega_1 \times \Omega_2$ *of groupoids, the subset* $\widetilde{\mathbf{G}}_1 = \{(\alpha_1, 1_{x_2}) \mid \alpha_1 \in \mathbf{G}_1, x_2 \in \Omega_2\} = \mathbf{G}_1 \times \mathbf{1}_{\Omega_2}$ *is a normal subgroupoid of* $\mathbf{G}_1 \times \mathbf{G}_2$ *(similarly with* $\widetilde{\mathbf{G}}_2 = \{(1_{x_1}, \alpha_2) \mid \alpha_2 \in \mathbf{G}_2, x_1 \in \Omega_1\} = \mathbf{1}_{\Omega_1} \times \mathbf{G}_2$*).*

2. *Consider the direct product* $\mathbf{G}(\Omega) \times G$*. Show that* $\widetilde{\mathbf{G}(\Omega)} = \mathbf{G}(\Omega)$ *and* $G \times \Omega = G \times \mathbf{1}_\Omega$ *is the fundamental groupoid of* $\mathbf{G}(\Omega) \times G$*. In particular,* $\mathbf{G}(\Omega)$ *is a connected normal subgroupoid of the product.*

We will end up this first glance to the notion of normal subgroupoids by observing that the connected components $\mathbf{G}_\mathcal{O}$ of a groupoid $\mathbf{G}$ are normal subgroupoids.

Proposition 6.2
Let $\mathbf{G} \rightrightarrows \Omega$ *be a groupoid and* $\mathbf{G} = \bigsqcup_{\mathcal{O} \in \Omega/\mathbf{G}} \mathbf{G}_\mathcal{O}$ *be its decomposition as a disjoint union of connected groupoids (Thm. 3.1). Then, each subgroupoid* $\mathbf{G}_\mathcal{O}$ *is normal and*

$$\mathbf{G}/\widetilde{\mathbf{G}}_\mathcal{O} = \bigsqcup_{\mathcal{O}' \in \Omega/\mathbf{G}, \mathcal{O}' \neq \mathcal{O}} \mathbf{G}_{\mathcal{O}'} . \tag{6.5}$$

Proof 6.3 Notice that $\alpha \circ \widetilde{\mathbf{G}}_\mathcal{O}$ is different from α only if $s(\alpha) \in \mathcal{O}$, in which case $\alpha \circ \widetilde{\mathbf{G}}_\mathcal{O} = \mathbf{G}_\mathcal{O}$ which is the same as $\widetilde{\mathbf{G}}_\mathcal{O} \circ \alpha$.

Finally, notice that $\mathbf{G}/\widetilde{\mathbf{G}}_\mathcal{O}$ is the set of cosets $\alpha \circ \widetilde{\mathbf{G}}_\mathcal{O}$, that is, they are the elements α themselves if they do not lie in $\mathbf{G}_\mathcal{O}$ or $\mathbf{G}_\mathcal{O}$ if $\alpha \in \mathbf{G}_\mathcal{O}$. Then, the space of cosets is the space $\mathbf{G}$ with the component $\mathbf{G}_\mathcal{O}$ collapsed to a point, that with a slight abuse of notation we write as in Eq. (6.5).

6.2 Simple groupoids

The previous theorem, Prop. 6.2, complements the first structure theorem of groupoids and allows us to understand their structure in terms of connected groupoids.

The fact that the fundamental (or isotropy) subgroupoid is normal too will be thoroughly discussed in the next section, for the moment we will just state:

Proposition 6.3
Let $\mathbf{G}$ *be a groupoid, then the fundamental subgroupoid* $\mathbf{G}_0$ *of* $\mathbf{G}$ *is normal.*

Proof 6.4 Notice that the isotropy groups of $\mathbf{G}_0$ are just the isotropy groups G_x of $\mathbf{G}$ and by construction $\alpha \circ G_x \circ \alpha^{-1} = G_y$, for any $\alpha : x \to y$.

Actually we can construct many disconnected subgroupoids of a given (connected) groupoid. It suffices to consider the restriction of the groupoid to any subset of its space of objetcs.

Proposition 6.4
Let $\mathbf{G} \rightrightarrows \Omega$ be a groupoid and $S \subset \Omega$ a subset. The subgroupoid $\mathbf{G}_S$ obtained by restriction of $\mathbf{G}$ to S, Def. 3.7, is normal.

Proof 6.5 The isotropy groups of $\mathbf{G}_S$ are the same as those of $\mathbf{G}$, then, if $x, y \in S$ and $\alpha : x \to y$, then $\alpha \in \mathbf{G}_S$ by definition. Moreover, $\alpha \circ G_x \circ \alpha^{-1} = G_y$ and $\mathbf{G}_S$ is normal.

Example 6.6

1. *The restriction of a groupoid to the subset $S = \{x\}$, $x \in \Omega$ is the isotropy group G_x, thus because of Prop. 6.4, G_x is a normal subgroupoid.*
2. *Let $\mathbf{G} \rightrightarrows \Omega$ be a groupoid and consider a partition of the space of objects Ω into two sets $\Omega = X \cup Y$, $X \cap Y = \emptyset$, and consider the disjoint union of the subgroupoids $\mathbf{G}_X \sqcup \mathbf{G}_Y$ obtained by restriction of $\mathbf{G}$ to either X and Y. Then $\mathbf{G}_X \sqcup \mathbf{G}_Y$ is a normal subgroupoid of $\mathbf{G}$.*

   Exercise 6.7 *Let $\mathbf{G} \rightrightarrows \Omega$ be a connected groupoid. Let $\Omega = \cup_{i=1}^{r} X_i$, $X_i \cap X_j = \emptyset$, $i \neq j$, be a partition of Ω Then the subgroupoid $\mathbf{H} = \bigsqcup_{i=1}^{r} \mathbf{G}_{X_i}$ is normal.*

3. *In particular, we may consider the partition given by the objects x of Ω themselves, then the groupoid described in (2) is just the fundamental subgroupoid $\mathbf{G}_0 = \bigsqcup_{x \in \Omega} G_x$.*

The previous discussions and examples show that a given groupoid has many natural normal subgroupoids, to begin with, its connected components. Even if the groupoid is connected there is always a totally disconnected normal subgroupoid, its fundamental subgroupoid, that has a natural trivial normal subgroupoid given by its units, and all normal subgroupoids (non-connected) that can be obtained by restricting the groupoid to a partition of its space of objects. Even if we consider just a subset of the space of objects, the restriction of the groupoid to this subset will not be connected if we consider its natural extension to the total space of objects. A simple way of not taking into consideration all those 'trivial' normal subgroupoids that any (connected) groupoid has, is to consider only connected subgroupoids, such that its space of objects is the total space, that is full connected subgroupoids. Taking all this into account it makes sense to extend the notion of simple groups to the class of groupoids.

Definition 6.3 We will say that the groupoid $\mathbf{G} \rightrightarrows \Omega$ is simple if it possesses no non-trivial full connected normal subgroupoids. We will call these normal subgroupoids proper.

Example 6.8 *1. Simple groups G are simple as groupoids too.*

2. The groupoid of pairs $\mathbf{G}(\Omega)$ is simple (it is connected and its isotropy group is $\{e\}$). Notice that the groupoid of pairs has many normal subgroupoids (any subset S of Ω defines a subgroupoid which is extended trivially to Ω) but they are not connected[1].

Proposition 6.5
The connected groupoid $\mathbf{G} = \mathbf{G}(\Omega) \times G$ has only the proper normal subgroupoid $\mathbf{G}(\Omega)$ if and only if G is simple.

Proof 6.6 Notice that if $\mathbf{G}$ had a proper normal subgroupoid $\mathbf{H}$, then H_x would be a normal subgroup of G, but G is simple, so either $H_x = \{e\}$ or $H_x = G$. In the second case, $\mathbf{H} = \mathbf{G}$, so that $H_x = \{e\}$ and $\mathbf{H} = \mathbf{G}(\Omega)$.

Conversely, if G is not simple, then there exists a proper normal subgroup H, then the subgroupoid $\mathbf{H} = \mathbf{G}(\Omega) \times H$ is normal.

Exercise 6.9 *Prove that if the groupoid $\mathbf{G}$ is simple, its isotropy groups G_x are simple.*

We will continue the discussion on simple groupoids after the discussion in the following section where we will obtain the second structure theorem for groupoids.

6.3 The structure of groupoids: Second structure theorem

Let us consider then a connected groupoid $\mathbf{G}$ over the object space Ω. As we know, the fundamental groupoid $\mathbf{G}_0$ is a totally disconnected normal subgroupoid of $\mathbf{G}$ with the obvious inclusion map $j \colon \mathbf{G}_0 \to \mathbf{G}$. Hence, as it was discussed in the previous section, Sect. 6.1, Prop. 6.1, because $\mathbf{G}_0$ is a normal subgroupoid of $\mathbf{G}$ we can form the corresponding quotient groupoid $\mathbf{G}/\mathbf{G}_0$. Then we can prove the following statement (recall Exercise 5.10):

Theorem 6.2
With the notations above, the groupoid $\mathbf{G}/\mathbf{G}_0$ is canonically isomorphic to the groupoid of pairs $\mathbf{G}(\Omega)$ of the object space Ω.

[1]That the groupoid of pairs is simple according to Def. 6.3 is particularly interesting as its algebra its simple too, which is not true for simple groups, whose algebras are just semisimple (see Chap. 10).

Proof 6.7 The quotient groupoid $\mathbf{G}/\mathbf{G}_0$ is defined as the set of classess $\alpha \circ \mathbf{G}_0 = \mathbf{G}_0 \circ \alpha$ (because $\mathbf{G}_0$ is normal), hence, the morphisms in the class $\alpha \circ \mathbf{G}_0$ have the form $\gamma_y \circ \alpha \circ \gamma_x$, with $\gamma_x \in G_x$, $\gamma_x \in G_x$.

It is clear then that there is a canonical identification between the quotient groupoid $\mathbf{G}/\mathbf{G}_0$ and the groupoid of pairs $\mathbf{G}(\Omega)$ provided by the assignment of the pair (y,x) to the coset $\alpha\mathbf{G}_0$ with $\alpha : x \to y$. Notice that the previous assignment defines an isomorphism between both groupoids because the composition of cosets is given by $(\alpha' \circ \mathbf{G}_0) \circ (\alpha \circ \mathbf{G}_0) = (\alpha' \circ \beta \circ \alpha) \circ \mathbf{G}_0$ with $\beta \in \mathbf{G}_0$, which implies that $\beta \in G_z$ for some $z \in \Omega$, then $\alpha : x \to z$, $\alpha' : z \to y$, the corresponding classes are identified with (z,x) and (y,z), respectively, and the composition with (y,x) showing that the indentifcation is a homomorphism. Checking that the identification is an isomorphism is trivial.

If we denote by $\pi : \mathbf{G} \to \mathbf{G}(\Omega)$ the map defined by the previous identification, $\pi(\alpha) = (y,x)$, provided $\alpha : x \to y$, then $\ker \pi = \mathbf{G}_0$ and the short sequence:

$$\mathbf{1}_\Omega \to \mathbf{G}_0 \to \mathbf{G} \stackrel{\pi}{\to} \mathbf{G}(\Omega) \to \mathbf{1}_* \,, \tag{6.6}$$

is exact, that is, $\ker \pi = \operatorname{ran} j$.

Therefore, in analogy with the corresponding terminology for groups, we will say that $\mathbf{G}$ is an extension of $\mathbf{G}(\Omega)$ by $\mathbf{G}_0$. We will call the short exact sequence of groupoids (6.6) the fundamental exact sequence of the groupoid $\mathbf{G} \rightrightarrows \Omega$ (or fundamental sequence for short). The structure of the fundamental sequence provides an alternative computation for the order of $|\mathbf{G}|$ to Cor. (4.1). Actually, we may prove:

Corollary 6.1
If $\mathbf{G} \rightrightarrows \Omega$ is a finite groupoid, then:

$$|\mathbf{G}| = \sum_{\mathcal{O} \in \Omega/\mathbf{G}} |G_x||\mathcal{O}|^2 \,. \tag{6.7}$$

where the sum is taken over the orbits $\mathcal{O}$ in the space of orbits $\Omega/\mathbf{G}$ of the groupoid $\mathbf{G}$, with G_x the isotropy group of any object x in $\mathcal{O}$, and $|\mathcal{O}|$ denotes the cardinal of the orbit $\mathcal{O}$. In particular, if $\mathbf{G} \rightrightarrows \Omega$ is a connected finite groupoid, then $|\mathbf{G}| = |G_x||\Omega|^2$.

Proof 6.8 The result is an immediate corollary of the first structure theorem,Thm. 3.1, Eq. 3.5 and the fact that for a connected groupoid, the normal subgroupoid $\mathbf{G}_0$ is full and its orbits are the objects in Ω. Then the space of orbits $\Omega/\mathbf{G}_0$ is Ω and $|\mathbf{G}/\mathbf{G}_0| = |\mathbf{G}(\Omega)| = |\Omega|^2$. Moreover because the fundamental sequence (6.6) is exact, $\ker \pi = \operatorname{ran} j$, thus for each $x \in \Omega$, $\pi^{-1}(x) = G_x$, then for each connected component of the groupoid we get $|\mathbf{G}_\mathcal{O}| = |\mathcal{O}|^2 |G_x|$.

Notice that the structure theorem above is void for groups as the fundamental subgroupoid of a group G is the group G itself and the corresponding quotient space consists of a single element.

The second structure theorem suggests that a groupoid is 'like' the product of the groupoid of pairs of its space of objects and the fundamental groupoid, actually an extension of $\mathbf{G}(\Omega)$ by $\mathbf{G}_0$. In the case of groups, the fact that a short exact sequence of groups $1 \to H \to G \to K = G/H \to 1$, H normal, splits implies that G is the semidirect product $H \rtimes_\omega K$ of K and H with respect the homomorphism $\omega\colon K \to \mathrm{Aut}(H)$ (Prop. 2.20).

We would expect a similar result in the case of groupoids but, because of the special structure of the extension defined by the fundamental sequence, we get more: We will show that if the fundamental sequence splits (recall Def. 5.7), the groupoid is the direct product of the groupoid of pairs and a group.

Theorem 6.3
Let $\mathbf{G} \rightrightarrows \Omega$ be a connected groupoid, then $\mathbf{G}$ is isomorphic to the direct product $\mathbf{G}(\Omega) \times G$, G a group, iff the fundamental short exact sequence (6.6) splits.

Proof 6.9 If $\mathbf{G} = \mathbf{G}(\Omega) \times G$, then the fundamental sequence splits because we can define the section $\sigma(y,x) = ((y,x),e)$ with e the neutral element of G. Clearly, the section σ is a groupoid homomorphism.

Suppose now that the fundamental sequence splits, that is, there is a section $\sigma\colon \mathbf{G}(\Omega) \to \mathbf{G}$ of the fundamental sequence which is a groupoid homomorphism. This means that $\sigma(x,y)\colon x \to y$, $\sigma(x,x) = 1_x$ and $\sigma(z,y) \circ \sigma(y,x) = \sigma(z,x)$. Moreover, because $\sigma(y,x)^{-1} \circ \sigma(y,x) = 1_x = \sigma(x,y) \circ \sigma(y,x)$, we get $\sigma(y,x)^{-1} = \sigma(x,y)$.

Consider the subset $\mathbf{H} = \{\sigma(x,y) | x,y \in \Omega\}$. Clearly $\mathbf{H}$ is a subgroupoid of $\mathbf{G}$ isomorphic to $\mathbf{G}(\Omega)$, the isomorphism provided by the section $\sigma\colon \mathbf{G}(\Omega) \to \mathbf{H}$ itself.

On the other hand, given the section σ and fixing a point $x_0 \in \Omega$, Prop. 5.7 shows that there is an isomorphism of groupoids $\varphi_{\sigma,x_0}\colon \mathbf{G}_0 \to \Omega \times G_{x_0}$, given by $\varphi_{\sigma,x_0}(\gamma_x) = (1_x, \sigma(x_0,x) \circ \gamma_x \circ \sigma(x,x_0))$.

Let us consider the map $\Lambda\colon \mathbf{G}(\Omega) \times G_{x_0} \to \mathbf{G}$ given by:

$$\Lambda((y,x),\gamma) = \sigma(y,x) \circ \varphi_{\sigma,x_0}^{-1}(1_x,\gamma),$$

with $x,y \in \Omega$ and $\gamma \in G_{x_0}$. We will show that the map Λ is an isomorphism of groupoids.

First we show that Λ is a homomorphism of groupoids:

$$\begin{aligned}
\Lambda((z,y),\gamma) \circ \Lambda((y,x),\gamma') &= \sigma(z,y) \circ \varphi_{\sigma,x_0}^{-1}(1_y,\gamma) \circ \sigma(y,x) \circ \varphi_{\sigma,x_0}^{-1}(1_x,\gamma') \\
&= \sigma(z,y) \circ \sigma(y,x_0) \circ \gamma \circ \sigma(x_0,y) \circ \sigma(y,x_0) \circ \gamma' \circ \sigma(x_0,x) \\
&= \sigma(z,x_0) \circ \gamma \circ \gamma' \circ \sigma(x_0,x) \\
&= \sigma(z,x_0) \circ \sigma(x_0,x) \circ \sigma(x,x_0) \circ \gamma \circ \gamma' \circ \sigma(x_0,x) \\
&= \sigma(z,x) \circ \varphi_{\sigma,x_0}^{-1}(1_x,\gamma \circ \gamma') \\
&= \Lambda((z,x),\gamma \circ \gamma') = \Lambda(((z,y),\gamma) \circ ((y,x),\gamma')).
\end{aligned}$$

Moreover, Λ is a monomorphism. In fact if $((y,x),\gamma) \in \ker\Lambda$, this means that $\sigma(y,x) \circ \varphi_{\sigma,x_0}^{-1}(\gamma) = 1_x$, then $\sigma(y,x) = \varphi_{\sigma,x_0}(\gamma) = \sigma(x,x_0) \circ \gamma \circ \sigma(x_0,x)$. In consequence $y = x$, and $1_x = \sigma(x,x_0) \circ \gamma \circ \sigma(x_0,x)$, hence, $\gamma = 1_{x_0}$ and $\ker\Lambda = \{((x,x),1_{x_0})\}$.

Finally we prove that Λ is an epimorphism. If we are given $\alpha : x \to y$, it is easy to check that $\gamma_{x_0} = \sigma(x_0,y) \circ \alpha \circ \sigma(x,x_0)$ satisfies $\Lambda((y,x),\gamma_{x_0}) = \alpha$:

$$\Lambda((y,x),\gamma_{x_0}) = \sigma(y,x) \circ \varphi_{\sigma,x_0}^{-1}(1_x,\gamma_{x_0}) = \sigma(y,x_0) \circ \gamma_{x_0} \circ \sigma(x_0,x) = \alpha\,.$$

Note that the key argument in the previous proof is the identification via the isomorphism φ_{σ,x_0}, provided by the chosen section σ, of the fundamental subgroupoid $\mathbf{G}_0$ with the product $\mathbf{G}(\Omega) \times G_{x_0}$. This also implies, as it will be discussed later on, that the analog of the homomorphism ω in the case of groups, recall Prop. 2.18, is trivial and the groupoid is a direct product.

Example 6.10 *1. The Loyd's groupoid $\mathfrak{L}_2$ is isomorphic to the product of groupoids $\mathbf{A}_2 \times \mathbb{Z}_3$. For that it is sufficient to check that its fundamental sequence splits. We do that by choosing the section $\sigma(2,1) = m_{1,2}$, $\sigma(3,2) = m_{2,3}$, $\sigma(4,3) = m_{3,4}$ and $\sigma(1,4) = (\sigma(2,1) \circ \sigma(3,2) \circ \sigma(4,3))^{-1} = m_{4,3}m_{3,2}m_{2,1}$.*

2. *Let Γ be a group and H a normal subgroup, such that Γ is isomorphic to the direct product $H \times \Gamma/H$. Consider the action groupoid $\mathbf{G}(\Gamma,\Gamma/H)$, Def. 4.7, corresponding to the natural action of the group Γ on the quotient space Γ/H. Then the fundamental sequence of the groupoid $\mathbf{G}(\Gamma,\Gamma/H)$ splits. In fact, it suffices to consider the section $\sigma(y,x) = (y,g,x)$ where $x,y \in G/H$, $g = (h,z)$, $h \in H$ and $y = zx$. Notice that given y,x in Γ/H, $z \in \Gamma/H$ is unique. Then, the decomposition $g = (h,z)$, $h \in H$ is unique and h is well defined. Then $\mathbf{G}(\Gamma,\Gamma/H) \cong H \times \mathbf{G}(\Gamma/H)$.*

Exercise 6.11 *Prove that all cyclic groupoids $\mathfrak{C}_n(r)$ split.*

Actually, as the previous examples suggest, it is easy to show that the fundamental sequence of a connected groupoid splits. For that, consider a section σ of the canonical epimorphism $\pi\colon \mathbf{G} \to \mathbf{G}(\Omega)$. Let us fix an object $x_0 \in \Omega$, and define the new section $\sigma'(y,x) = \sigma(y,x_0) \circ \sigma(x,x_0)^{-1}$. Then, we check easily that $\sigma'(x,x) = 1_x$ for all $x \in \Omega$, and:

$$\sigma'(z,y) \circ \sigma'(y,x) = \sigma(z,x_0) \circ \sigma(y,x_0)^{-1} \circ \sigma(y,x_0) \circ \sigma(x,x_0)^{-1} = \sigma'(z,x)\,,$$

which shows that σ' is a homomorphism. Then, Thm. 6.3 implies:

Corollary 6.2

Any connected groupoid $\mathbf{G} \rightrightarrows \Omega$, is isomorphic, albeit not canonically, to the direct product groupoid $\mathbf{G}(\Omega) \times G$, with G the isotropy group of $\mathbf{G}$.

We can take back the discussion we had on simple groupoids in Sect. 6.2. The next theorem shows that we can characterize them quite thoroughly:

Theorem 6.4
A groupoid $\mathbf{G} \rightrightarrows \Omega$ is simple iff it is connected, its isotropy groups are simple and the fundamental exact sequence does not split.

Proof 6.10 Because of Prop. 6.2, Exercise 6.9 and Thm. 6.3, the conditions in the theorem are necessary. But they are sufficient too.

Let $\mathbf{H} \subset \mathbf{G}$ be a connected normal subgroupoid. The isotropy groups H_x of $\mathbf{H}$ must be normal subgroups of G_x, but if G_x are simple, then H_x is either 1_x or G_x. If $H_x = 1_x$ for some $x \in \Omega$, then $H_y = 1_y$ for all $y \in \Omega$ (because we are assuming that $\mathbf{G}$ is connected). Alternatively if $H_x = G_x$ for some $x \in \Omega$, again the connectness condition implies that $H_y = G_y$ for all $y \in \Omega$, then we conclude that because $\mathbf{H}$ is connected it is either $\mathbf{G}$ or the fundamental sequence of $\mathbf{G}$ splits, because if $H_x = \{e\}$ for every $x \in \Omega$ implies that there is just one element in each $\mathbf{H}(x,y)$ for every $x,y \in \Omega$. Denote it by $\alpha_{x,y}$. Then $\mathbf{H} = \cup_{x,y\in\Omega}\mathbf{H}(x,y) = \{\alpha_{x,y} \colon y \to x \mid x,y \in \Omega\}$. Then there is a section $\sigma \colon \mathbf{G}(\Omega) \to \mathbf{G}$ defined as $\sigma(x,y) = \alpha_{x,y}$. But $\mathbf{H}$ is a subgroupoid which means that $\alpha_{x,y} \circ \alpha_{y,z} \in \mathbf{H}(x,z) = \{\alpha_{x,z}\}$, hence $\alpha_{x,y} \circ \alpha_{y,z} = \alpha_{x,z}$ and $\sigma(x,y) \circ \sigma(y,z) = \sigma(x,z)$, which implies that σ is a homomorphism of groupoids and the fundamental exact sequence splits.

Example 6.12 1. *Simple groups G are also simple as groupoids.*

2. *The Loyd's groupoid $\mathfrak{L}_2$ is not simple. It is connected, its isotropy group $\mathbb{Z}_3$ is simple but its fundamental sequence splits. It is the direct product of $\mathbf{A}_2$ and $\mathbb{Z}_3$.*

3. *The direct product $\mathbf{G}(\Omega) \times G$ is never simple because the fundamental sequence splits (it always has a proper normal subgroupoid, the subgroupoid $\mathbf{G}(\Omega)$).*

Then, because of Cor. 6.2, we conclude:

Corollary 6.3
The only simple groupoids are simple groups G, or groupoids of pairs $\mathbf{G}(\Omega)$.

6.4 Classification of groupoids up to order 20

One of the major achievements of any theory, not just in Mathematics, is to provide a classification of the subjects of their study. In our case, a classification of groupoids. Because groups are particular instances of groupoids, a classification of groupoids must contain a classification of groups. This is not a minor task, but one that has reached extraordinary success.

We will not try to reproduce the results already known on the classification of groups nor we will try to attempt a classification theory of finite groupoids. What we will offer in this section, however, as a closing point to the first part of this work, is a classification of 'small' groupoids. Of course 'small' is not well-defined and depends on the context. In our case, we have decided to discuss the classification of groupoids up to order 20 because, first, it can be done by using just elementary methods and, secondly, because it already exhibits a variety of groups and groupoids that summarises many of the ideas developed so far and opens the door to keep exploring this fascinating subject.

The following theorem will contain a list of groupoids (and groups) up to isomorphisms. Because of the first structure theorem for groupoids, only connected groupoids will be shown (because any groupoid is a disjoint union of connected groupoids). The first column of the table below will display the order of the groupoid $\mathbf{G}$ and the second column the order of its space of objects Ω. The third column will describe the isotropy groupoid $\mathbf{G}_0$ of the groupoid, that is, the group itself G if there is just one object, or the model isotropy group G - as all the isotropy groups G_x are isomorphic, $G_x \cong G$. Then, finally, the fourth column will list the groupoids of a given order (only if the object space has more than one point).

The proof of the theorem will split between the case of groups and groupoids with non-trivial object space.

Theorem 6.5

Classification of groupoids up to order 20. Up to isomorphisms, there are 72 different connected groupoids of order less than 20. Every connected groupoid of order $|\mathbf{G}| \leq 20$ is isomorphic to one of the list below.

The notations for groups in the table have mostly been introduced before: Abelian groups and semidirect products were discussed in Sects. 2.5.2 and 2.5.3, respectively. For the specific notations already introduced in the main text we have: D_n is the dihedral group of order 2n, S_n is the nth symmetric group, A_n is the alternating group of n indices. For the groups in the list not referred to before (see also Appendix B.3): $\mathrm{Dic}_n = Q_{4n}$ is the dicyclic group of order 4n and QD_n, is the quasi-dihedral group of order 2^n.

The notation for the groupoids in the last column is: $\mathbf{G}(n) = \mathbf{A}_n$ is the groupoid of pairs of n elements, $\mathfrak{C}_n(m)$ is the cyclic groupoid with n objects and shift of order m.

Table 6.1: Classification of groupoids up to order 20.

$\|\mathbf{G}\|$	$\|\Omega\|$	$\mathbf{G}_0(G)$	$\mathbf{G}$
1	1	$\mathbb{Z}_1$	
2	1	$\mathbb{Z}_2$	
3	1	$\mathbb{Z}_3$	
4	1	$\mathbb{Z}_4$, $\mathbb{Z}_2 \oplus \mathbb{Z}_2 = V_4$	
	2	$\mathbb{Z}_1$	$\mathbf{G}(2) = \mathbf{A}_2$
5	1	$\mathbb{Z}_5$	
6	1	$\mathbb{Z}_6$, $D_3 = S_3$	
7	1	$\mathbb{Z}_7$	
8	1	$\mathbb{Z}_8$, $\mathbb{Z}_4 \oplus \mathbb{Z}_2$, $\mathbb{Z}_2 \oplus \mathbb{Z}_2 \oplus \mathbb{Z}_2$, D_4, $\mathrm{Dic}_2 = Q_8$	
	2	$\mathbb{Z}_2$	$\mathbf{G}(2) \times \mathbb{Z}_2 = \mathfrak{C}_2(4)$
9	1	$\mathbb{Z}_9$, $\mathbb{Z}_3 \oplus \mathbb{Z}_3$	
	3	$\mathbb{Z}_1$	$\mathbf{G}(3) = \mathbf{A}_3$
10	1	$\mathbb{Z}_{10}$, D_5	
11	1	$\mathbb{Z}_{11}$	
12	1	$\mathbb{Z}_{12}$, $\mathbb{Z}_6 \oplus \mathbb{Z}_2$, D_6, A_4, $\mathrm{Dic}_3 = Q_8$	
	2	$\mathbb{Z}_3$	$\mathbf{G}(2) \times \mathbb{Z}_3 = \mathfrak{C}_2(3)$
13	1	$\mathbb{Z}_{13}$	
14	1	$\mathbb{Z}_{14}$, D_7	
15	1	$\mathbb{Z}_{15}$	
16	1	$\mathbb{Z}_{16}$, $\mathbb{Z}_4 \oplus \mathbb{Z}_4$, $\mathbb{Z}_8 \oplus \mathbb{Z}_2$, $\mathbb{Z}_4 \oplus \mathbb{Z}_2 \oplus \mathbb{Z}_2$, $\mathbb{Z}_2 \oplus \mathbb{Z}_2 \oplus \mathbb{Z}_2 \oplus \mathbb{Z}_2$, $(\mathbb{Z}_2 \oplus \mathbb{Z}_2) \rtimes \mathbb{Z}_4$, $(\mathbb{Z}_4 \oplus \mathbb{Z}_2) \rtimes \mathbb{Z}_2$, $\mathbb{Z}_4 \rtimes \mathbb{Z}_4$, $\mathbb{Z}_8 \rtimes \mathbb{Z}_2$, D_8, $\mathrm{Dic}_4 = Q_{16}$, $D_4 \times \mathbb{Z}_2$, QD_4, $Q_8 \times \mathbb{Z}_2$	
	2	$\mathbb{Z}_4$	$\mathbf{G}(2) \times \mathbb{Z}_4 = \mathfrak{C}_2(8)$
		$V_4 = \mathbb{Z}_2 \oplus \mathbb{Z}_2$	$\mathbf{G}(2) \times V_4$
	4	$\mathbb{Z}_1$	$\mathbf{G}(4) = \mathbf{A}_4$
17	1	$\mathbb{Z}_{17}$	
18	1	$\mathbb{Z}_{18}$, $\mathbb{Z}_6 \oplus \mathbb{Z}_3$, $(\mathbb{Z}_3 \oplus \mathbb{Z}_3) \rtimes \mathbb{Z}_2$, $\mathbb{Z}_3 \times S_3$, D_9	
	3	$\mathbb{Z}_2$	$\mathbf{G}(3) \times \mathbb{Z}_2 = \mathfrak{C}_3(2)$
19	1	$\mathbb{Z}_{19}$	
20	1	$\mathbb{Z}_{20}$, $\mathbb{Z}_{10} \times \mathbb{Z}_2$, D_{10}, Dic_5, $\mathrm{Hol}(\mathbb{Z}_5) = \mathrm{Aut}(\mathbb{Z}_5) \ltimes \mathbb{Z}_5$	
	2	$\mathbb{Z}_5$	$\mathbf{G}(2) \times \mathbb{Z}_5 = \mathfrak{C}_2(5)$

Because of the second structure theorem for groupoids, the order of a groupoid is of the form $n^2 r$, where n is the order of Ω and r is the order of the isotropy group. Thus, in the list, there will be groupoids with object space possessing more than one point in the case of orders $|\mathbf{G}| = 2^2 r$ $(r = 1,2,3,4,5)$; $|\mathbf{G}| = 3^2 r$ $(r = 1,2)$; and $|\mathbf{G}| = 4^2$, that is, corresponding to orders $|\mathbf{G}| = 4,8,9,12,16,18,20$. Thus, all groupoids appearing in the list which are not groups (i.e., with object space with more than one point) have isotropy groups of order less or equal than five, consequently all of them are Abelian. The first groupoid with non-trivial object space and non-Abelian isotropy group will have order 24 $(= 2^4 \times 6)$.

Proof 6.11 As indicated above, the proof will split into the case of groups and groupoids with non-trivial object space. Because of Cor. 6.2 the proof for groupoids with non-trivial space of objects is trivial.

The proof for groups will consider separately the case of Abelian groups and non-Abelian ones. The classification of Abelian groups follows as consequence of the Fundamental Theorem of Finite Abelian groups, Thm. 2.13. Groups of prime order are cyclic (see Exercise 2.17), this determines orders 2, 3, 5, 7, 11,13, 17 and 19.

The non-Abelian cases follow from various theorems and *ad hoc* results. The proof up to order 15 can be done without relying on them, see, for instance, the recent paper by G. Thompson [62], where the structure of the conjugation classes of the group is exploited. A similar approach is used in [25], where groups of order less than 32 are classified by emphasizing the role of the conjugation classes of the group.

6.5 Groupoids with Abelian isotropy group

Even if not strictly necessary for the proof of the classification theorem above, Thm. 6.5, we will conclude this chapter by introducing two notions that are relevant in the study of topological groupoids, the modular invariant and the canonical cohmology class, that can be easily described with the tools introduced so far.

6.5.1 The modular invariant of the fundamental sequence

First, inspired by the discussion on the structure of groups in Sect. 2.5.3, we will discuss the natural extension of the modular invariant ω of an extension of groups, to groupoids.

We will consider the action of the groupoid $\mathbf{G} \rightrightarrows \Omega$ on its fundamental groupoid $\mathbf{G}_0$ by conjugation, that is, given $\alpha \colon x \to y \in \mathbf{G}$, we consider the map $\omega(\alpha) \colon G_x \to G_y$, given by:

$$\omega(\alpha)(\gamma) = \alpha \circ \gamma \circ \alpha^{-1}, \qquad \gamma \in G_x .$$

That is, ω is a functor from $\mathbf{G}$ to $\mathrm{Aut}(\mathbf{G}_0)$ with $\mathrm{Aut}(\mathbf{G}_0)$ the groupoid of automorphisms of the groupoid $\mathbf{G}_0$ (Sect. 5.3.4, Def. 5.9).

Let $\sigma \colon \mathbf{G}(\Omega) \to \mathbf{G}$ be a cross-section of the canonical functor $\pi \colon \mathbf{G} \to \mathbf{G}(\Omega)$, that is, σ maps the element $(y,x) \in \mathbf{G}(\Omega)$ into a morphism $\alpha = \sigma(y,x) \colon x \to y \in \mathbf{G}$, i.e., $\sigma(y,x) \in \mathbf{G}(y,x)$. However, in general, we cannot guarantee that σ is a functor, that is in general $\sigma((z,y) \circ (y,x)) \neq \sigma(z,y) \circ \sigma(y,x)$. What is true instead, is that, both $\sigma(z,x)$ and $\sigma(z,y) \circ \sigma(y,x)$ belong to $\mathbf{G}(z,x)$, then there would exist $\lambda(z,y,x) \in G_z$ such that:

$$\sigma(z,y) \circ \sigma(y,x) = \lambda(z,y,x) \circ \sigma(z,x) .$$

We may always choose the section σ, such that $\sigma(x,x) = 1_x$ for all $x \in \Omega$, then we get that $\lambda(x,x,x) = 1_x$. It would also be convenient to choose the section σ such that $\sigma(x,y) = \sigma(y,x)^{-1}$ for every $x,y \in \Omega$. Moreover, because $1_x = \sigma(x,x) = \sigma(x,y) \circ \sigma(y,x) = \lambda(x,y,x)\sigma(x,x)$, we get the normalization conditions for λ:

$$\lambda(x,y,x) = e , \qquad \lambda(x,x,y) = e , \qquad \lambda(y,x,x) = e , \tag{6.8}$$

where e is the neutral element of the corresponding isotropy group, and the remaining equalities follow from $\sigma(y,x) \circ \sigma(x,x) = \sigma(y,x)$ and $\sigma(y,y) \circ \sigma(y,x) = \sigma(y,x)$. Moreover, notice that $(\sigma(z,y) \circ \sigma(y,x))^{-1} = \sigma(x,y) \circ \sigma(y,z) = \lambda(x,y,z)\sigma(x,z) = \lambda(x,y,z)\sigma(z,x)^{-1}$. Then we get:

$$\sigma(x,z) \circ \lambda(z,y,x)^{-1} = \lambda(x,y,z) \circ \sigma(x,z), \qquad \forall x,y,z \in \Omega. \tag{6.9}$$

Now, we will restrict the map ω to the graph of σ, i.e., to morphisms α of the form $\sigma(x,y)$. Then we get:

$$\omega(\sigma(y,x))(\gamma_x) = \sigma(y,x) \circ \gamma_x \circ \sigma(x,y) \in G_y,$$

for every $\gamma_x \in G_x$. Denoting by ω^σ the composed map $\omega \circ \sigma$, we get

$$\begin{aligned} \omega^\sigma(z,y) \circ \omega^\sigma(y,x)(\gamma_x) &= \sigma(z,y) \circ \sigma(y,x) \circ \gamma_x \circ \sigma(x,y) \circ \sigma(y,z) \\ &= \lambda(z,y,x) \circ \sigma(z,x) \circ \gamma_x \circ \sigma(x,z) \circ \lambda(z,y,x)^{-1} \\ &= \mathrm{Int}(\lambda(z,y,x))(\omega^\sigma(z,x)(\gamma_x)). \end{aligned}$$

Notice that $\mathrm{Int}(\lambda(z,y,x))$ is the inner automorphism of the group G_z defined by the element $\lambda(z,y,x)$ acting on the element $\omega^\sigma(z,x)(\gamma_x) \in G_z$, then we get:

$$\omega^\sigma(z,y) \circ \omega^\sigma(y,x) = \mathrm{Int}(\lambda(z,y,x))\omega^\sigma(z,x),$$

and ω^σ defines a homomorphism of groupoids from $\mathbf{G}(\Omega)$ to the groupoid obtained by quotienting the groupoid of automorphisms by the normal subgroupoid of inner automorphisms, i.e., $\mathrm{Aut}(\mathbf{G}_0)/\mathrm{Int}(\mathbf{G}_0)$, where $\mathrm{Int}(\mathbf{G}_0)$ is the normal subgroupoid of inner automorphisms of $\mathbf{G}_0$ (see Sect. 5.3.4). The groupoid $\mathrm{Aut}(\mathbf{G}_0)/\mathrm{Int}(\mathbf{G}_0)$ will be called, as in the case of groups, the groupoid of outer automorphisms of $\mathbf{G}_0$ and denoted accordingly $\mathrm{Out}(\mathbf{G}_0)$.

Let us suppose that we choose a different section σ' of π, then $\sigma'(y,x) = \mu(y,x) \circ \sigma(y,x)$, with $\mu(y,x) \in G_y$. Then, repeating the argument before we obtain the map $\omega^{\sigma'} : \mathbf{G}(\Omega) \to \mathrm{Aut}(\mathbf{G}_0)$ as:

$$\begin{aligned} \omega^{\sigma'}(y,x)(\gamma) &= \sigma'(y,x) \circ \gamma \circ \sigma'(x,y) = \mu(y,x) \circ \sigma(y,x) \circ \gamma \circ \sigma(x,y) \circ \mu(y,x)^{-1} \\ &= \mathrm{Int}(\mu(y,x))(\omega^\sigma(y,x)(\gamma)), \end{aligned}$$

and then we conclude that ω^σ and $\omega^{\sigma'}$ define the same homomorphism of groupoids $\tilde{\omega} \colon \mathbf{G}(\Omega) \to \mathrm{Out}(\mathbf{G}_0)$.

Definition 6.4 The homomorphism of groupoids $\tilde{\omega} \colon \mathbf{G}(\Omega) \to \mathrm{Out}(\mathbf{G}_0)$ is called the modular invariant of the short exact sequence $\mathbf{1} \to \mathbf{G}_0 \to \mathbf{G} \stackrel{\pi}{\to} \mathbf{G}(\Omega) \to \mathbf{1}$.

Notice that the previous definition of the modular invariant works for any short exact sequence $\mathbf{1} \to \mathbf{H} \to \mathbf{G} \stackrel{\pi}{\to} \mathbf{K} = \mathbf{G}/\mathbf{H} \to \mathbf{1}$ where $\mathbf{H}$ is a normal subgroupoid of $\mathbf{G}$.

Exercise 6.13 *Define and compute the modular invariant of the short exact sequence defined by a normal subgroupoid of the groupoid of pairs of a set.*

Fortunately, the modular invariant of the fundamental sequence can be computed easily and it turns to be trivial, as it should.

Theorem 6.6
Let $\mathbf{G} \rightrightarrows \Omega$ be a finite connected groupoid, then the modular invariant $\tilde{\omega}\colon \mathbf{G}(\Omega) \to \mathrm{Out}(\mathbf{G}_0)$ of the fundamental sequence of $\mathbf{G}$ is trivial, i.e, $\tilde{\omega} = 1$.

Proof 6.12 It is an obvious consequence of the definitions. Notice that $\mathrm{Aut}(\mathbf{G}_0) \cong \mathbf{G}(\Omega) \times \mathrm{Aut}(G_x)$, x a fixed element in Ω. The isomorphism can be constructed by using a section $\sigma\colon \mathbf{G}(\Omega) \to \mathbf{G}$ of the canonical projection. Then, we define the isomorphism $A\colon \mathbf{G}_0 \to \Omega \times G_x$, where G_x is the isotropy group of a given fixed point $x \in \Omega$, $A(\gamma_y) = (y, \sigma(x,y) \circ \gamma_y \circ \sigma(y,x))$. Clearly, $A(\gamma_y) \in G_x$ for any $\gamma_y \in G_y$. Because $\mathbf{G}_0$ is totally disconnected there are no morphisms joining any two points in Ω. The groupoid structure on $\Omega \times G_x$ is the trivial one. Then A is an isomorphism of groupoids.

Using A, we may define the isomorphism $\mathrm{Aut}(\mathbf{G}_0) \cong \mathbf{G}(\Omega) \times \mathrm{Aut}(G_x)$. Consider $\varphi \in \mathrm{Aut}(\mathbf{G}_0)$, this means that φ is a group isomorphism from G_y to G_z for some $y,z \in \Omega$. Now, define the element $\tilde{\varphi}$ in $\mathrm{Aut}(\Omega \times G_x)$ as $\tilde{\varphi} = A \circ \varphi \circ A^{-1}$. That is, $\tilde{\varphi}(\gamma) = \sigma(x,z) \circ \varphi(\sigma(y,x) \circ \gamma \circ \sigma(x,y)) \circ \sigma(z,x)$ and $\tilde{\varphi}$ is a group homomorphism of the isotropy group at y of $\Omega \times G_x$ (which is G_x).

But now, when we compute the modular invariant using the previous isomorphism $\varphi = \omega^\sigma(z,y)$, we get

$$\begin{aligned} A \circ \omega^\sigma(z,y) \circ A^{-1}(\gamma) &= \sigma(x,z) \circ \omega^\sigma(z,y)(\sigma(y,x) \circ \gamma \circ \sigma(x,y)) \circ \sigma(z,x) \\ &= \lambda(z,y,x) \circ \gamma \circ \lambda(x,y,z) = \mathrm{Int}(\lambda(z,y,x))(\gamma). \quad (6.10) \end{aligned}$$

In the particular instance that the isotropy group G_x of $\mathbf{G}$ is Abelian, Eq. (6.10) shows that ω^σ is actually trivial. Notice that, in such case, $\mathrm{Out}(\mathbf{G}_0) = \mathrm{Out}(\Omega \times G_x) = \Omega \times \mathrm{Out}(G_x) = \Omega \times \mathrm{Aut}(G_x)$.

6.5.2 The canonical cohomology class

In what follows, we will concentrate on the case that the isotropy group of our groupoid $\mathbf{G}$ is Abelian, this is actually the case that concerns us regarding the proof of the classification theorem[2]. So, we will consider a finite connected groupoid $\mathbf{G} \rightrightarrows \Omega$ with isotropy group the finite Abelian group A. Thus, we identify the fundamental groupoid $\mathbf{G}_0$ with the groupoid $\Omega \times A$. Then, given a section σ, we may consider the function $\lambda(z,y,x)$ associated to it as a map $\lambda\colon \Omega \times \Omega \times \Omega \to A$ satisfying the

[2]The theory can be developed in similar ways for non-Abelian isotropy groups but it becomes more involved and it will not be discussed in this work.

properties (6.8) and (6.9). In addition, because of the associativity property of the composition law of groupoids, we get:

$$(\sigma(u,z)\circ\sigma(z,y))\circ\sigma(y,x)=\sigma(u,z)\circ(\sigma(z,y)\circ\sigma(y,x)), \tag{6.11}$$

for any four points x,y,z,u in Ω. Then we obtain from the r.h.s. of (6.11):

$$(\sigma(u,z)\circ\sigma(z,y))\circ\sigma(y,x)=\lambda(u,z,y)\circ\sigma(u,y)\circ\sigma(y,x)=\lambda(u,z,y)\lambda(u,y,x)\circ\sigma(u,x).$$

On the other hand, from the l.h.s. of (6.11) we get:

$$\sigma(u,z)\circ(\sigma(z,y)\circ\sigma(y,x))=\sigma(u,z)\circ\lambda(z,y,x)\circ\sigma(z,x)=\lambda(z,y,x)\lambda(u,z,x)\circ\sigma(u,x),$$

where we have use the fact that $\omega^\sigma=1$ for Abelian isotropy groups. Then, we obtain the following equation for λ:

$$\lambda(u,z,y)\lambda(u,y,x)=\lambda(z,y,x)\lambda(u,z,x), \qquad \forall x,y,z,u\in\Omega. \tag{6.12}$$

Definition 6.5 A function $\lambda:\Omega\times\Omega\times\Omega\to A$, A an Abelian group, satisfying (6.8), (6.9) and (6.12) will be called a 2-cocycle on the groupoid of pairs $\mathbf{G}(\Omega)$ with values in the Abelian group A. Equation (6.12) will be also called the cocycle condition.

However the function λ depends on the choice of a section σ. What happens if we change the section? Suppose that we pick up another section σ' (satisfying the same properties $\sigma'(x,x)=e$ and $\sigma'(x,y)=\sigma'(y,x)^{-1}$ as σ). Then we have:

$$\sigma'(y,x)=\mu(y,x)\circ\sigma(y,x), \qquad \mu(y,x)\in G_y.$$

With the identification $\mathbf{G}_0\cong\Omega\times A$ discussed before, the function $\mu:\Omega\times\Omega\to A$, satisfies $\mu(x,x)=e$, for all $x\in\Omega$. Then, we compute on one side $\sigma'(z,y)\circ\sigma'(y,x)=\lambda'(z,y,x)\circ\sigma'(z,x)=\lambda'(z,y,x)\mu(z,x)\sigma(z,x)$ and on the other: $\sigma'(z,y)\circ\sigma'(y,x)=\mu(z,y)\circ\sigma(z,y)\circ\mu(y,x)\circ\sigma(y,x)=\mu(z,y)\mu(y,x)\circ\sigma(z,y)\circ\sigma(y,x)=\mu(z,y)\mu(y,x)\lambda(z,y,x)\circ\sigma(z,x)$. Hence, we conclude:

$$\lambda'(z,y,x)=\mu(z,y)\mu(y,x)\mu(z,x)^{-1}\lambda(z,y,x). \tag{6.13}$$

Then, we will say that two functions λ', λ related by (6.13) are equivalent. Notice that since all functions λ, λ' and μ take values in the Abelian group A we are keeping the multiplicative notation.

Giving a function $\mu:\Omega\times\Omega\to A$, we define a map ∂ taking μ into a function $\partial\mu:\Omega\times\Omega\times\Omega\to A$, given by $\partial\mu(z,y,x)=\mu(z,y)\mu(y,x)\mu(z,x)^{-1}$. Clearly $\partial\mu(y,x,x)=e$, and if we choose $\mu(x,y)=\mu(y,x)^{-1}$, then $\partial\mu(x,y,x)=e$. Finally, notice that:

$$\begin{aligned}\partial\mu(u,z,y)\partial\mu(u,y,x) &= \mu(u,z)\mu(z,y)\mu(y,u)\mu(u,y)\mu(y,x)\mu(x,u)\\ &= \mu(u,z)\mu(z,y)\mu(y,x)\mu(x,u),\end{aligned}$$

while

$$\begin{aligned} \partial\mu(u,z,x)\partial\mu(z,y,x) &= \mu(u,z)\mu(z,x)\mu(x,u)\mu(z,y)\mu(y,x)\mu(x,z) \\ &= \mu(u,z)\mu(z,y)\mu(y,x)\mu(x,u) . \end{aligned}$$

Then, we get

$$\partial\mu(u,z,y)\partial\mu(u,y,x) = \partial\mu(u,z,x)\partial\mu(z,y,x) ,$$

and $\partial\mu$ satisfies the cocycle condition (6.12).

We will now introduce some terminology that is standard in the cohomology theory of groups in the specific form that is taken in our present context.

Definition 6.6 The maps $\mu : \Omega\times\Omega \to A$ satisfying $\mu(x,x) = e$, $\mu(x,y) = \mu(y,x)^{-1}$ will be called 1-cochains and they form an Abelian group with respect to the natural multiplication defined by A. They will be denoted by $C^1(\Omega,A)$.

Similarly, the set of maps $\lambda : \Omega\times\Omega\times\Omega \to A$, satisfying $\lambda(x,y,x) = \lambda(x,x,y) = \lambda(y,x,x) = e$ and $\lambda(x,y,z) = \lambda(z,y,x)^{-1}$ for all $x,y,z \in \Omega$ will be called 2-cochains. They will be denoted by $C^2(\Omega,A)$. The set of 2-cochains have a natural structure of Abelian group.

The map $\partial : C^1(\Omega,A) \to C^2(\Omega,A)$, $\partial\mu(z,y,x) = \mu(z,y)\mu(y,x)\mu(x,z)$ will be called the coboundary operator.

The set of 2-cochains satisfying the cocycle condition will be called 2-cocycles and denoted by $Z^2(\Omega,A)$. The set of 2-cocycles of the form $\partial\mu$ will be called 2-coboundaries and will be denoted by $B^2(\Omega,A)$.

The set of coboundaries $B^2(\Omega,A) \subset Z^2(\Omega,A)$ define a normal subgroup and the quotient Abelian group $Z^2(\Omega,A)/B^2(\Omega,A)$ will be called the second cohomology group of Ω with coefficients in A and will be denoted by $H^2(\Omega,A)$.

Elements in the second cohomology group $H^2(\Omega,A)$ will be denoted by $[\lambda]$, that is $[\lambda] = \{\lambda'(z,y,x) = \partial\mu(z,y,x)\lambda(z,y,x) \mid \mu \in C^1(\Omega,A)\}$. Elements in $H^2(\Omega,A)$ will be called cohomology classes, and $[\lambda] \in H^2(\Omega,A)$, will be said to be the cohomology class containing λ.

With this terminology, the conclusion of the discussion above is that, given a finite connected groupoid $\mathbf{G} \rightrightarrows \Omega$ with Abelian isotropy group A, it has associated a cohomology class $[\lambda] \in H^2(\Omega,A)$. The importance of the 2-cocycle λ lies in the following statement.

Theorem 6.7

Let $\mathbf{G}$ *be a finite connected groupoid over* Ω *with Abelian isotropy group. Then,* $\mathbf{G}$ *is isomorphic to the direct product* $\mathbf{G}(\Omega)\times A$ *iff the cohomology class* $[\lambda]$ *is trivial.*

Proof 6.13 If the groupoid $\mathbf{G}$ is the direct product $\mathbf{G}(\Omega)\times A$, then we may choose a section $\sigma(y,x) = ((y,x),e)$, and $\lambda(z,y,x) = e$ for all $x,y,z \in \Omega$ and $[\lambda] = 0$.

To prove the converse, we will observe first that if the cohomology class $[\lambda]$ of the fundamental sequence vanishes, we may choose a section σ which is a homomorphism, that is, such that $\lambda(z,y,x) = e$ for all $x,y,z \in \Omega$ and $\sigma(z,y) \circ \sigma(y,x) = \sigma(z,x)$, in other words, the fundamental sequence splits[3]. Then, given $\alpha : x \to y \in \mathbf{G}$, we may write:

$$\alpha = \gamma(\alpha)_y \circ \sigma(y,x) , \qquad \alpha : x \to y , \gamma(\alpha)_y \in G_y . \tag{6.14}$$

Note that the previous factorization of α is unique and only depends on the section σ. However, we may identify the fundamental groupoid $\mathbf{G}_0$ with $\Omega \times A$ using an arbitrary section σ and a reference point $x_0 \in \Omega$, then, using the section above, we will identify the isotropy group G_y with $G_{x_0} = A$ as $\gamma_y \mapsto \sigma(x_0,y) \circ \gamma_y \circ \sigma(y,x_0) \in A = G_{x_0}$. Thus, given the element $\gamma(\alpha)_y \in G_y$ we will denote by $\gamma(\alpha) \in A$ the element:

$$\gamma(\alpha) = \sigma(x_0,y) \circ \gamma(\alpha)_y \circ \sigma(y,x_0) . \tag{6.15}$$

Then, we define the following map: $\varphi : \mathbf{G} \to A \times \mathbf{G}(\Omega)$, $\varphi(\alpha) = (\gamma(\alpha),(y,x))$, $\alpha : x \to y$, where $\gamma(\alpha)$ is given according to the decomposition (6.14) and the identification (6.15). Notice that $1_x = \gamma(1_x) \circ \sigma(x,x) = \gamma(1_x)$ (because we choose the section σ such that $\sigma(x,x) = 1_x$. Then $\varphi(1_x) = (1_x,(x,x))$.

Moreover $\varphi(\beta \circ \alpha) = (\gamma(\beta \circ \alpha),(z,x))$, provided that $\alpha : x \to y$ and $\beta : y \to z$. On the other hand $\beta \circ \alpha = \gamma(\beta \circ \alpha)_z \circ \sigma(z,x) = \gamma(\beta \circ \alpha)_z \sigma(z,y)\sigma(y,x)$. Then, because

$$\beta \circ \alpha = \gamma(\beta)_z \circ \sigma(z,y) \circ \gamma(\alpha)_y \circ \sigma(y,x) = \gamma(\beta)_z \circ \sigma(z,y) \circ \gamma(\alpha)_y \circ \sigma(y,z) \circ \sigma(z,y) \circ \sigma(y,x) ,$$

we conclude:

$$\gamma(\beta \circ \alpha)_z = \gamma(\beta)_z \circ \sigma(z,y) \circ \gamma(\alpha)_y \circ \sigma(y,z) .$$

Then, a simple computation shows:

$$\begin{aligned}
\gamma(\beta \circ \alpha) &= \sigma(x_0,z) \circ \gamma(\beta \circ \alpha)_z \circ (z,x_0) \\
&= \sigma(x_0,z) \circ \gamma(\beta)_z \circ \sigma(z,y) \circ \gamma(\alpha)_y \circ \sigma(y,z) \circ (z,x_0) \\
&= (\sigma(x_0,z) \circ \gamma(\beta)_z \circ \sigma(z,x_0)) \circ (\sigma(x_0,y) \circ \gamma(\alpha)_y \circ \sigma(y,z) \circ (z,x_0)) \\
&= (\sigma(x_0,z) \circ \gamma(\beta)_z \circ \sigma(z,x_0)) \circ (\sigma(x_0,y) \circ \gamma(\alpha)_y \circ \sigma(y,x_0)) \\
&= \gamma(\beta)\gamma(\alpha) ,
\end{aligned}$$

where we have used the fact that σ is a homomorphism at various places along the computation. Thus, we have proved that φ is a functor, i.e., a groupoid homomorphism. Now, clearly it is surjective and injective, then it defines an isomorphism of groupoids.

[3]Consequently, because the modular invariant $\tilde{\omega}$ is trivial then, in the spirit of the corresponding theorem for groups, the conclusion would follow, however, we will describe the decomposition explicitly.

6.5.3 Groupoids with Abelian isotropy group and $|\Omega| = 2,3,4$

We can apply the results obtained in the previous section, mainly Thm. 6.7, to groupoids with 'small' space of objects, 'small' in this context means $|\Omega| = 2,3,4$.

The case $|\Omega| = 2$

If $|\Omega| = 2$, the set Ω has just two elements, x, y say. Then, it is obvious that any 2-cochain λ is trivial, i.e., $\lambda = e$, because all terms in the cocycle equation (6.12) are of the form $\lambda(x,y,x)$, $\lambda(x,x,y)$, etc., i.e., there is always a repeated entry. Then, the second cohomology group $H^2(\Omega, A)$ is trivial and the groupoid is isomorphic to the direct product $\mathbf{G}(2) \times A$.

The case $|\Omega| = 3$

Now the set Ω has three elements x,y,z. The cocycle condition is obtained by choosing four points u,z,y,x in Ω. Then, there will always be at least one repeated, let it be, without loosing any generality, $u = z$. Then, the cocycle condition becomes just $\lambda(z,y,x) = e$ and we conclude that any cocycle is trivial $\lambda = e$, then again $H^2(\Omega, A) = \{0\}$, and the second cohomology group is trivial.

This exhausts the cases in the proof of the classification theorem.

The case $|\Omega| = 4$

Now, the situation is more interesting. The space of objects will have four objects x,y,z,u, then the only non-trivial case, that is, the only situation where there are no repetition of the objects x,y,z,u, is precisely the cocycle condition (6.12), i.e,.

$$\lambda(u,z,y)\lambda(u,y,x) = \lambda(z,y,x)\lambda(u,z,x).$$

Hence, given three terms in the equation, the fourth is determined by them. Thus, we may independently choose $\lambda(u,y,x)$, $\lambda(z,y,x)$ and $\lambda(u,z,x)$, while $\lambda(u,z,y)$ will be determined by the others. If the isotropy group A has order $a = |A|$, then there will be $3a$ possible 2-cocycles. We must determine the number of 2-coboundaries, or, in other words, if given a 2-cocycle λ, we may find μ such that $\lambda = \partial\mu$. Then, for $\lambda(u,y,x)$ we choose $\mu(u,y)$, $\mu(y,x)$ and $\mu(x,u)$, such that $\lambda(u,y,x) = \mu(u,y)\mu(y,x)\mu(x,u)$. We repeat the same for $\lambda(z,y,x)$ and $\lambda(u,z,x)$, hence, we get the set of conditions:

$$\lambda(u,y,x) = \mu(u,y) + \mu(y,x) + \mu(x,u) \quad (6.16)$$

$$\lambda(z,y,x) = \mu(z,y) + \mu(y,x) + \mu(x,z) \quad (6.17)$$

$$\lambda(u,z,x) = \mu(u,z) + \mu(z,x) + \mu(x,u). \quad (6.18)$$

where we have switched to additive notation in order to make it easier to visualize. Thus, we have a system of 3 equations with 6 unknowns $\mu(x,y)$, $\mu(x,z)$, $\mu(x,u)$, $\mu(y,z)$, $\mu(y,u)$, $\mu(z,u)$, with matrix of coefficients:

$$M = \begin{bmatrix} -1 & 0 & 1 & 0 & -1 & 0 \\ -1 & 1 & 0 & -1 & 0 & 0 \\ 0 & -1 & 1 & 0 & 0 & -1 \end{bmatrix}.$$

The matrix M has rank 3 and, consequently, the system always has a solution. Then, we conclude that the cohomology class defined by any 2-cocycle λ is trivial. The first groupoid with four objects and nontrivial Abelian group $A = \mathbb{Z}_2$ appearing in the classification would be the groupoid $\mathbf{G}(4) \times \mathbb{Z}_2$ that has order $4^2 \times 2 = 32$.

II

LINEAR REPRESENTATIONS OF FINITE GROUPS AND GROUPOIDS

Chapter 7

Linear Representations of Groups

We will start the second part of this work by discussing the theory of linear representations of finite groups in detail. The purpose of this is twofold. On one hand we will offer a succinct review of a well-known theory that will help readers without previous knowledge to grasp the main ideas and to put it to use immediately. On the other hand, we offer to the knowledgeable reader a smooth transition from the standard presentations of the theory of linear representations to the language that will help us very much in developing the theory of representations for groupoids in subsequent chapters.

As in the previous chapters of this book, we will keep the exposition elementary, meaning that only basic linear algebra will be used apart, of course, from the notions and methods about groups and categories developed in the first part. Thus, the notions of linear representations of groups, their equivalence and the main questions regarding their structure will be addressed. In addition, the notion of unitary representations, with the corresponding background of Hilbert spaces (finite-dimensional mostly) will be discussed. The chapter will end with the statement of Schur's lemmas, that will become a recurrent theme in this part of the book.

We will leave the proofs of the main theorems for the next chapter, Chap. 8, where the powerful notion of characters of representations will be discussed in detail. Ideas that will be exploited again in Chapter 12 when discussing the structure of representations of groupoids.

7.1 Linear and unitary representations of groups

7.1.1 *Why linear representations?*

First of all, we should devote a few lines to say why linear representations. Beyond historical reasons, that include, among other things, many inputs from physical problems –which had their origins in the developments taking place in the foundational moments of Quantum Mechanics (see for instance [69], [68], [71], Sect. 8.9 and Appendix C in this work)– we should stress that the main idea behind the theory of linear representations, and behind any theory of representations, as it was already emphasized in the presentation of the notion of functor (recall Sect. 5.1.2), is that of realising abstract objects (and morphisms) in terms of elements of linear algebra (a summary of Linear Algebra notions is offered in Appendix A): Vectors, matrices, etc., that we (presumably) understand very well and that we can translate into concrete applications very quickly.

Certainly, other concrete mathematical objects, rather than linear spaces and linear maps, can be chosen to 'represent' our abstract objects. For instance, we may use polynomials (which also share a linear algebraic structure), or we might use other objects with a rich inner structure, like finite fields or rings of pseudo-differential operators. Actually, this could be convenient depending on the problem we would be dealing with. For instance, if we were dealing with abstract coding theory, it could be relevant to use representations over finite fields or, in the example of databases discussed in Sect. 5.2.2, it was shown that filling a database with data was nothing other than 'representing' the database scheme in the category of sets.

However it just happens that the structures offered by linear algebra, in particular the algebra of matrices, are rich enough to accommodate in a natural way the abstract relations encoded by categorical structures. Furthermore, representing our abstract objects in terms of linear operators (or matrices) allows us to use the wealth of wonderful theorems and (numerical) algorithms accumulated during more than two hundred years in order to solve difficult problems presented in this way.

7.1.2 *Linear representations of groups*

We will repeat now the standard, set theoretical, definition of a linear representation of a group (see Sect. 5.1.2).

Definition 7.1 A linear representation of the group G with support on the linear space V is a map μ that assigns to any $g \in G$ a linear map $\mu(g) \colon V \to V$ (also denoted as $\mu_g \colon V \to V$ following the conventions with actions of groups, Dcf. 4.1) and such that:

$$\mu(e) = \mathrm{id}_V \, ; \qquad \mu(gg') = \mu(g)\mu(g') \, ; \qquad \forall g, g' \in G \, . \tag{7.1}$$

If the linear space V has dimension n, we will say that μ has dimension n. The linear space V will be called the support of the representation. We will denote the linear representation μ of G on the linear space V by (μ, V) or just μ (or even V if there is no risk of confusion). Lowercase greek letters μ, ν, ρ,..., will be used to denote linear representations of groups.

Notice that because $gg^{-1} = e$ for any $g \in G$, then from (7.1) we get immediately that $\mu(g)\mu(g^{-1}) = \mathrm{id}_V$ and the linear maps $\mu(g)$ associated to elements in the group G must be invertible:

$$\mu(g^{-1}) = \mu(g)^{-1}, \qquad \forall g \in G. \tag{7.2}$$

The idea, then, is to 'represent' each element g of the abstract group G by a linear map $\mu(g)$ that, once we fix a basis of the linear space V, becomes a matrix. Let us exhibit a few examples that will prove to be relevant in what follows.

Example 7.1 *Let us consider the "exchange group"* $\mathbb{Z}_2 = \{e, \tau\}$ *with* $\tau^2 = e$. *Let* V *be a linear space of dimension* 2 *and take a basis on it whose vectors will be denoted by* e_1, e_2. *Now define the representation* λ *of* $\mathbb{Z}_2$ *on* V *by means of:*

$$\lambda(\tau)e_1 = e_2, \qquad \lambda(\tau)e_2 = e_1.$$

(Note that, by definition, the linear map associated to e *is the identity). Then, with respect to the given basis, the matrices associated to the elements of the group are:*

$$[\lambda(e)] = \begin{bmatrix} 1 & 0 \\ 0 & 1 \end{bmatrix}, \qquad [\lambda(\tau)] = \begin{bmatrix} 0 & 1 \\ 1 & 0 \end{bmatrix}.$$

This is just a fancy way of representing the abstract elements e, τ *by means of* 2×2 *matrices, but this is not by any means the only way to do it.*

We could define another representation μ_1, *this time one-dimensional, given by:* $\mu_1(e) = 1, \mu_1(\tau) = -1$. *We may even consider another representation, called the trivial representation, given by* $\mu_0(e) = \mu_0(\tau) = 1$.

Example 7.2 *The previous example invites us to consider not just the group* $\mathbb{Z}_2$ *but the groups* $\mathbb{Z}_n$ *of congruences of integers mod n studied in Sect. 2.1, Example 2.2.2. Just recall that elements* $\mathbf{k}$ *in* $\mathbb{Z}_n$ *represent classes of integers mod n, that is* $\mathbf{k} = k + n\mathbb{Z}$, $k = 0, 1, \ldots, n-1$. *We may define a representation* ρ *of* $\mathbb{Z}_n$ *on a n-dimensional linear space by assigning a* $n \times n$ *matrix* $\rho(\mathbf{1})$ *to the element* $\mathbf{1}$ *and then defining* $\rho(\mathbf{k}) = \rho(\mathbf{1})^k$.

Let us consider, for instance, $n = 4$, *then we assign the matrix:*

$$\rho(\mathbf{1}) = \begin{bmatrix} 0 & 1 & 0 & 0 \\ 0 & 0 & 1 & 0 \\ 0 & 0 & 0 & 1 \\ 1 & 0 & 0 & 0 \end{bmatrix}, \tag{7.3}$$

to the generator $\mathbf{1}$ *and we get:*

$$\rho(\mathbf{2}) = \begin{bmatrix} 0 & 0 & 1 & 0 \\ 0 & 0 & 0 & 1 \\ 1 & 0 & 0 & 0 \\ 0 & 1 & 0 & 0 \end{bmatrix}, \rho(\mathbf{3}) = \begin{bmatrix} 0 & 0 & 0 & 1 \\ 1 & 0 & 0 & 0 \\ 0 & 1 & 0 & 0 \\ 0 & 0 & 1 & 0 \end{bmatrix},$$

$$\rho(\mathbf{0}) = \lambda(\mathbf{4}) = \begin{bmatrix} 1 & 0 & 0 & 0 \\ 0 & 1 & 0 & 0 \\ 0 & 0 & 1 & 0 \\ 0 & 0 & 0 & 1 \end{bmatrix}.$$

The previous examples can be extended without difficulty to $\mathbb{Z}_n$ represented on an n-dimensional linear space.

In addition to this matrix representation, there are simpler ones, actually, in the case of $\mathbb{Z}_4$, there are exactly four different one-dimensional representations (which are the building blocks of all others, as we shall see). These representations are given by:

1. *μ_0 is the trivial representation (all elements take the value 1).*
2. *$\mu_1(\mathbf{1}) = \mathrm{i}$ (hence $\mu_1(\mathbf{2}) = -1$, $\mu_1(\mathbf{3}) = -\mathrm{i}$).*
3. *$\mu_2(\mathbf{1}) = -1$ (hence $\mu_2(\mathbf{2}) = 1$, $\mu_2(\mathbf{3}) = -1$).*
4. *$\mu_3(\mathbf{1}) = -\mathrm{i}$ (hence $\mu_3(\mathbf{2}) = -1$, $\mu_3(\mathbf{3}) = \mathrm{i}$).*

The reader will realise the use of the roots of unity: $1, \mathrm{i}, -1, -\mathrm{i}$, because if μ is a one-dimensional representation of $\mathbb{Z}_4$ then $\rho(\mathbf{1})^4 = 1$, hence, $\rho(\mathbf{1}) = \sqrt[4]{1}$.

Exercise 7.3 *Show that the group $\mathbb{Z}_n$ has n different one-dimensional representations and list them.*

7.1.3 Equivalence of representations: Abstract versus concrete

The notion of linear representation introduced above is abstract, that is, both the group G and the linear spaces V involved are abstract mathematical objects. However, in the previous examples, we have seen that, by fixing linear basis, we can describe the linear representations by using matrices. Actually, this is what we do in concrete applications. We associate concrete matrices $D(g)$ to the elements of the group satisfying the representation property, that is $D(e) = I$ and $D(g)D(g') = D(gg')$, $g, g' \in G$. The order of the matrices $D(g)$ will be called the dimension of the matrix representation.

Sometimes the assignment $g \mapsto D(g)$ is called a matrix representation of the group. Clearly, given any (abstract) linear representation μ, the choice of a linear basis $\{e_i\}$ in the space V supporting the representation, will associate a matrix representation $D(g) = [\mu(g)]$, where $[\mu(g)]$ is the matrix associated with the linear map $\mu(g) \colon V \to V$ in the linear basis $\{e_i\}$ (see Sect. A.4.2 for more details):

$$\mu(g)e_j = \sum_{i=1}^{n} D(g)_{ij} e_i . \tag{7.4}$$

Conversely, given a matrix representation $D(g)$ of the group G, and given a linear space V of dimension the dimension of the matrix representation, for any linear basis $\{e_i\}$ of V we may define an abstract linear representation by the formula above (7.4).

Thus, in the examples above, Examples 7.1 and 7.2, the linear representations of the groups $\mathbb{Z}_n$ were given as matrix representations and the abstract representation can be constructed by choosing a basis on the given linear space.

Moreover, we observe that, dealing with representations of groups, there is always the possibility of changing the basis that we use to describe them. Changing basis is nothing but establishing isomorphisms between the corresponding linear spaces (see Chap. A.4.2). Thus, it is only natural to establish that two linear representations are equivalent if they 'differ' by an isomorphism, more precisely:

Definition 7.2 Given a group G and two linear representations μ, ν on the linear spaces V, W, respectively, we will say that the two representations μ, ν are equivalent, and we will denote it by $\mu \sim \nu$, if there exists an isomorphism $\phi\colon V \to W$ such that:

$$\phi \circ \mu(g) = \nu(g) \circ \phi\,, \qquad \forall g \in G.$$

(Compare this definition with the equivalence notion introduced by means of natural transformations, Example 5.17.2).

Clearly the relation among linear representations defined by the equivalence of representations in the previous definition is an equivalence relation and the collection of all linear representations of finite groups decomposes in equivalence classes of representations. Equivalence classes are customarily denoted by enclosing a given representative among brackets[1], that is $[\mu] = \{\mu' \mid \mu' \sim \mu\}$. However, we will seldom use this notation.

It is clear that any property satisfied by a linear representation (as defined on an abstract linear space) is shared by all linear representations equivalent to it. This is not true, however, for properties depending on a specific choice of basis (like possessing a specific value in a given position, $(1,1)$ for instance, in the specific matrix representation), or depending on a concrete realisation of the representation using a specific linear space (for instance, using a linear space of polynomials and the property of the representation related to the properties of their roots).

Selecting one representative of an equivalence class of linear representations determines all the others because they are all related by isomorphisms (that is, they differ only on the selections of linear basis in the corresponding spaces).

Thus, from now, unless we want to describe a property depending on the specific choice of the linear representation μ, we will assume that we are considering the properties common to the equivalence class to which μ belongs, and we will avoid the cumbersome statement "let μ be a representative of the equivalence class of linear represenations $[\mu]$".

Even on each equivalence class of linear representations there are plenty of matrix representations. In fact, once we choose a basis, the given linear representation μ is equivalent to the matrix representation D determined by the choice of basis, thus, in what follows, we will not insist on the difference between abstract and ma-

[1] Sometimes we will also use bold fonts like in the case of elements in $\mathbb{Z}_n$.

trix representations, that is, we will describe the results always in terms of abstract representations and in specific situations, along the proofs or in examples, we will use appropriate basis to write the corresponding matrix representations.

Before continuing studying linear representations, we will discuss two fundamental examples where the questions raised above will become apparent. The first one is extremely important and will be discussed again in depth later on. For the moment, we will use it to illustrate how the higher-dimensional linear representations exhibited in Examples 7.1 and 7.2 were obtained.

7.1.4 The regular representations

Given a group G, there are always two canonical representations of the group called the left and right-regular representations. The support space for both representations is the space of functions on G, that is $V := \mathcal{F}(G) = \{\psi \colon G \to \mathbb{C}\}$, which is a concrete complex linear space with the natural pointwise addition of functions and multiplication by scalar operations.

If the group G we are considering is finite, there is no restriction on the class of functions we use[2] and the vector space $V = \mathcal{F}(G)$ is finite-dimensional. In such case, there is a natural basis on V given by the delta-like functions $\delta_g \colon G \to \mathbb{C}$ defined as:

$$\delta_g(g') = \begin{cases} 1 & \text{if } g = g' , \\ 0 & \text{otherwise} . \end{cases}$$

An arbitrary function ψ on G can be written as:

$$\psi = \sum_{g \in G} \psi(g) \delta_g .$$

Obviously the family of functions $\{\delta_g \mid g \in G\}$ is linearly independent. The space of functions V has dimension the order of G: $\dim V = |G|$. The left and right regular representations of G on V, λ and ρ, are defined respectively as:

$$(\lambda(g)\psi)(g') = \psi(g^{-1}g') , \qquad (\rho(g)\psi)(g') = \psi(g'g) , \quad \forall g, g' \in G .$$

Notice that

$$\lambda(g)\delta_{g'} = \delta_{gg'} , \qquad \rho(g)\delta_{g'} = \delta_{g'g^{-1}} \quad \forall g, g' \in G . \tag{7.5}$$

If we label the elements of the group as $G = \{e = g_1, g_2, \ldots, g_n\}$, $n = |G|$, then the matrix $D(g_i)$ associated to the element g_i by the left regular representation in the canonical basis δ_{g_i}, will have an entry 1 in the position (k, j) if $g_k = g_i g_j$.

Thus, if we consider the multiplication table of the group G, the ith row will correspond to multiplication by the element g_i on the left. Hence, if g_k is the (i, j) entry of the table, that is, it is the result of multiplying g_i with g_j, then there is "1" in the (k, j) entry of the matrix $D(g_i)$. If we abstract the table of the group by just

[2]If the group were not finite but a locally compact topological we could restrict to the space of square integrable functions with respect to the Haar measure defined on it.

writing the indexes i corresponding to the elements g_i of the group, the numbers k of the ith row indicate in which rows the ones in the matrix corresponding to the given element are located.

Now, it is clear that the representations exhibited in the previous examples were just the matrix representations of the right regular representations of the groups $\mathbb{Z}_2$ and $\mathbb{Z}_4$. In fact, if we label by $1,2,3,4$ the elements **0**, **1**, **2**, **3**, of the group $\mathbb{Z}_4$, respectively, and we compute the matrix associated to the element $\mathbf{1}(=g_2)$ in left-regular representation in the natural basis δ_i, then we get that the (k,j) entry of the matrix is 1 if $g_2 g_j = g_k$, which translates in $k = j+1$, $j = 1,2,3$, and $k = 1$ if $j = 4$. Hence, we get the matrix:

$$\lambda(\mathbf{1}) = \begin{bmatrix} 0 & 0 & 0 & 1 \\ 1 & 0 & 0 & 0 \\ 0 & 1 & 0 & 0 \\ 0 & 0 & 1 & 0 \end{bmatrix},$$

We proceed similarly for the right-regular representation. In such case, we use formulas (7.5) again in order to show that the (k,j) entry of the matrix $D(g_i)$ will be one if $g_i g_j^{-1} = g_k$, but this translates (using the same convention as before to label the elements of the group $\mathbb{Z}_4$) to $k = 4$ if $j = 1$ and $k = j-1$, if $j = 2,3,4$. Hence, we obtain that the matrix $\rho(\mathbf{1})$ is exactly (7.3).

Exercise 7.4 *Write down the matrix representation of the left and right regular representations of the group $D_3 = S_3$ and D_4 in the standard basis.*

7.1.5 The defining representation of the symmetric group

We will end this introduction by describing another family of representations that will turn out to be important in what follows: The "defining representation" of the group of permutations.

We will start by consider the permutation group S_3. Recall that $|S_3| = 6$. Consider a linear space of dimension 3 and a given basis on it $\{e_1, e_2, e_3\}$. Then, there is a representation of S_3 on V defined as follows:

$$\mu(\sigma)e_i = e_{\sigma(i)}, \qquad i = 1,2,3, \quad \sigma \in S_3 .$$

Thus, the matrix associated to the transposition $\tau = (12)$ is:

$$[\mu(\tau)] = \begin{bmatrix} 0 & 1 & 0 \\ 1 & 0 & 0 \\ 0 & 0 & 1 \end{bmatrix},$$

etc.

Let S_n be the symmetric group of n elements. Consider a linear space V of dimension n and a linear basis $\{e_k\}$ on it. We will define a linear representation of S_n on V as follows (now we use the notation ρ_σ instead of $\rho(\sigma)$ in order to avoid too many parenthesis in the formulas):

$$\rho_\sigma(v) = \sum_k v_k e_{\sigma(k)}; \qquad \forall \sigma \in S_n, \quad v = \sum_k v_k e_k \in V.$$

Clearly $\rho(e) = \mathrm{id}_V$. Moreover: $\rho_\sigma(e_j) = \sum_k \delta_{kj} e_{\sigma(k)} = \sum_l \delta_{\sigma^{-1}(l),j} e_l$, or, more easily:

$$\rho_\sigma(e_j) = e_{\sigma(j)}, \qquad \forall j = 1, \ldots, n.$$

Then,

$$\begin{aligned} \rho_{\sigma_1}(\rho_{\sigma_2} e_j) &= \rho_{\sigma_1}\left(\sum_l \delta_{\sigma_2^{-1}(l),j} e_l\right) = \sum_l \delta_{\sigma_2^{-1}(l),j} \rho_{\sigma_1} e_{l_1} \\ &= \sum_{l_1,l_2} \delta_{\sigma_2^{-1}(l_1)j} \delta_{\sigma_1^{-1}(l_2)l_1} e_{l_2} = \sum_l \delta_{\sigma_2^{-1}\sigma_1^{-1}(l),j} e_l = \rho_{\sigma_1\sigma_2}(e_j). \end{aligned}$$

Notice that the matrix associated to the linear map $\rho(\sigma)$ with respect to the basis $\{e_k\}$, is given by:

$$\rho_\sigma(e_j) = \sum_k P(\sigma)_{kj} e_k,$$

then:

$$P(\sigma)_{kj} = \delta_{\sigma^{-1}(k)j},$$

that is, $P(\sigma)_{kj} = 1$ if $\sigma^{-1}(k) = j$ (or if $\sigma(j) = k$), and it will be zero otherwise, in other words, there is 1 on the row $\sigma(j)$ and zeros at all other positions.

Of course, the previous representation depends on the choice of a basis. To make it 'more canonical', we may consider the canonical basis in the linear space V_n freely generated by the numbers $1, 2, \ldots, n$. That is, we will consider the numbers $1, 2, \ldots$ as abstract symbols. To emphasise this abstraction we will denote them as $|1\rangle$, $|2\rangle, \ldots$, then we will consider the set V_n of formal linear combinations

$$v = c_1|1\rangle + c_2|2\rangle + \cdots + c_n|n\rangle, \qquad c_k \in \mathbb{C}. \tag{7.6}$$

The sum and the multiplication by scalars are defined in the obvious way:

$$\sum_k c_k|k\rangle + \sum_k c'_k|k\rangle = \sum_k (c_k + c'_k)|k\rangle; \qquad \lambda \sum_k c_k|k\rangle = \sum_k (\lambda c_k)|k\rangle.$$

It is a simple exercise to show that V_n with the previous operations is a complex linear space of dimension n with canonical basis given by the vectors $|k\rangle$ themselves.

With this notation, we see that $\rho_\sigma|j\rangle = |\sigma(j)\rangle$, and the permutation σ can be read directly from the transformed vectors. For instance the matrix representation of the permutation $\sigma = (123)$ in S_3 is given by the matrix:

$$P(\sigma) = \begin{bmatrix} 0 & 0 & 1 \\ 1 & 0 & 0 \\ 0 & 1 & 0 \end{bmatrix},$$

that corresponds to the transformation of basis vectors $|1\rangle \mapsto |2\rangle$, $|2\rangle \mapsto |3\rangle$ and $|3\rangle \mapsto |1\rangle$. We will find many more examples in the coming chapters, where we will obtain the basic theorems that will allow us to construct all of them efficiently.

7.2 Irreducible representations

We try to identify the 'simplest' representations expecting that they will be the blocks that will allow us to construct any other representation. Because of this, it appears natural to consider the situation where a representation does not contain non-trivial subrepresentations.

7.2.1 *Reducible and irreducible representations*

Definition 7.3 A linear representation μ of the group G on the linear space V is said to be reducible if there exists a proper subspace $W \subset V$ (that is, different from the trivial subspace $\{\mathbf{0}\}$ and V) such that $\mu_g(W) \subset W$ for all $g \in G$.

A linear subspace W, such that $\mu_g(W) \subset W$ for all $g \in G$, will be said to be G-invariant (or just invariant if there is no risk of confusion) and the restriction of the linear representation μ to it will be called a subrepresentation of (μ, V). We will sometimes indicate it simply by $W \subset V$ if there is no risk of confusion.

Notice that, if the representation (ρ, V) is reducible, we may consider smaller subspaces to support it, hence, we may use smaller matrices to represent the elements of G.

Definition 7.4 A linear representation μ which is not reducible is said to be irreducible. More explicitly, a linear representation (ρ, V) is irreducible if there are not proper subspaces W of V such that $\mu_g(W) \subset W$, for all $g \in G$.

Example 7.5 *Let us provide a few examples of reducible and irreducible representations.*

1. *One-dimensional representations are obviously irreducible. In particular the one-dimensional representations of the groups $\mathbb{Z}_n$ exhibited in Examples 7.1, 7.2 are irreducible.*

   Exercise 7.6 *Consider the dihedral group D_3. Show that the left and right-regular representations are reducible.*

2. *The defining representation of the permutation group is reducible. In fact, you may check that the subspace generated by the vector $(1, 1, \ldots, 1)$ is invariant and defines, obviously, an irreducible representation.*

3. *We may consider the more general situation of a group Γ acting transitively on a finite set Ω. Then, there is a natural representation μ_Γ of Γ in the space of functions defined on Ω, that is, we consider the space $\mathcal{H}_\Omega = \mathcal{F}(\Omega, \mathbb{C})$ as the support of the representation and the representation is defined as: $(\mu_\Gamma(g)\psi)(x) = \psi(g^{-1}x)$, for all $\psi \in \mathcal{H}_\Omega$, $g \in \Gamma$, $x \in \Omega$.*

This situation generalizes the previous example with $\Gamma = S_n$, $\Omega = \{1,2,\ldots,n\}$ *and the standard action of* S_n *on* Ω.

The representation $(\mu_\Gamma, \mathcal{H}_\Omega)$ *is reducible. It is enough to observe that the subspace defined by constant functions is invariant (which corresponds to the subspace spanned by the vector* $(1,1,\ldots,1)$ *in the previous example).*

Let (μ, V) denote an irreducible representation of G. Let $v \in V$ denote an arbitrary non-zero vector and consider the linear span W of the collection of vectors $\{\mu_g(v) \mid g \in G\}$. Notice that the set $\{\mu_g(v) \mid g \in G\}$ is the orbit of v with respect to the action of the group G on the linear space V.

It is clear that W is G-invariant because $\mu_g\left(\sum_l c_l \mu(g_l)v\right) = \sum_l c_l \mu_{gg_l}(v) \in W$. Hence, because $W \neq \{\mathbf{0}\}$, it must be V. A vector $v \in V$, such that the linear span of its orbit under the group G is the total space, is called a cyclic vector (for the representation). Thus, we have proved that, if V supports an irreducible representation, any non-zero vector is cyclic. The converse is also true.

Proposition 7.1
Let μ *be a representation of the group* G *in the linear space* V*, then the representation* μ *is irreducible iff any non-zero vector in* V *is cyclic.*

Proof 7.1 The direct implication was proved above. To prove the converse, we proceed by contradiction. Assume that V is not irreducible, then there exists an invariant proper subspace $W \subset V$. Clearly, any vector in W is not cyclic.

It follows from the previous observation that any irreducible representation of a finite group must be finite-dimensional.

Proposition 7.2
If (μ, V) *is an irreducible representation of a finite group* G*, then* V *has finite dimension.*

Proof 7.2 If V is irreducible, any non-zero vector v is cyclic, hence, the family $S = \{\mu_g(v) \mid g \in G\}$ generates V, but if G is finite, the generating system S is finite and the dimension of V must be finite.

The argument used above to show that the linear span of the orbit of a vector under an irreducible representation is the total space will be used again and again and it is the essence of the instrumental result known as Schur's Lemmas, Sect. 7.4.

7.2.2 Operations with representations and complete reducibility

There are a number of natural operations that can be done with linear representations which are induced by the operations we can do with linear spaces and linear maps (see Appendix A.3, for the description of the basic operations). Given two linear representations (μ_a, V_a), $a = 1, 2$, we can construct the direct sum and tensor product of them:

1. Direct sum: The direct sum $(\mu_1 \oplus \mu_2, V_1 \oplus V_2)$ is defined as the linear representation of G on the linear space $V_1 \oplus V_2$ given by $(\mu_1 \oplus \mu_2)(g) = \mu_1(g) \oplus \mu_2(g)$ for all $g \in G$, i.e, $(\mu_1 \oplus \mu_2)_g(v_1 \oplus v_2) = (\mu_1)_g(v_1) \oplus (\mu_2)_g(v_2)$, for all $v_1 \oplus v_2 \in V_1 \oplus V_2$.

2. Tensor product: The tensor product $(\mu_1 \otimes \mu_2, V_1 \otimes V_2)$ is defined as the linear representation of G on the linear space $V_1 \otimes V_2$ given by $(\mu_1 \otimes \mu_2)(g) = \mu_1(g) \otimes \mu_2(g)$ for all $g \in G$. Hence, on monomials $v_1 \otimes v_2$, $(\mu_1 \otimes \mu_2)(g)$ acts as $(\mu_1 \otimes \mu_2)_g(v_1 \otimes v_2) = (\mu_1)_g(v_1) \otimes (\mu_2)_g(v_2)$.

3. Quotient: Let (μ, V) be a representation and (μ, W) be a subrepresentation, that is, W is an invariant subspace of V. Then, we defined the quotient representation $(\mu, V/W)$ as the linear representation $\mu_g(v + W) = \mu_g(v) + W$.

4. Dual: The dual representation (μ^*, V^*) is defined as $(\mu_g^*(\alpha))(v) = \alpha(\mu_{g^{-1}}(v))$ for all covectors $\alpha \in V^*$, $g \in G$ and $v \in V$.

Exercise 7.7 *Check that for the dual representation we actually have that $\mu^*(gg') = \mu^*(g)\mu^*(g')$. Observe that if we were defining the dual representation 'without the -1', e.g. as $(\tilde{\mu}_g(\alpha))(v) = \alpha(\mu_g(v))$, we would get $\tilde{\mu}(gg') = \tilde{\mu}(g')\tilde{\mu}(g)$ with the wrong order in the factors.*

The natural identifications that apply for linear spaces, Prop. A.9, become equivalences of representations. For instance, $\mu_1 \oplus \mu_2 \sim \mu_2 \oplus \mu_1$, $\mu_1^* \oplus \mu_2^* \sim (\mu_1 \oplus \mu_2)^*$, etc.

We can be more specific now about our previous considerations on the possibility of constructing any linear representation by using 'simple' blocks. For instance, we may ask if it is possible to construct any finite-dimensional representation μ of a finite group as a direct sum of irreducible representations. The answer is positive and the following theorem provides the first fundamental description of a finite-dimensional representation of a finite group.

Theorem 7.1

Let (μ, V) be a linear representation of the finite group G in the finite-dimensional linear space V, then there exists a family W_a, $a = 1, \ldots, r$, of invariant subspaces supporting irreducible representations μ_a of G, such that $\mu = \mu_1 \oplus \cdots \oplus \mu_r$, or, in even simpler terms, such that the support space decomposes as the direct sum of W_a:

$$V = W_1 \oplus \cdots \oplus W_r \,. \tag{7.7}$$

Proof 7.3 The proof is by induction on n the dimension of V. If $n = 1$, the representation is irreducible.

Let us suppose that the hypothesis is true for $n > 1$. Let (μ, V) be a linear representation of G with $\dim V = n+1$. Choose an irreducible invariant subspace W of V with $0 < \dim W < \dim V$. If such subspace does not exist, the representation μ is irreducible and we are done.

Suppose then, that such subspace W exists. Select an invariant supplementary subspace U of W. We may construct such subspace as follows. Select an arbitrary supplementary subspace U' of W (Thm. A.1). Denote by P' the projector on U' along W (Def. A.9). Then define a new projector by the formula:

$$P = \frac{1}{|G|} \sum_{g \in G} \mu(g) \circ P' \circ \mu(g)^{-1}.$$

Exercise 7.8 *Prove that P is a projector, i.e., $P^2 = P$, and* $\ker P = W$.

Because P is a projector and $\ker P = W$, we have $V \cong W \oplus U$ with $U = \mathrm{ran}(P)$.

Exercise 7.9 *Prove that $\mu(g) \circ P = P \circ \mu(g)$, for all $g \in G$. Conclude that* $\ker P$ *and* $\mathrm{ran}\, P$ *are invariant subspaces.*

Then $U = \mathrm{ran}(P)$ is invariant, and it is the support of a representation of G, the restriction of the representation μ to U. But now, $\dim U < n$, hence, by the induction hypothesis $U = W_1 \oplus \cdots \oplus W_r$, with W_i invariant and irreducible. Then $V = W \oplus W_1 \oplus \cdots \oplus W_r$ with all factors invariant and irreducible.

Remark 7.1 Much more can be proved. For instance, it is possible to show that the decomposition (7.7) is essentially unique in the sense that, given any other decomposition, $V = W_1' \oplus \cdots \oplus W_{r'}'$ of V with W_k' irreducible invariant subspaces, then $r = r'$ and there exists an isomorphism $T \colon V \to V$ such that $T(W_l) = W'_{\sigma(l)}$ (with σ a permutation of the indices). We will postpone the proof of this fact until next chapter (see Thm. 8.6).

In a more abstract form, the previous result is a consequence of the Jordan-Hölder theorem. We will establish it for groupoids (Thm. 9.5) and it will be presented in its more general form for representations of algebras, Thm. 11.4. ∎

If a linear representation (ρ, V) satisfies the decomposition property stated in Prop. 7.1 we will say that it is completely reducible or semisimple. In the case of finite groups and finite dimensional linear spaces, all linear representations are completely reducible. However this is not the case, in general, for infinite-dimensional representations of infinite groups (even if they are discrete). The general discussion on semisimplicity will be postponed until Chapter 11.

7.3 Unitary representations of groups

It is a common situation in almost any application of the theory that the linear spaces we are using are equipped with an inner product. Not only that, in the theory of finite-dimensional representations of finite-groups and groupoids, all support linear spaces are equipped with inner products. Thus, in the following paragraphs we will offer a succinct description of the main properties of linear spaces with inner product (that will hold mostly in the infinite-dimensional situation).

7.3.1 *A succinct description of Hilbert spaces*

Definition 7.5 An inner product on a complex linear space V is a map $\langle \cdot, \cdot \rangle : V \times V \to \mathbb{C}$, such that:

1. $\langle v, v \rangle \geq 0$ for all $v \in V$ and $\langle v, v \rangle = 0$ iff $v = 0$.
2. $\langle v, w \rangle = \overline{\langle w, v \rangle}$.
3. $\langle v + u, w \rangle = \langle v, w \rangle + \langle u, w \rangle$.
4. $\langle \lambda v, w \rangle = \lambda \langle v, w \rangle$, for all $\lambda \in \mathbb{C}$.

An inner product allows to define a norm by the formula:

$$||v|| = +\sqrt{\langle v, v \rangle}\,.$$

Actually, it is easy to show that:

$$||v|| = 0 \quad \text{iff} \quad v = 0\,,$$

and

$$||\lambda v|| = |\lambda|\,||v||\,; \qquad \forall \lambda \in \mathbb{C}, v \in V\,.$$

However, to prove that the triangle inequality holds, that is:

$$||v + w|| \leq ||v|| + ||w||\,,$$

we need to prove first the Cauchy-Schwartz inequality:

$$|\langle v, w \rangle| \leq ||v||\,||w||\,, \qquad \forall v, w \in V\,, \tag{7.8}$$

(where the equality happens only if v and w are proportional).

For that, we will consider the nonnegative function on s: $||v - sw||^2 = \langle v - sw, v - sw \rangle \geq 0$, for all $s \in \mathbb{C}$. Substituting $s = \langle v, w \rangle / ||w||^2$, and expanding it by using the properties listed in the definition of an inner product (7.5), we get:

$$0 \leq ||v||^2 - \frac{|\langle v, w \rangle|^2}{||w||^2}\,, \tag{7.9}$$

from which the fundamental Cauchy-Schwarz inequality (7.8) follows.

Two vectors v, w, such that $\langle v, w \rangle = 0$, are called orthogonal.

Exercise 7.10 *Prove that two non-null vectors which are orthogonal are linearly independent.*

Definition 7.6 A linear map $T\colon V \to W$ with V and W linear spaces with inner products $\langle \cdot, \cdot \rangle_V$ and $\langle \cdot, \cdot \rangle_W$, respectively, is called an isometry if $\langle Tu, Tv \rangle_W = \langle u, v \rangle_V$ for all $u, v \in V$.

Exercise 7.11 *Prove that T is an isometry if and only if $||Tu||_W = ||u||_V$ for all $u \in V$.*

Notice that, if T is an isometry, it must be an injective map. In fact, if $Tu = 0$, then $\langle u, u \rangle_V = \langle Tu, Tu \rangle_W = 0$ and $u = 0$. In general, an isometry does not have to be surjective. We may consider $W = V \oplus F$ and T the canonical inclusion map $V \to V \oplus F$, $v \mapsto v + 0$. Then, defining the inner product $\langle u + f, v + g \rangle_W = \langle u, v \rangle_V + \langle f, g \rangle_F$ for any inner product $\langle \cdot, \cdot \rangle_F$ in F, the map T is an isometry but it is not surjective.

Sometimes this notion of linear map preserving the inner products is called a partial isometry and the name isometry is reserved for linear maps which are both injective and surjective. We will call such isometries unitary maps (or unitary operators). As it will be discussed below, choosing appropriate basis (i.e., orthonormal), the matrices U associated to unitary maps are unitary matrices, that is $U^\dagger U = UU^\dagger = I$, where $U^\dagger$ denotes the conjugate transpose of the matrix U.

7.3.2 Geometry of spaces with inner products

Given $S \subset V$ any subset of a linear space with inner product, we will define its orthogonal subspace as $S^\perp = \{v \in V \mid \langle v, s \rangle = 0, \forall s \in S\}$. Some pieces of the geometry of linear spaces with inner products are provided by the following results:

Proposition 7.3

Let $S \subset V$ be any subset of a finite-dimensional linear space with inner product. Then:

1. *$S^\perp$ is a linear subspace of V.*
2. $(S \cap R)^\perp = S^\perp + R^\perp$, $(S \cup R)^\perp = S^\perp \cap R^\perp$.
3. *$(S^\perp)^\perp$ is the linear span of S.*
4. *If S is a linear subspace then, $V = S \oplus S^\perp$.*

The proof of the previous statements are simple exercises that we leave to the reader.

Remark 7.2 When dealing with finite dimensional spaces, the nuances introduced by the topology induced by the norm $|| \cdot ||$ are not relevant, however, when dealing with infinite dimensional spaces, they become fundamental and the proper notion of support for a linear representation is that of Hilbert spaces (that is, linear spaces with inner product such that they are complete as normed spaces). Such issues will not be discussed further on and, in most of what follows, all linear spaces will be assumed to be finite dimensional. ■

Spaces with inner products have distinguished bases.

Proposition 7.4
Any finite-dimensional linear space V of dimension n with an inner product $\langle\cdot,\cdot\rangle$ has an orthonormal basis, that is a basis u_k, $k = 1,\ldots,n$, such that $\langle u_j, u_k\rangle = \delta_{jk}$.

Proof 7.4 The proof is standard and proceeds as follows. Let $v \neq 0$ be any vector on V. Define the unitary vector $u_1 = v/||v||$ and consider the linear subspace $V' = \{u_1\}^{\perp}$. Choose another non-zero vector $v' \in V'$. Define $u_2 = v'/||v'||$ and iterate the process which is obviously finite. The set of vectors $\{u_k\}$ obtained in this way forms an orthonormal basis.

As a corollary of the previous existence theorem we conclude that all finite dimensional linear spaces with inner product are isometrically isomorphic or unitarily isomorphic.

Corollary 7.1
Two finite-dimensional linear spaces V and W of the same dimension with inner products are unitarily equivalent.

Proof 7.5 Let $\{u_k\}$, $\{w_k\}$ be orthonormal basis of V and W, respectively, then we define a linear map $U\colon V \to W$ as $U(\sum_k c_k u_k) = c_k \sum_k w_k$. It is a trivial computation to show that U is unitary.

Remark 7.3 Notice that, if two linear spaces are unitarily equivalent, that does not mean that there is a natural choice for such isomorphism. Actually, the proof of Corollary 7.1 shows that such identification depends on the choice of two orthonormal bases, so it is not canonical at all. ■

Remark 7.4 The same result holds true, and the ones that follow, even if harder to prove, for complex separable Hilbert spaces: Any two complex separable Hilbert spaces are isometrically isomorphic. Again, the proof is based on the existence of countable orthonormal basis. ■

Exercise 7.12 *Prove that, given a finite-dimensional Hilbert space $\mathcal{H}$ and an orthonormal basis $\{v_k\}$, any vector $w \in \mathcal{H}$ can be written as: $w = \sum_k \langle v_k, w\rangle v_k$.*

Then, we may prove that:

Proposition 7.5
Let $\{v_k\}$ a family of non-zero orthogonal vectors on a Hilbert space $\mathcal{H}$, then the family $\{v_k\}$ defines a basis if and only if the family is complete, that is any vector $w \in \mathcal{H}$, such that $\langle v_k, w\rangle = 0$ for all k, vanishes.

Proof 7.6 We prove it in finite dimension. First note that the family $\{v_k\}$ is linearly independent. If fact, if $\lambda_1 v_1 + \ldots + \lambda_k v_k = 0$, then, taking the inner product with the vector v_l in the previous identity, we get $\lambda_l ||v_l||^2 = 0$, then $\lambda_l = 0$.

Now, if $\{v_k\}$ is a basis, we can redefine the vectors by dividing them by their norms, so that they define an orthonormal basis that we will denote with the same symbols $\{v_k\}$. Then, for any $w \in \mathcal{H}$, $w = \sum_k \langle v_k, w\rangle v_k$, thus, if $\langle v_k, w\rangle = 0$ for all k, then $w = 0$.

Conversely, let us assume that the system of orthogonal vectors $\{v_k\}$ is complete, then consider a vector w which is not in the subspace W generated by them. The Hilbert space decomposes as $\mathcal{H} = W \oplus W^\perp$, then $w = w' \oplus w''$, where $w' \in W$ and $w'' \in W^\perp$. But $\langle v_k, w''\rangle = 0$ for all k, then $w'' = 0$ and $w = w' \in W$.

Exercise 7.13 *Prove that, given any family of orthogonal vectors $\{v_k\}$, it is possible to complete it to obtain an orthonormal basis.*

7.3.3 Adjoint operators and orthogonal projectors

Given a finite-dimensional[3] Hilbert space $\mathcal{H}$ and a linear map $T : \mathcal{H} \to \mathcal{H}$, there is a canonical map $T^\dagger : \mathcal{H} \to \mathcal{H}$ associated to it defined by:

$$\langle T^\dagger v, w\rangle = \langle v, Tw\rangle ,$$

for all $v, w \in \mathcal{H}$. It is a simple exercise to check that $T^\dagger$ is well defined. The map $T^\dagger$ is called the adjoint of the map T.

If we select an orthonormal basis $\{u_k\}$ and T_{jk} denote the coefficients of the matrix associated with the linear map T on this basis, then $(T^\dagger)_{jk} = \overline{T}_{kj}$ that is, the matrix associated with $T^\dagger$ is the transpose conjugate to the matrix associated with T.

Exercise 7.14 *Prove that: $(T^\dagger)^\dagger = T$, $(T_1 + T_2)^\dagger = T_1^\dagger + T_2^\dagger$ and $(T_1 \circ T_2)^\dagger = T_2^\dagger \circ T_1^\dagger$.*

Definition 7.7 A linear map $T : \mathcal{H} \to \mathcal{H}$ such that $T^\dagger = T$ is called self-adjoint or Hermitean.

If $W \subset \mathcal{H}$ is a subspace of a Hilbert space $\mathcal{H}$, then 7.3.4 shows that there is a canonical projector P_W associated to it, the projector onto W along $W^\perp$. The projector

[3]The same result holds for bounded operators on infinite-dimensional Hilbert spaces.

P_W is defined as $P_W(v) = w$ for $v = w \oplus w' \in W \oplus W^\perp$. In addition, the projector P_W satisfies $P_W = P_W^\dagger$, i.e., it is a self-adjoint operator.

Definition 7.8 Let $\mathcal{H}$ be a Hilbert space. A linear map $P\colon \mathcal{H} \to \mathcal{H}$, such that $P^2 = P = P^\dagger$, is called an orthogonal projector.

Proposition 7.6
Let $\mathcal{H}$ be a Hilbert space. There is a one-to-one correspondence between subspaces $W \subset \mathcal{H}$ and orthogonal projectors.

Proof 7.7 We have already seen that, given a subspace W, there is an orthogonal projector P_W associated to it. Conversely, if P is an orthogonal projector, then consider $W = \operatorname{ran} P$. Then, it is trivial to check that $P = P_W$ (notice that $\ker P = W^\perp$ because $\langle u, Pv\rangle = \langle Pu, v\rangle$ for all $u, v \in \mathcal{H}$).

7.3.4 Unitary representations

In what follows, a finite-dimensional complex linear space carrying an inner product will be called a Hilbert space and, unless stated otherwise, Hilbert spaces will be assumed to be finite-dimensional. We will denote them by using script letters like $\mathcal{H}$ and, unless there is risk of confusion, the inner product $\langle \cdot, \cdot \rangle$ will be omitted.

Definition 7.9 Let G be a finite group and $\mathcal{H}$ be a Hilbert space. A unitary representation of the group G on $\mathcal{H}$ is a map U assigning to each element $g \in G$ a unitary operator $U(g)$ on $\mathcal{H}$ such that:

$$U(e) = \mathrm{id}_{\mathcal{H}}; \qquad U(gg') = U(g)U(g'), \quad \forall g, g' \in G.$$

In other words, a unitary representation is an ordinary linear representation by unitary operators on the supporting Hilbert space.

Remark 7.5 Because we are assuming that the group G is discrete (actually) finite, there is no mention to any continuity property. If we were considering topological groups instead, a continuity condition on the unitary representation U would have been desirable, since it would make the topology on G and the topology on the supporting space $\mathcal{H}$ compatible. ∎

Example 7.15 *The defining representation of the symmetric group S_n is obviously unitary, Sect. 7.1.5.*

The subspace spanned by the vector $(1, 1, \ldots, 1)$ defines an irreducible representation of S_n and its orthogonal complement $W = (1, \ldots, 1)^\perp$ defines an invariant subspace that supports another irreducible representation of S_n of dimension $n-1$.

Exercise 7.16 *Consider the linear representation μ_Γ defined in Example 7.6, associated to a group Γ acting transitively on a finite set Ω. Prove that μ_Γ is a unitary representation when $\mathcal{H}_\Omega$ is equipped with the standard inner product $\langle \psi, \phi \rangle = \sum_{x\in\Omega} \overline{\psi(x)}\phi(x)$.*

Because of the nature of unitary maps, unitary representations have extra properties. In fact, one of the reasons that unitary representations play a fundamental role in the theory of linear representations is the following theorem which shows that, in any equivalence class of linear representations, there is a unitary one. In other words, all we need in order to understand the linear representations of a finite group are the unitary ones.

Theorem 7.2

Any linear representation of a finite group G on a linear space V is equivalent to a unitary one.

Proof 7.8 Let μ be a linear representation of G on the linear space V. Let us consider an inner product $\langle \cdot, \cdot \rangle_0$ on V. Notice that such inner product always exists as it suffices to consider an arbitrary linear basis $\{u_i\}$ and to declare it to be an orthonormal basis for the inner product, that is, we define $\langle u_i, u_j \rangle_0 = \delta_{ij}$. Now, consider the bilinear form defined as:

$$\langle u, v \rangle_1 = \frac{1}{|G|} \sum_{g\in G} \langle \mu(g)u, \mu(g)v \rangle_0 .$$

It is a routine computation to check that the bilinear form $\langle \cdot, \cdot \rangle_1$ defines an inner product on V.

It is also a routine computation to show that the inner product $\langle \cdot, \cdot \rangle_1$ is such that: $\langle \mu(g)u, \mu(g)v \rangle_1 = \langle u, v \rangle_1$, for all $g \in G$, $u, v \in V$, that is, μ is unitary on the Hilbert space $(V, \langle \cdot, \cdot \rangle_1)$.

$$\begin{aligned} \langle \mu(g)u, \mu(g)v \rangle_1 &= \frac{1}{|G|} \sum_{g'\in G} \langle \mu(g')\mu(g)u, \mu(g')\mu(g)v \rangle_0 \\ &= \frac{1}{|G|} \sum_{g'\in G} \langle \mu(g'g)u, \mu(g'g)v \rangle_0 \\ &= \frac{1}{|G|} \sum_{g''\in G} \langle \mu(g'')u, \mu(g'')v \rangle_0 = \langle u, v \rangle_1 . \end{aligned} \tag{7.10}$$

The last step in (7.10) is a standard trick that will be used over and over. It consists of relabelling the sum $\sum_g$ by taking advantage that the elements g' and $g'g$ are in one-to-one correspondence (the operation of multiplying by $g \in G$ is invertible, since G is a group!).

Now, we may interpret this result in two ways. First, we may simply say that the representation μ is unitary on the Hilbert space $(V, \langle\cdot,\cdot\rangle_1)$ or, in a more convoluted way, say that the representation μ is equivalent to the unitary representation $\nu(g) = T^{-1} \circ \mu(g) \circ T$ on the Hilbert space $(V, \langle\cdot,\cdot\rangle_0)$, where T is the isomorphism that transforms an orthonormal basis $\{e_i\}$ for the inner product $\langle\cdot,\cdot\rangle_0$ into an orthonormal basis $\{u_j\}$ for the inner product $\langle\cdot,\cdot\rangle_1$. We immediately check that $\langle \nu(g)u, \nu(g)v\rangle_0 = \langle \mu(g)Tu, \mu(g)Tv\rangle_1 = \langle Tu, Tv\rangle_1 = \langle u, v\rangle_0$ (because, by definition, $\langle Tu, Tv\rangle_1 = \langle u, v\rangle_0$, and (7.10)), hence, ν is unitary with respect to $\langle\cdot,\cdot\rangle_0$.

Actually, we have from the definition of ν that $T \circ \nu(g) = \mu(g) \circ T$, hence, μ and ν are equivalent.

Because of Thm. 7.2, any finite-dimensional representation of a finite group is equivalent to a unitary one, then, we may take advantage of this to simplify the arguments in the previous paragraphs. Thus, if we assume that $(\mu, \mathcal{H})$ is a unitary representation of the group G on the finite-dimensional Hilbert space $\mathcal{H}$, then it is completely reducible, that is there is a family of irreducible invariant subspaces W_i such that $\mathcal{H} = \oplus_{i=1}^r W_i$. Moreover the subspaces W_i are orthogonal to each other.

The proof simplifies with respect to the proof of Thm. 7.1 because, given an invariant irreducible subspace W, it is immediate to check that the orthogonal subspace $W^\perp$ is invariant too, actually, because $\mu(g)$ is unitary, then if $w \in W$ and $v \in W^\perp$, we get:

$$\langle w, \mu(g)v\rangle = \langle \mu(g^{-1})w, v\rangle = 0,$$

for any $g \in G$, because W is invariant and $\mu(g^{-1})w \in W$.

Then we continue by selecting another invariant irreducible subspace of $W^\perp$ and repeating the argument. Thus, the Hilbert space $\mathcal{H}$ decomposes as an orthogonal sum of subspaces W_i. The corresponding orthogonal projectors $P_i \colon \mathcal{H} \to W_i$, $i = 1, \ldots, r$, recall (7.8), satisfy: $P_i P_j = 0$, $i \neq j$, and $\mathrm{id}_{\mathcal{H}} = P_1 \oplus \cdots \oplus P_r$. Such family of orthogonal projectors $\{P_i\}$ is called a decomposition of the identity.

7.4 Schur's lemmas for groups

In this section, we will explore another aspect of the notion or irreducible representation. The result described below, even if very easy to prove, lies at the heart of the classical representation theory and is known as Schur's Lemma. It will appear again later on and it will be established in a general form in the context of representations of algebras, Sect. 10.4.4.

Theorem 7.3 Schur's Lemma

Let (μ_a, V_a), $a = 1, 2$, be two irreducible representations of the group G and let $\varphi \colon V_1 \to V_2$ be a linear map such that

$$\varphi \circ \mu_1(g) = \mu_2(g) \circ \varphi, \qquad \forall g \in G, \tag{7.11}$$

then either $\varphi = 0$ or φ is an isomorphism and both representations are equivalent.

Proof 7.9 Let $\varphi\colon V_1 \to V_2$ satisfying (7.11). Then, it is clear that $\ker\varphi$ is an invariant subspace, because if $v \in \ker\varphi$, then $\varphi(\mu_1(g)v) = \mu_2(g)(\varphi(v)) = 0$. Hence, because (μ_1, V_1) is irreducible then, either $\ker\varphi = \mathbf{0}$ and φ is injective, or $\ker\varphi = V_1$ in which case $\varphi = 0$.

It is also true that $\varphi(V_1)$ is an invariant subspace of V_2. In fact, if $w = \varphi(v) \in \varphi(V_1)$, then $\mu_2(g)w = \mu_2(g)(\varphi(v)) = \varphi(\mu_1(g)v) \in \varphi(V_1)$. Then, because (μ_2, V_2) is irreducible, either $\varphi(V_1) = \mathbf{0}$, that is $\varphi = 0$, or $\varphi(V_1) = V_2$ and φ is surjective. But if $V_1 \neq \mathbf{0}$, and $\varphi \neq 0$, then it must be injective and $\varphi(V_1) \neq \mathbf{0}$, hence, φ is surjective too, then an isomorphism.

Taking advantage of the fact that when we consider complex linear spaces, the scalars of the theory are complex numbers, we get as an immediate consequence of Schur's Lemma that any linear map $\varphi\colon V \to V$, such that $\varphi \circ \mu(g) = \mu(g) \circ \varphi$ with μ irreducible, that is, a linear map commuting with an irreducible representation, must be a multiple of the identity.

Theorem 7.4
Let (ρ, V) be a finite-dimensional representation of the finite group G. The representation (ρ, V) is irreducible iff the only linear maps $\varphi\colon V \to V$, such that $\mu(g) \circ \varphi = \varphi \circ \mu(g)$, $\forall g \in G$ are homotecies, i.e., multiples of the identity, $\varphi(v) = \lambda v$, with $\lambda = \mathrm{Tr}\,\varphi / \dim V$.

Proof 7.10 Suppose that μ is irreducible. Let $\varphi\colon V \to V$ be a linear map such that $\mu(g) \circ \varphi = \varphi \circ \mu(g)$, $\forall g \in G$. Let λ be an eigenvalue of φ. Define the map $\varphi' = \varphi - \lambda\, \mathrm{id}_V$, then φ' is not invertible and $\mu(g) \circ \varphi' = \varphi' \circ \mu(g)$, then because of Schur's Lemma, Thm. 7.3, $\varphi' = 0$, that is, $\varphi = \lambda\, \mathrm{id}_V$. The result $\lambda = \mathrm{Tr}\,\varphi / \dim V$ follows from the properties of the trace (Prop. A.14).

Conversely, the argument will proceed by contradiction. Suppose that the representation μ is reducible, then, because of Thm. 7.1, the representation is completely reducible, and there exists a decomposition of V as the direct sum of (irreducible) invariant subspaces $V = W_1 \oplus \cdots \oplus W_r$, $r > 1$. Then any projector $P_k\colon V \to W_k$, $k = 1, \ldots, r$, commutes with the representation μ,

$$(\mu(g) \circ P_k)(v) = \mu(g)v_k = P_k(\mu(g)(v)) = (P_k \circ \mu(g))(v),$$

for all $v \in V$, $g \in G$, where $v = v_1 \oplus \cdots \oplus v_k \oplus \cdots \oplus v_r$, $v_k \in W_k$, and $\mu(g)v_k \in W_k$, for all $k = 1, \ldots, r$. However, P_k is not a multiple of the identity which contradicts the assumption that the only linear maps commuting with μ are multiple of the identity.

Definition 7.10 Given two linear representations (μ_a, V_a), $a = 1, 2$ of the group G, a linear map $\varphi\colon V_1 \to V_2$ such that: $\varphi \circ \mu_1(g) = \mu_2(g) \circ \varphi$, for all $g \in G$, is called an intertwiner (or equivariant linear map). The space of intertwiners between V_1 and V_2 will be denoted by $\mathrm{Hom}_G(V_1, V_2)$.

Clearly, the space of intertwiners $\mathrm{Hom}_G(V_1,V_2)$ is a subspace of $\mathrm{Hom}(V_1,V_2)$, the linear space of all linear maps from V_1 to V_2. The use of the intertwining space $\mathrm{Hom}_G(V_1,V_2)$ provides an alternative way of thinking about irreducibility that will be exploited later on when dealing with groupoids and algebras. In fact, we can rewrite Schur's Lemmas as:

Theorem 7.5
Let (μ_a,V_a), $a=1,2$, be two irreducible representations of the group G, then:

1. $\mathrm{Hom}_G(V_1,V_2)=\mathbf{0}$ *iff* $\mu_1 \nsim \mu_2$.
2. $\mathrm{Hom}_G(V_1,V_2)\cong \mathbb{C}$ *iff* $\mu_1 \sim \mu_2$.

Another striking consequence of Schur's Lemmas is that irreducible representations of Abelian groups must be one-dimensional.

Proposition 7.7
Let A be an Abelian group, then any irreducible representation of A is one-dimensional.

Proof 7.11 Let (μ,V) be an irreducible representation of A. Since A is Abelian, we have $\mu(a)\circ\mu(a')=\mu(a')\circ\mu(a)$ for all $a'\in A$, and $\mu(a)\in \mathrm{Hom}_A(V,V)$ for any $a\in A$. But μ is irreducible, then $\mu(a)=\mu_a$ is a multiple of the identity, and there exists a complex number $\lambda(a)\in\mathbb{C}$, such that $\mu_a(v)=\lambda(a)v$, hence, the space generated by v is a proper invariant subspace of V. Thus, the space V must be one-dimensional.

It is convenient to write down the consequences of the previous theorems in terms of the corresponding matrix representations.

Corollary 7.2
Let D and D' be matrix representations of two non-equivalent irreducible representations of the group G, then, if A is a matrix, such that $D'(g)A=AD(g)$ for all $g\in G$, then $A=0$. In particular, if the dimensions of the two matrix representations are different, then $A=0$.

Corollary 7.3
Let D and D' be matrix representations associated with two irreducible representations of the group G of the same dimension. If A is a singular matrix, such that $D'(g)A=AD(g)$ for all $g\in G$, then $A=0$.

Corollary 7.4
Let A be a square matrix that commutes with all matrices of an irreducible representation of the same dimension, then A is a multiple of the unit matrix.

Chapter 8

Characters and Orthogonality Relations

In this chapter, we will introduce the notion of the character of a representation and the orthogonality relations satisfied by them. Characters will prove to be invaluable tools to understanding the structure of representations. We will use them again when working with representations of groupoids where the properties of characters of group representations will be heavily exploited.

The use of characters will lead us easily to the fundamental theorems of the theory of linear representations of finite groups. This theory is well-known and there are already very good references available, however, we will discuss it in some detail in order to prepare the reader for the more advanced topics that will be introduced in subsequent chapters and also as a way of presenting a number of notations and examples that will be used extensively afterwards.

8.1 Orthogonality relations

As it was stated at the end of the previous chapter, Schur's Lemmas have far-reaching consequences. In this section, we will state some of them, known as orthogonality relations. They will be instrumental in classifying the linear representations of finite groups.

Theorem 8.1
Let (μ_i, V_i), $i = 1, 2$, be two irreducible representations of the finite group G. For each linear map $f : V_1 \to V_2$, let ϕ_f be the linear map $\phi_f : V_1 \to V_2$ defined as:

$$\phi_f = \frac{1}{|G|}\sum_{g\in G}\mu_2(g^{-1})\circ f\circ\mu_1(g)\,.$$

Then:

i) If μ_1 and μ_2 are not equivalent, then $\phi_f = 0$.

ii) If $V_1 = V_2 = V$ and $\mu_1 \sim \mu_2$, then ϕ_f is a multiple of the identity:

$$\phi_f = (\mathrm{Tr}\, f/\dim V)\,\mathrm{id}_V\,.$$

Proof 8.1 It is a direct application of Schur's Lemmas 7.3 and 7.4.

i) Using the standard trick of relabelling the indexes inside a sum (recall, for instance, the comments after the proof of Thm. 7.2), it is an easy exercise to check that $\forall g \in G$, $\mu_2(g)\circ\phi_f = \phi_f\circ\mu_1(g)$. Hence, because Thm. 7.3, if μ_1 and μ_2 are not equivalent, then $\phi_f = 0$.

ii) If $\mu_1 \sim \mu_2$, then, because Thm. 7.4, ϕ_f is a homotecy, i.e., a multiple of the identity map, $\phi_f = \lambda\,\mathrm{id}_V$, and because $\mathrm{Tr}\,\phi_f = \mathrm{Tr}\, f$, then $\lambda = \mathrm{Tr}\, f/\dim V$.

Remark 8.1 Notice that $f \in \mathrm{Hom}\,(V_1,V_2) \cong V_1^*\otimes V_2$. There is a natural representation of G on $V_1^*\otimes V_2$ given by $\mu_1^*\otimes\mu_2$ (recall Sec. 7.2.2). Then ϕ_f is nothing but the average of f with respect to the linear representation $\mu_1^*\otimes\mu_2$ of G on $\mathrm{Hom}\,(V_1,V_2)$.

■

The relevance of the previous result becomes apparent when using matrix representations. In fact, it leads immediately to the following formulas.

Corollary 8.1

(Orthogonality relations) Let D_1 and D_2 be two irreducible matrix representations of the finite group G. Then:

i) If D_1 and D_2 are inequivalent,

$$\sum_{g\in G}[D_2(g^{-1})]_{pq}[D_1(g)]_{ij} = 0\,,$$

for every $1\le i,j\le n_1 = \dim D_1$, $1\le p,q\le n_2 = \dim D_2$.

ii) If $D_1 = D_2 = D$, and n is the dimension of D,

$$\frac{1}{|G|}\sum_{g\in G}[D(g^{-1})]_{pq}[D(g)]_{ij} = \frac{1}{n}\delta_{pj}\,\delta_{iq}\,.$$

for any $1\le i,j,p,q\le n$.

iii) If D_1 and D_2 are equivalent irreducible matrix representations of dimension n, and M an invertible matrix such that $D_2(g) = M^{-1}D_1(g)M$, then,

$$\frac{1}{|G|}\sum_{g\in G}[D_2(g^{-1})]_{pq}[D_1(g)]_{ik} = \frac{1}{n}M_{iq}M^{-1}_{pk}, \tag{8.1}$$

for every $1 \leq i,k,p,q \leq n$.

Proof 8.2 i) Let (μ_1,V_1) and (μ_2,V_2) be two linear representations of G associated to D_1 and D_2, respectively, i.e., D_1 and D_2 are the matrix representations associated to μ_1 and μ_2 in some chosen bases of V_1 and V_2, respectively. Let A be the matrix associated in the chosen bases to an arbitrary linear map $f\colon V_1 \to V_2$. Then, because Thm. 8.1.i we get that $\phi_f = 0$ and, computing the element (p,j) in the given basis we get:

$$\sum_{g\in G}\sum_{q,i}[D_2(g^{-1})]_{pq}A_{qi}[D_1(g)]_{ij} = 0, \quad 1 \leq p \leq n_2, 1 \leq j \leq n_1,$$

for any matrix A and the conclusion follows.

ii) With the same conventions, we apply now Thm. 8.1.ii to get:

$$\frac{1}{|G|}\sum_{g\in G}\sum_{jl}[D(g^{-1}]_{ij}A_{jl}[D(g)]_{lk} = \frac{1}{n}\delta_{ik}\sum_{r}A_{rr} = \frac{1}{n}\delta_{ik}\sum_{lj}\delta_{lj}A_{lj}.$$

Again, because the equality holds for any matrix A we get the desired conclusion.

iii) Applying the previous result, Cor. 8.1.ii, to D_2 we get:

$$\frac{1}{|G|}\sum_{g\in G}[D_2(g^{-1})]_{pq}[D_2(g)]_{ik} = \frac{1}{n}\delta_{pk}\,\delta_{qi},$$

but, because $[D_2(g)]_{ik} = \sum_{j,l}[M^{-1}]_{ij}[D_1(g)]_{jl}M_{lk}$:

$$\frac{1}{m}\sum_{j,l}\sum_{g\in G}[D_2(g^{-1})]_{pq}[M^{-1}]_{ij}[D_1(g)]_{jl}M_{lk} = \frac{1}{n}\delta_{pk}\,\delta_{qi}.$$

Multiplying both sides of the previous equation by $M_{ri}[M^{-1}]_{ks}$ and summing on i and k we get the formula (8.1) in (iii).

Clearly, if the matrix representations are unitary, then $[D(g^{-1})]_{ij} = \overline{[D(g)]}_{ji}$ and Cor. 8.1.i takes the particular form:

$$\sum_{g\in G}\overline{[D_2(g)]}_{pq}[D_1(g)]_{ij} = 0,$$

for all $1 \leq p,q \leq \dim D_2$, $1 \leq i,j \leq \dim D_1$ provided that D_1 and D_2 are disequivalent.

Moreover, if $D_1 = D_2 = D$, then 8.1.ii becomes

$$\frac{1}{|G|}\sum_{g\in G}\overline{[D(g)]_{ij}}[D(g)]_{kl} = \frac{1}{n}\delta_{ik}\,\delta_{jl}\,,$$

for all $1 \leq i,j,k,l \leq \dim D$ and, finally, 8.1.iii becomes:

$$\frac{1}{|G|}\sum_{g\in G}\overline{[D_2(g)]_{ij}}[D_1(g)]_{kl} = \frac{1}{n}\overline{[U]_{lj}}U_{ki},$$

if D_1 are D_2 are equivalent, and U is a unitary matrix such that $D_2(g) = U^\dagger D_1(g)U$, for all $g \in G$.

The previous relations can be written in a slightly more compact form. Denoting by μ_α an (representative of an equivalence class of an) irreducible unitary representation of G, and D^α the corresponding matrix representation on an orthonormal basis, the previous relations can be written as:

$$\frac{1}{|G|}\sum_{g\in G}\overline{[D^\alpha(g)]_{ij}}[D^\beta(g)]_{kl} = \frac{1}{n_\alpha}\,\delta_{\alpha\beta}\,\delta_{ik}\,\delta_{jl}, \tag{8.2}$$

where n_α is the dimension of D^α.

These relations are known as *orthogonality relations*. A clever use of the orthogonality relations will allow us to show that there is just a finite number of different irreducible representations of any finite group.

We will denote by $\widehat{G}$ the set of equivalence classes of irreducible representations of the finite group G (all of them finite-dimensional because of Prop. 7.2). Thus, an element in $\widehat{G}$ will be an equivalence class of irreducible representations $[\mu^\alpha]$, and we can choose a representative $(\mu^\alpha, \mathcal{H}^\alpha)$ for such class or just use the index α to indicate it. We will use any of these conventions in what follows.

Theorem 8.2
The set $\widehat{G}$ of equivalence classes of irreducible representations of the finite group G is finite. If n_α is the dimension of the irreducible representation μ^α, then:

$$\sum_{\alpha\in\widehat{G}} n_\alpha^2 \leq |G|\,.$$

Proof 8.3 Let us consider the set $\mathcal{F}(G,\mathbb{C})$ of complex valued function on the group G. It is clear that $\mathcal{F}(G,\mathbb{C})$ is a complex linear space of finite dimension $|G|$ (see Sect. 7.1.4). We can also consider an inner product (Sect. 7.3, Def. 7.5) on $\mathcal{F}(G,\mathbb{C})$ defined as:

$$\langle f_1, f_2\rangle = \sum_{g\in G}\overline{f_1(g)}\,f_2(g)\,. \tag{8.3}$$

Notice that the functions δ_g form an orthonormal basis, i.e., $\langle\delta_g,\delta_{g'}\rangle = \delta_{gg'}$.

Hence, once we have selected a matrix representation D^α on each equivalence class, for each triple (i,j,α) with $1 \leq i,j \leq n_\alpha$, we define the function $f_{ij\alpha}(g) = [D^\alpha(g)]_{ij}$. The orthogonality relations (8.2) show that the functions $f_{ij\alpha}$ are mutually orthogonal, hence, linearly independent. Because for each α there exists n_α^2 of them as the indices i,j run from 1 to n_α, the total number of them will be $\sum_{\alpha\in\widehat{G}} n_\alpha^2$, which must be finite as they cannot exceed the dimension of $\mathcal{F}(G,\mathbb{C})$, that is, $|G|$.

We will show later on that the previous inequality is, in fact, an equality. We can summarise the information we have so far on the family $\widehat{G}$ of irreducible representations of a finite group.

Corollary 8.2
Let G be a finite group. Each irreducible representation of G is finite-dimensional and equivalent to a unitary representation. The number of inequivalent representations is finite and $\sum_{\alpha\in\widehat{G}} n_\alpha^2 \leq |G|$, hence, $|\widehat{G}| \leq |G|$.

8.2 Characters

A basic idea when attempting to classify a certain family of abstract objects is to associate numbers to them that do not depend on the equivalence relations defined among them, that is, such that they are the same for any pair of equivalent objects. Hence, we could ascertain that two objects will not be equivalent if the numbers associated to them are different.

We often call such numbers 'invariants' (with respect to the given equivalence relation) and a good strategy to classify the objects in the given family will be to find enough invariants such that equivalence classes will be totally determined by them, that is, two objects will be equivalent iff they possess the same invariants.

Regarding linear representations, this idea will be implemented attaching to any linear representation (μ,V) a family of numbers, one for each linear map $\mu(g)$, such that they will coincide in equivalent representations. A simple way to obtain such numbers is exploiting properties of the algebra of matrices and realising that the traces of matrices are invariant under changes of basis. So, we may define our first set of invariants for linear representations, called characters, as follows.

Definition 8.1 Given a finite-dimensional linear representation (μ,V) of the group G, we will define the character of μ at $g\in G$ as the the trace of the linear map $\mu(g)\colon V\to V$, that is, $\chi(g) = \operatorname{Tr}\mu(g)$, where $\operatorname{Tr}\mu(g)$ is the trace of any matrix representation $D(g)$ associated to μ (recall the definition of the trace of a linear map, Def. A.17).

We immediately get:

Proposition 8.1
The character of the neutral element e is the dimension of the representation.

Proof 8.4 Obvious, because $D(e) = I_n$, n the dimension of the linear space supporting the representation, then $\mathrm{Tr}\, D(e) = n$.

We state explicitly the fundamental fact that characters depend just on the equivalence classes of linear representations.

Proposition 8.2
The characters of two equivalent representations are the same.

Proof 8.5 If μ' is equivalent to μ there exists an isomorphism $\varphi\colon V \to V'$ such that $\mu'(g) = \varphi \circ \mu(g) \circ \varphi^{-1}$, $\forall g \in G$. Let D' be a matrix representation associated to μ' and D a matrix representation associated to μ, that means that there exist linear bases $\{u_i\}$, $\{u'_j\}$ on V and V', respectively, such that $D(g)$ is the matrix associated to $\mu(g)$ in the basis $\{u_i\}$ (and $D'(g)$ is the matrix associated to $\mu'(g)$ in the basis $\{u'_j\}$). But then $D'(g) = AD(g)A^{-1}$ and the characters of $\mu'(g)$ and $\mu(g)$ coincide.

Proposition 8.3
Let (μ, V) be a finite-dimensional linear representation of the group G. The map $\chi : G \to \mathbb{C}$, defined as: $\chi : g \to \chi(g)$, satisfies $\chi(gg') = \chi(g'g)$ for all $g, g' \in G$, hence, χ is a function which is constant on each conjugacy class of G. Functions satisfying this property will be called class functions or central functions.

Proof 8.6 That $\chi(gg') = \chi(g'g)$ is trivial because it follows immediately from the trace circular property $\mathrm{Tr}(AB) = \mathrm{Tr}(BA)$ for all A, B (A.14.1). If g_2 y g_1 are conjugate in G, then there exists $g \in G$ such that $g_2 = g^{-1} g_1 g$, in consequence, $D(g_2) = D(g^{-1} g_1 g) = D(g^{-1}) D(g_1) D(g)$ and, $\chi(g_2) = \mathrm{Tr}\, D(g_2) = \mathrm{Tr}\, D(g_1) = \chi(g_1)$.

Proposition 8.4
The set $\mathcal{C}(G)$ of central functions is a subspace of $\mathcal{F}(G, \mathbb{C})$. Its dimension is equal to the number of conjugacy classes in the group G.

Proof 8.7 It is clear that the sum of two central functions is again central as well as the product by a scalar. Moreover, because central functions are constant on conjugacy classes, the set of characteristic functions form a basis of $\mathcal{C}(G)$.

We will finish this introduction to the properties of characters using, for the first time, and obtaining a non-trivial consequence, the fact that any linear representation of a finite group is equivalent to a unitary one.

Proposition 8.5
Let (μ, V) be a finite-dimensional linear representation of the group G. The characters of an element g and its inverse g^{-1} are related by $\chi(g^{-1}) = \overline{\chi(g)}$.

Proof 8.8 Any representation is equivalent to a unitary representation (Thm. 7.2), then, choosing a matrix representation for a unitary representation equivalent to μ (in an orthonormal basis) we get $D(g^{-1}) = D^\dagger(g)$ and then $\mathrm{Tr}\, D(g^{-1}) = \mathrm{Tr}\,(D(g)^\dagger) = \overline{\mathrm{Tr}\, D(g)} = \overline{\chi(g)}$ (because of A.14.3).

8.3 Orthogonality relations of characters

In this section, the main orthogonality relations satisfied by characters will be obtained. They will lead immediately to the main theorems of the theory. In what follows, we will consider the standard normalized inner product in the space of functions on the finite group G, that is:

$$\langle \psi, \phi \rangle = \frac{1}{|G|} \sum_{g \in G} \overline{\psi(g)} \phi(g) . \tag{8.4}$$

Theorem 8.3
Let μ_1 and μ_2 be two irreducible representations of the finite group G. If they are not equivalent then their characters χ_1 and χ_2 are orthogonal with respect to the standard inner product on $\mathcal{F}(G,\mathbb{C})$, i.e., $\langle \chi_1, \chi_2 \rangle = 0$.

Proof 8.9 Let D_1 and D_2 be two matrix representations corresponding to the representations μ_1, μ_2. From the orthogonality relations (8.2), with pairwise identified indices, we get: $\sum_{g \in G} \overline{[D_2(g)]_{ii}} [D_1(g)]_{jj} = 0$, and summing up over the indices i and j we get: $\sum_{g \in G} \overline{\chi_2(g)}\, \chi_1(g) = 0$.

Theorem 8.4
If χ is the character of an irreducible linear representation (μ, V) of the finite group G, then $\langle \chi, \chi \rangle = 1$.

Proof 8.10 We will again use the orthogonality relations (8.2). Choosing a matrix representation D for μ, we get:

$$\frac{1}{|G|} \sum_{g \in G} \overline{[D(g)]_{ij}} [D(g)]_{kl} = \frac{1}{n} \delta_{ik}\, \delta_{jl} .$$

Identifying the indices $i = j$, $k = l$, and summing over i and k, we get:

$$\frac{1}{|G|}\sum_{g\in G}\overline{\chi(g)}\,\chi(g) = 1\,.$$

Let $\{C_i\}_{i\in I}$ be the set of conjugacy classes of the group G, i.e., $C_i = \{gg_ig^{-1} \mid g \in G\}$, and I a collection of indices labelling the different conjugacy classes of G. We will denote by m_j the number of elements in the class C_j, $m_j = |C_j|$. Then $\sum_{i\in I} m_i = m = |G|$.

Because characters are class functions, the inner product $\langle\cdot,\cdot\rangle$ can be expressed as:

$$\langle\psi,\varphi\rangle = \frac{1}{|G|}\sum_{g\in G}\overline{\psi(g)}\varphi(g) = \frac{1}{|G|}\sum_{i\in I} m_i\overline{\psi}_i\varphi_i\,, \qquad \psi,\varphi\in\mathcal{C}(G)\,, \tag{8.5}$$

where ψ_i,φ_i are the common values that the class functions ψ, φ take on the corresponding conjugacy class C_i.

Then, the previous theorems can be restated as:

Corollary 8.3
Let μ_1 and μ_2 be two inequivalent irreducible linear representations. If χ_i^a is the value taken by the character corresponding to the representation μ_a, $a = 1,2$, in the class C_i, $m_i = |C_i|$, then:

$$\sum_{i\in I} m_i\,\overline{\chi_i^2}\,\chi_i^1 = 0\,.$$

Corollary 8.4
Let χ be the character of an irreducible representation of the finite group G. If χ_i is the common value of χ on the elements of the conjugacy class C_i, then:

$$\frac{1}{|G|}\sum_{i\in I} m_i|\chi_i|^2 = 1\,.$$

The orthogonality relations of characters obtained before can be summarized as:

$$\langle\chi_\alpha,\chi_\beta\rangle = \delta_{\alpha\beta}\,, \tag{8.6}$$

where χ_α,χ_β are the characters of the irreducible representations μ_α, μ_β.

Proposition 8.6
The number of equivalence class of irreducible representations of the finite group G is smaller or equal than the number of conjugacy classes of G.

Proof 8.11 We may consider the set of central functions $\mathcal{C}(G)$ equipped with the inner product obtained by restriction of the standard inner product $\langle\cdot,\cdot\rangle$ on $\mathcal{F}(G,\mathbb{C})$. The orthogonality relations (8.6) imply that characters of irreducible representations are orthogonal, hence, independent, but $\mathcal{C}(G)$ is a linear space of dimension the number of conjugacy classes, then, because non-equivalent irreducible representations possess orthogonal characters, they are independent and the conclusion follows.

8.4 Inequivalent representations and irreducibility criteria

In this section, we will put together the information provided by the orthogonality relations of characters, to close the previous chain of arguments and reach the main results of the theory.

Definition 8.2 Characters of irreducible representations of G will be called simple.

Example 8.1

1. *Clearly, the characters of one-dimensional representations are simple. For instance, the representation of the symmetric group S_n on the subspace spanned by the vector $(1,\ldots,1)$ is irreducible. Its character is given by $\chi(\sigma) = \mathrm{Tr}(D(\sigma)) = 1$. Then $\langle\chi,\chi\rangle = 1$.*

2. *The character of the representation $(\mu_\Gamma, \mathcal{H}_\Omega)$ naturally associated to a group Γ acting transitively on a set Ω with more than one element is given by: $\chi(e) = |\Omega|$, $\chi(g) = 0$, $g \neq e$. Then, $\langle\chi,\chi\rangle = |\Omega| > 1$ and the character is not simple.*

Proposition 8.7
The character χ of a linear representation (μ, V) is a linear combination with non-negative integer coefficients of simple characters:

$$\chi = \sum_{\alpha\in\widehat{G}} d_\alpha \chi_\alpha \,. \tag{8.7}$$

Proof 8.12 Let (μ, V) be a linear representation of G, then, because of Thm. 7.1, the representation μ is completely reducible and

$$V = V_1 \oplus \cdots \oplus V_r \,, \tag{8.8}$$

with V_α invariant subspaces supporting irreducible representations of G. We may choose a basis $\mathcal{B}$ of V adapted to the previous decomposition, and the corresponding matrix representation will have the form of a block diagonal matrix:

$$D(g) = \begin{pmatrix} D_1(g) & 0 & \cdots & 0 \\ 0 & D_2(g) & \cdots & 0 \\ \vdots & \vdots & \ddots & \vdots \\ 0 & 0 & \cdots & D_r(g) \end{pmatrix},$$

then $\chi(g) = \sum_{\beta=1}^{r} \chi_\beta(g)$, $\forall g \in G$. Some of the representations that appear in the decomposition above can be equivalent, hence, they will have the same character (Prop. 8.2), then, once we group equivalent irreducible representations in the decomposition (8.8) together, we will get: $\chi(g) = \sum_{\beta\in\widehat{G}} d_\beta \chi_\beta(g)$, $\forall g \in G$, d_β non-negative integers.

To understand in what sense the decomposition of a character as sum of simple ones is characteristic of it, we need to understand better the space of class functions $\mathcal{C}(G)$ of the group.

Proposition 8.8
Let $f \in \mathcal{C}(G)$ be a class function and (μ, V) an irreducible representation of the finite group G with character χ, then the endormphism $\mu_f : V \to V$ defined by

$$\mu_f = \sum_{g\in G} f(g)\mu(g)$$

is a homothety of ratio $\lambda = \langle \bar{f}, \chi \rangle / \dim V$.

Proof 8.13 We show easily that for any $g \in G$, $\mu(g^{-1}) \circ \mu_f \circ \mu(g) = \mu_f$, in fact:

$$\begin{aligned} \mu(g^{-1}) \circ \mu_f \circ \mu(g) &= \sum_{g'\in G} f(g')\mu(g^{-1}g'g) \\ &= \sum_{g''\in G} f(gg''g^{-1})\mu(g'') = \sum_{g'\in G} f(g')\mu(g') = \mu_f . \end{aligned}$$

But μ is irreducible, then because of Schur's Lemma, Thm. 7.4, μ_f is a homothety of ratio $\operatorname{Tr}\mu_f/\dim V$. But $\operatorname{Tr}\mu_f = \sum_{g\in G} f(g)\chi(g) = \langle \bar{f}, \chi \rangle$.

Theorem 8.5
Simple characters form an orthonormal basis of the space of class functions $\mathcal{C}(G)$.

Proof 8.14 We already know that simple characters are class functions and orthonormal because of the orthogonality relations, Thm. 8.3. To prove that the family $\{\chi_\alpha\}$ of simple characters is a basis, it is sufficient to show that the only function $f \in \mathcal{C}(G)$ orthogonal to all of them $\langle f, \chi_\alpha \rangle = 0$, for all $\alpha \in \widehat{G}$, is zero (recall Prop. 7.5).

Thus, let us assume that $\langle f, \chi_\alpha \rangle = 0$, for all $\alpha \in \widehat{G}$. Then, if $\mu = \mu_\alpha$ is an irreducible representation of G, μ_f vanish. Actually, $\mu_{\bar{f}}$ is the homothety with ratio $\langle f, \chi_\alpha \rangle / \dim V$, but $\langle f, \chi_\alpha \rangle = 0$ by hypothesis, then $\mu_{\bar{f}} = 0$.

If μ is reducible, we choose a decomposition of μ as a sum of irreducibles $\mu = \bigoplus_\alpha d_\alpha \mu_\alpha$. Then, $\mu_f = \sum_{g \in G} f(g)\mu(g) = \sum_{g \in G} f(g) \sum_\alpha d_\alpha \mu_\alpha = \sum_\alpha d_\alpha \sum_{g \in G} f(g)\mu_\alpha(g)$. But each of the factors $\mu_{\alpha,f} = \sum_{g \in G} f(g)\mu_\alpha(g)$ vanish because of the previous argument and we conclude that $\mu_f = 0$.

Consider now the particular instance with $\mu = \lambda$, the left regular representation of G. Then, on one side, applying the definition, we get $\lambda_{\bar{f}}(e) = \sum_{g \in G} \overline{f(g)} \lambda_g(e) = \sum_{g \in G} \overline{f(g)} g$, but we have proved that $\lambda_{\bar{f}} = 0$, then $\overline{f(g)} = 0$, for all $g \in G$, and we conclude that $\bar{f} = 0$, hence, $f = 0$.

Theorem 8.6
Let (μ, V) be a finite dimensional linear representation of the finite group G. The decomposition of (μ, V) as the sum of irreducible representations, $\mu = \bigoplus_\alpha d_\alpha \mu_\alpha$, $V = \bigoplus_\alpha d_\alpha V_\alpha$, is essentially unique in the sense that the multiplicity d_α of the factors V_α is unique; its character is given by $\chi = \sum_{\alpha \in \widehat{G}} d_\alpha \chi_\alpha$ and the multiplicity d_α in μ is given by:

$$d_\alpha = \langle \chi_\alpha, \chi \rangle ,$$

Proof 8.15 Let $V = d_1 V_1 \oplus \cdots \oplus d_r V_r$ be a decomposition of the representation (μ, V) with (μ_α, V_α) irreducible representations of G. Let $\chi = \sum_{\beta \in \widehat{G}} d_\beta \chi_\beta$ (Prop. 8.7), the corresponding decomposition of the character χ of μ. Then, because of (8.6), we get:

$$\langle \chi_\alpha, \chi \rangle = \sum_{\beta \in \widehat{G}} d_\beta \langle \chi_\alpha, \chi_\beta \rangle = d_\alpha .$$

If $V = d'_1 V'_1 \oplus \cdots \oplus d'_{r'} V'_{r'}$ is another decomposition in irreducibles of the representation μ. Then, if μ_α is an irreducible representation, we get $d_\alpha = \langle \chi_\alpha, \chi \rangle = d'_\alpha$, hence, $d_\alpha = d'_\alpha$ (note that if an irreducible representation is not in the decomposition $d_\alpha = 0$), then, the factors that appear in both decompositions are the same.

Definition 8.3 The non-negative integer d_α that appears in the decomposition of the character χ in terms of simple ones will be called the frequency or the multiplicity of the class represented by the irreducible representation μ_α in μ.

Remark 8.2 The previous theorem makes explicit the observation about the uniqueness of the decomposition made after the proof that any finite-dimensional representation of a finite group is completely reducible, Thm. 7.1. ▮

The previous results allow us to describe linear representations entirely in terms of their corresponding characters. We already know that if two representations are equivalent their characters coincide. The converse is also true.

Corollary 8.5
If the characters χ, χ' of two linear representations (μ, V) and (μ', V') coincide, $\chi = \chi'$, then μ and μ' are equivalent.

Proof 8.16 Consider the decompositions of $V = \bigoplus_\alpha d_\alpha V_\alpha$ and $V' = \bigoplus_\alpha d'_\alpha V_\alpha$ as direct sum of irreducible representations. Then, because of the uniqueness theorem, Thm. 8.6, we get

$$d_\alpha = \langle \chi_\alpha, \chi \rangle = \langle \chi_\alpha, \chi' \rangle = d'_\alpha ,$$

and the representations (μ, V) and (μ', V') are equivalent.

Corollary 8.6
The linear representation (μ, V) of the finite group G is irreducible iff its character χ satisfies $\langle \chi, \chi \rangle = 1$.

Proof 8.17 Let $\chi = \sum_{\alpha \in \widehat{G}} d_\alpha \chi_\alpha$, be the decomposition of the character χ in terms of simple characters. Then, because simple characters define an orthonormal basis of the space of class functions, we get:

$$\langle \chi, \chi \rangle = \sum_{\beta \in \widehat{G}} d_\beta^2 . \tag{8.9}$$

Then, if μ is irreducible, only one of the coefficients d_α is one and the others vanish.

Conversely, $\langle \chi, \chi \rangle = 1$, then $\sum_{\beta \in \widehat{G}} d_\beta^2 = 1$ and there exists only one non-vanishing coefficient $d_\alpha = 1$.

Finally, we obtain an expression for the dimension of the space of intertwiners $\mathrm{Hom}_G(V_1, V_2)$ between two representations (μ_1, V_1) and (μ_2, V_2).

Theorem 8.7
Let (μ_1, V_1), (μ_2, V_2) be two finite-dimensional representations of the finite group G with characters χ_1, χ_2, respectively. Then:

1. $\langle \chi_1, \chi_2 \rangle = \dim \mathrm{Hom}_G(V_1, V_2)$.
2. *If μ_1, μ_2 are irreducible, then $\langle \chi_1, \chi_2 \rangle = 1$ iff they are equivalent and zero otherwise.*

Proof 8.18 The second statement is a direct consequence of Thm. 8.5 and Cor. 8.6.

To prove the first formula, we consider decompositions of the representations μ_1 and μ_2 as a sum of irreducibles: $V_1 = \bigoplus_\alpha d_\alpha V_\alpha$ and $V_2 = \bigoplus_\alpha f_\alpha V_\alpha$. Then, we have that $\mathrm{Hom}(V_1, V_2) = \bigoplus_{\alpha,\beta} d_\alpha f_\beta \mathrm{Hom}(V_\alpha, V_\beta)$ (Appendix A, Eq. (A.2)). Then, $\mathrm{Hom}_G(V_1, V_2) = \bigoplus_{\alpha,\beta} d_\alpha f_\beta \mathrm{Hom}_G(V_\alpha, V_\beta)$, but because of Schur's Lemmas, if the irreducible representations V_α and V_β are not equivalent, then $\mathrm{Hom}_G(V_\alpha, V_\beta) = 0$ and $\mathrm{Hom}_G(V_\alpha, V_\alpha) = \mathbb{C}$. Then, we obtain:

$$\mathrm{Hom}_G(V_1, V_2) = \bigoplus_\alpha d_\alpha f_\alpha \mathrm{Hom}_G(V_\alpha, V_\alpha) = \bigoplus_\alpha \mathbb{C}^{d_\alpha f_\alpha},$$

and, consequently, $\dim \mathrm{Hom}_G(V_1, V_2) = \sum_\alpha d_\alpha f_\alpha$. On the other hand

$$\langle \chi_1, \chi_2 \rangle = \sum_{\alpha,\beta} d_\alpha f_\beta \langle \chi_\alpha, \chi_\beta \rangle = \sum_\alpha d_\alpha f_\alpha,$$

and the desired formula follows.

We will end this section by proving the following theorem which provides a definite answer to the question of how many equivalence classes of irreducible representations a finite group has.

Theorem 8.8
The number of equivalence classes of irreducible representations of a finite group equals the number of its conjugation classes.

Proof 8.19 An orthonormal basis of $\mathcal{C}(G)$ consists of characteristic functions on conjugacy classes C_i, because any function $f \in \mathcal{C}(G)$ can be written as $f = \sum_{i \in I} f(C_i)\xi_{C_i}$, where $\xi_{C_i}(g) = 1$ if $g \in C_i$ and 0 otherwise.

Then, because both sets, simple characters and characteristic functions, are orthonormal bases, they have the same number of elements.

8.5 Decomposition of the regular representation

It is instructive to analyse the decomposition in irreducible representations of the regular representation.

Theorem 8.9
The regular representation of the finite group G (left or right) is completely reducible. In the decomposition of such a representation, there appear all irreducible representations μ_α of the group, each one with multiplicity its dimension d_α.

Proof 8.20 Let ρ be the right regular representation, then $\rho = \sum_{\alpha \in \widehat{G}} d_\alpha \mu_\alpha$. In particular, the character χ_ρ of the regular representation is $\chi_\rho(e) = |G|$ at the neutral element of G and zero at any other element of the group, then we get:

$$\langle \chi_\rho, \chi_\alpha \rangle = \frac{1}{|G|} \sum_g \overline{\chi_\rho(g)} \chi_\alpha(g) = \frac{1}{|G|} |G| \dim \mu_\alpha = d_\alpha,$$

and we conclude:

$$|G| = \chi_\rho(e) = \sum_{\alpha\in\widehat{G}} n_\alpha \chi^\alpha(e) = \sum_{\alpha\in\widehat{G}} n_\alpha^2 .$$

As an immediate consequence of the previous theorem we get:

Corollary 8.7
(Burnside Theorem) The dimensions n_α of the irreducible representations μ_α of the finite group G verify:

$$\sum_{\alpha\in\widehat{G}} n_\alpha^2 = |G| .$$

8.6 Tensor products of representations of groups

We will briefly discuss the tensor product of representations, one of the basic operations with representations discussed in Sect. 7.2.2 that have not been analyzed.

An interesting feature of the tensor product is that it is compatible with the direct product of groups, that is, if (μ, V) is a linear representation of the group G and (μ', V') is a linear representation of the group G', then we can define the tensor product of both represenations as a representation of the group $G\times G'$, that is, we define $(\mu\otimes\mu')_{(g,g')}(v\otimes v') = \mu_g(v)\otimes\mu_{g'}(v')$, for all $(g,g')\in G\times G'$, $v\in V$, $v'\in V'$. Note that the dimension of the tensor product of two representations is the product of the corresponding dimensions.

Remark 8.3 When the groups G,G' are the same, $G = G'$, then we can restrict the representation to the diagonal subgroup G of $G\times G$, that is, the subgroup whose elements are (g,g), $g\in G$, and we recover the definition of the tensor product of representations defined in Sect. 7.2.2. ■

Proposition 8.9
The character $\chi_{\mu\otimes\mu'}$ of the tensor product of two finite-dimensional representations is the product of the corresponding characters, that is: $\chi_{\mu\otimes\mu'} = \chi_\mu\,\chi_{\mu'}$.

Proof 8.21 The proof is a direct consequence of the mutiplicative property of the trace with respect to the tensor product, Prop. A.14.6. Then, $\chi_{\mu\otimes\mu'}(g,g') = \mathrm{Tr}\,(\mu\otimes\mu'(g,g')) = \mathrm{Tr}\,(\mu_g)\mathrm{Tr}\,(\mu'_{g'}) = \chi_\mu(g)\chi_{\mu'}(g')$.

The tensor product preserves irreducibility.

Theorem 8.10
Let (μ,V), (μ',V') be linear representations of the groups G, G', respectively. Then, μ and μ' are irreducible if and only if their tensor product $\mu\otimes\mu'$ is irreducible.

Proof 8.22 To prove the statement, we will take advantage of the fact that the character of the tensor product behaves multiplicatively, Prop. 8.9. On the other hand, we consider the space of functions on the direct product $G\times G'$, and the corresponding inner product, and we get:

$$\langle\psi,\phi\rangle_{G\times G'} = \frac{1}{|G\times G'|}\sum_{(g,g')\in G\times G'} \overline{\psi(g,g')}\phi(g,g') .$$

Thus, if ψ and ϕ are functions of the form $\psi(g,g') = \psi_1(g)\psi_2(g')$ or, using a consistent notation $\psi = \psi_1\otimes\psi_2$, then:

$$\begin{aligned}\langle\psi_1\otimes\psi_2,\phi_1\otimes\phi_2\rangle_{G\times G'} &= \frac{1}{|G||G'|}\sum_{g\in G, g'\in G'} \overline{\psi_1(g)\psi_1(g')}\phi_1(g)\phi_2(g') \\ &= \langle\psi_1,\phi_1\rangle_G\langle\psi_2,\phi_2\rangle_{G'} .\end{aligned}$$

Then, we have that for the characters χ and χ' of the representations μ and μ', respectively, $\chi_{\mu\otimes\mu'} = \chi\otimes\chi'$, and:

$$\langle\chi_{\mu\otimes\mu'},\chi_{\mu\otimes\mu'}\rangle_{G\times G'} = \langle\chi,\chi\rangle_G\langle\chi',\chi'\rangle_{G'} .$$

We conclude that $\langle\chi_{\mu\otimes\mu'},\chi_{\mu\otimes\mu'}\rangle_{G\times G'} = 1$ iff $\langle\chi,\chi\rangle_G$ and $\langle\chi',\chi'\rangle_{G'} = 1$ and then, because of Corollary 8.6, $\mu\otimes\mu'$ is irreducible iff μ and μ' are irreducible.

We may conclude this section by observing that if a group G is a direct product of groups H and K, its irreducible representations are the tensor products of irreducible representations of H and K.

Proposition 8.10
The irreducible representations of $G = H\times K$ are the tensor product $\mu_H\otimes\mu_K$ of irreducible representations μ_H and μ_K of H and K, respectively.

Proof 8.23 We have shown that $\mu_H\otimes\mu_K$ is irreducible iff both μ_H and μ_K are irreducible. To prove that they are all possible irreducible representations of $G = H\times K$, we use the decomposition of the regular representation of G, Thm. 8.7, then we get $|G| = |H||K| = \left(\sum_{\alpha\in\widehat{H}} d_\alpha^2\right)\left(\sum_{\beta\in\widehat{K}} d_\beta^2\right) = \sum_{\alpha\otimes\beta} d_{\alpha\otimes\beta}^2$ with $\alpha\in\widehat{H}$ and $\beta\in\widehat{K}$, d_α, d_β the dimensions of the irreducible representations of H and K, respectively, and $\alpha\otimes\beta\in\widehat{H\times K}$, $d_{\alpha\otimes\beta} = d_\alpha d_\beta$ the dimension of the tensor product of representations.

Remark 8.4 We would like to stress that the situation for the semidirect product is very different. For instance, the irreducible representations of the direct product $\mathbb{Z}_3 \otimes \mathbb{Z}_2$ are all one-dimensional and tensor products of irreducible representations of $\mathbb{Z}_3$ and $\mathbb{Z}_2$, while the semidirect product (the dihedral group D_3) $\mathbb{Z}_3 \rtimes \mathbb{Z}_2$ has a 2-dimensional irreducible representation (see Sect. 8.7.3) which is not the tensor product of any pair of representation of $\mathbb{Z}_3$ and $\mathbb{Z}_2$ (that would all be one-dimensional)[1]. ■

8.7 Tables of characters

It is very useful to construct the table of characters of the group G. This table is a square array where each of the r rows consists of r numbers given by the r values (on each conjugacy class) of a simple character. The orthogonality relation between the characters show that row vectors are orthogonal to each other with respect to the inner product induced by the standard inner product in the space of class functions, while the norm of each any such vector is one. On the other hand, it is also true that two different column vectors are also orthogonal.

Proposition 8.11
(Second orthogonality relation of characters). The family of simple characters $\{\chi_\alpha\}$ satisfies the orthogonality relation:

$$\sum_{\alpha \in \widehat{G}} \overline{\chi_i^\alpha} \chi_j^\alpha = \frac{|G|}{m_i} \delta_{ij},$$

with χ_j^α the value of the character χ_α on the conjugacy class C_i and $m_i = |C_i|$.

Proof 8.24 Any class function f could be expressed as a linear combination of simple characters, then picking up the characteristic function ξ_{C_i} of the class C_i, we get: $\xi_{C_i} = \sum_{\alpha \in \widehat{G}} x_\alpha \chi_\alpha$, with $x_\alpha = \langle \chi_\alpha, \xi_{C_i} \rangle$, because simple characters χ_α form an orthonormal basis. That is,

$$x_\alpha = \frac{1}{|G|} \sum_{g \in G} \overline{\chi_\alpha}(g) \xi_{C_i(g)=} \frac{m_i}{|G|} \overline{\chi_i^\alpha},$$

then, $\xi_{C_i} = \frac{m_i}{|G|} \sum_{\alpha \in \widehat{G}} \overline{\chi_i^\alpha} \chi_\alpha$, and evaluating the previous expression at an element $g \in C_k$, we get the desired expression.

[1] G.W. Mackey [49] developed a beautiful theory to describe the irreducible representations of semidirect product that we will not cover here.

8.7.1 Some simple examples: Abelian groups

Example 8.2 *The group $\mathbb{Z}_2$. The situation with Abelian groups is easy: All their irreducible representations are one-dimensional, Prop. 7.7, and there are as many as elements of the groups because conjugacy classes coincide with individual elements, Thm. 8.8.*

In the particular instance of $\mathbb{Z}_2$ there will be two one-dimensional irreducible representations, μ_0 and μ_1. Their characters define an orthonormal basis of the space of class functions, that can be identified with functions on the group itself. Thus, the values of the characters can be arranged as a 2×2 unitary matrix (because of the normalization we have chosen) whose first row is $(1\,,1)$.

Example 8.3 *The group $\mathbb{Z}_3$. The situation is similar to the case of $\mathbb{Z}_2$. There will be three one-dimensional non-equivalent irreducible representations μ_0, μ_1, μ_2, whose characters could be arranged as a 3×3 unitary matrix whose first row (corresponding to the trivial representation μ_0) is $(1\,1\,1)$. Then, we obtain the values displayed in Table 8.1.*

The general situation is easy to describe. Find the characters of irreducible representations of the group $\mathbb{Z}_n = \langle a \mid a^n = e \rangle$, in Tab. 8.2.

Table 8.1: Table of characters of the group $\mathbb{Z}_3$.

$\mathbb{Z}_3$	e	(123)	(132)
χ_0	1	1	1
χ_1	1	ω	ω^2
χ_2	1	ω^2	ω

The character table of the group $\mathbb{Z}_3$, $\omega^3 = 1$.

Table 8.2: Table of characters of the group $\mathbb{Z}_n$.

$\mathbb{Z}_n$	C_1	C_2	C_3	$\cdots$	C_n
χ_0	1	1	1	$\cdots$	1
χ_1	1	ω	ω^2	$\cdots$	ω^{n-1}
χ_2	1	ω^2	ω^4	$\cdots$	ω^{2n-2}
$\vdots$	$\vdots$	$\vdots$	$\vdots$	$\ddots$	$\vdots$
χ_{n-1}	1	ω^{-1}	ω^{2n-2}	$\cdots$	$\omega^{(n-1)^2}$

Table of characters of $\mathbb{Z}_n$, $\omega = e^{2\pi i/n}$.

Notice that, in the regular representation, the generator a of the group is represented by the matrix (recall Eq. (7.3)):

$$D(a) = \begin{bmatrix} 0 & 1 & 0 & 0 & \cdots & 0 & 0 \\ 0 & 0 & 1 & 0 & \cdots & 0 & 0 \\ 0 & 0 & 0 & 1 & \cdots & 0 & 0 \\ \vdots & \vdots & \vdots & \ddots & \ddots & \vdots & \vdots \\ 0 & 0 & 0 & 0 & \cdots & 1 & 0 \\ 0 & 0 & 0 & 0 & \cdots & 0 & 1 \\ 1 & 0 & 0 & 0 & \cdots & 0 & 0 \end{bmatrix}$$

8.7.2 Irreducible representations of Abelian groups

The previous examples provide many hints on the general situation with the space of irreducible representations of Abelian groups. Consider a finite Abelian group A, its irreducible representations are one-dimensional. Then, if μ is an irreducible representation, $\mu(a) \in \mathbb{C}$ for all $a \in A$, and the character χ and the representation μ are the same, so we will use the same notation for both. Moreover, in this particular instance, we have that characters $\chi\colon A \to \mathbb{C}$ of irreducible representations satisfy:

$$\chi(a+a') = \chi(a)\chi(a'), \qquad \chi(-a) = \overline{\chi(a)}, \quad \forall a, a' \in A.$$

Thus in the case of Abelian groups, characters are multiplicative functions.

Definition 8.4 One-dimensional unitary representations of Abelian groups are called characters. They are actually the characters of irreducible representations of Abelian groups.

In addition, because $\chi(0) = 1$, then $|\chi(a)| = 1$, and we can write characters of Abelian groups as $\chi(a) = \exp(i\varphi(a))$, with $\varphi\colon A \to A$ a group homomorphism $\varphi(a+b) = \varphi(a) + \varphi(b)$. On the other hand, if $a \in A$ has order r, then $ra = 0$, and $\chi(ra) = \chi(a)^r = 1$, in consequence $r\varphi(a) = 2\pi k$, then $\varphi(a) = 2\pi k/r = \omega_r^k$ and $\chi(a)$ is a rth-root of unity (see Table 8.2).

We will denote by $\widehat{A}$ the space of characters of the Abelian group A (because they do represent the equivalence classes of irreducible representations of A). Then, the space $\widehat{A}$ is an Abelian group with respect to the tensor product of representations. Clearly, if χ_1, χ_2 are two irreducible representations, then $\chi_1 \otimes \chi_2$ is a representation with support on $\mathbb{C} \otimes \mathbb{C} = \mathbb{C}$, hence, it is a one-dimensional representation again, a character of A. The tensor product $\otimes$ of representation induces a natural multiplication operation on $\widehat{A}$ given by $\chi_1 \otimes \chi_2(a) = \chi_1(a)\chi_2(a)$, that is the standard product of functions.

If we consider now the Abelian group $(\widehat{A}, \otimes)$, we may wonder about its irreducible representations again. Then, we obtain the celebrated Pontryagin's duality theorem[2].

Theorem 8.11
Let A be a finite Abelian group and $\widehat{A}$ the Abelian group of its characters. There is a canonical isomorphism between the group of characters of $\widehat{A}$ and A. This isomorphism induces a natural unitary map from the Hilbert space of functions on A and the Hilbert space of functions on $\widehat{A}$ called the Fourier transform of A.

Proof 8.25 Let $a \in A$ and $\chi \in \widehat{A}$, we define the map $\mathrm{ev}_a : \widehat{A} \to \mathbb{C}$, as $\mathrm{ev}_a(\chi) = \chi(a)$. Clearly ev_a is a character: $\mathrm{ev}_a(\chi^{-1}) = \chi^{-1}(a) = \overline{\chi(a)} = \overline{\mathrm{ev}_a(\chi)}$, and $\mathrm{ev}_a(\chi\chi') = (\chi\chi')(a) = \chi(a)\chi'(a) = \mathrm{ev}_a(\chi)\mathrm{ev}_a(\chi')$.

The map $A \to \widehat{\widehat{A}}$, given by $a \mapsto \mathrm{ev}_a$ is a monomorphism of groups as $a + a'$ is mapped into $\mathrm{ev}_{a+a'}$ that turns out to be $\mathrm{ev}_{a+a'}(\chi) = \chi(a+a') = \chi(a)\chi(a') = \mathrm{ev}_a(\chi)\mathrm{ev}_{a'}(\chi)$. If $\mathrm{ev}_a = 1$, then $\chi(a) = 1$, for all characters χ, but then $a = 0$. Because both A and $\widehat{A}$ have the same order, the previous map is an isomorphism.

8.7.3 Irreducible representations of the dihedral group

We can construct the character tables of the dihedral groups working from the presentation $D_n = \langle r, \tau \mid r^n = \tau^2 = e, \tau r \tau^{-1} = r^{-1} \rangle$ (Sect. 8.7.3) and D_n is the semidirect product $D_n = \mathbb{Z}_n \rtimes \mathbb{Z}_2$ (Example 2.45).

The structure of the conjugacy classes of D_n is different depending on whether n is even or odd. Computing the conjugacy classes, we get the conjugacy class of e denoted by $C_1 = \{e\}$. The conjugacy class of r will be denoted by $C(r) = \{r, \tau r \tau^{-1} = r^{-1}\}$. Similarly, $C(r^k) = \{r^k, r^{-k}\}$. If $n = 2m+1$, then there are $m+1$ different classes: C_1, $C(r) = \{r, r^{-1}\}$, ..., $C(r^m) = \{r^m, r^{m+1}\}$, while if $n = 2m$ is even there are $m+1$ different classes: C_1, $C(r) = \{r, r^{-1}\}$, ..., $C(r^{m-1}) = \{r^{m-1}, r^{m+1}\}$, and $C(r^m) = \{r^m\}$.

Now, the class $C(\tau) = \{\tau, r\tau r^{-1} = r^2\tau, \ldots\}$ gives all elements $r^k\tau$ if n is odd, but conjugation by r gives two different classes if n is even, $C(\tau)$ and $C(r\tau)$.

We easily compute the one-dimensional representations μ of D_n. Because $\tau^2 = 1$, $\mu(\tau) = \pm 1$, and because $r^n = e$ and $\tau r \tau^{-1} = r^{-1}$, $\mu(r)^n = \mu(r)^2 = 1$. Then, if n is odd, $\mu(r) = 1$. Thus, for n odd there are only two one-dimensional representations. On the other hand, the decomposition of the regular representations turns out: $2n = 1 + 1 + (n-1)/2 \times 2^2$, with $(n-1)/2 = m$ irreducible representations of dimension 2. To construct them, we represent r by a diagonal unitary matrix with diagonal entries $e^{2\pi i k/n}$ and $e^{-2\pi i k/n}$, $k = 1, \ldots, m$, and the corresponding character character $\chi_k(r) = 2c_k = 2\cos(2\pi k/n)$, while $\chi_k(\tau) = 0$, as shown in Table 8.3.

[2]The theorem can be stated for locally compact Abelian groups even if, for obvious reasons, we will only discuss it here in the case of finite Abelian groups.

Table 8.3: Table of characters of the group D_n.

D_{2m+1}	C_1	$2C(r)$	$2C(r^2)$	$\cdots$	$2C(r^m)$	$nC(\tau)$
χ_0	1	1	1	$\cdots$	1	1
χ_1	1	1	1	$\cdots$	1	-1
χ_2	2	$2c_1$	$2c_2$	$\cdots$	$2c_m$	0
χ_3	2	$2c_2$	$2c_4$	$\cdots$	$2c_{2m}$	0
$\vdots$	$\vdots$	$\vdots$	$\vdots$	$\ddots$	$\vdots$	$\vdots$
χ_{m+1}	2	$2c_m$	$2c_{2m}$	$\cdots$	$2c_{m^2}$	0

Table of characters of D_n with n odd, $n = 2m+1$.

We will leave the reader to figure out the table of characters for the case of even n.

8.7.4 Irreducible representations of the symmetric group S_3

The theory of representations of of the symmetric group is a well studied topic and will not be discussed here. We will present, as a simple example, some aspects of this theory in relation to the symmetric group S_3, which includes part of the most significant properties of the general case.

As we have discussed in previous chapters, the symmetric group S_3 is a non-Abelian group of order 6. The character table can be easily constructed, following the rules we have discussed above. The six elements are grouped into three conjugacy classes, the identity, the transpositions and the third order cycles (as it is well known, conjugacy classes are composed of elements with the same cycle structure). The number of nonequivalent irreducible representations is equal to the number of classes, that is, 3. Then, from Burnside's Theorem:

$$n_1^2 + n_2^2 + n_3^2 = 6$$

where n_i is the dimension of the i-th irreducible representation. The solution of this equation is easy to find ($n_1 = 1$ for the identity representation):

$$n_1 = 1,\ n_2 = 1,\ n_3 = 2$$

Using the orthogonality properties, and taking into account that the characters of the identity representation are equal to 1, and the character of the unit element is the dimension of the representation, we get the results in Table 8.4.

The representation μ_1 is the identity representation, all elements of S_3 are assigned to 1. This is the symmetric representation, associated to symmetric tensors of rank three. The representation μ_2 is the antisymmetric representation. The elements of S_3 are represented by their parity, $+1$ for the identity and the third order cycles and -1 for the transpositions (second order cycles). This representation is the basis for skewsymmetric tensors. These two one-dimensional representations appear in all symmetric groups S_n. Finally, the two-dimensional representation μ_3 is related to

Table 8.4: Table of characters of the group S_3.

S_3	C_1	C_2	C_3
χ_1	1	1	1
χ_2	1	-1	1
χ_3	2	0	-1

mixed-symmetry tensors (skewsymmetric in two indices and with a linear constraint on the components). If $n > 3$, there are other representations, related to these kinds of mixed symmetries.

A practical way to understand the structure of the irreducible representations of the symmetric group are the Young (or Ferrers) diagrams and tableaux introduced in Sect. 2.4.1 to analyse the structure of the symmetric group. We restrict the discussion to S_3 but it is easily extended to any n. To construct the Young diagrams of S_3 (recall Eq. 2.6), draw three boxes and arrange them in a rectangular scheme, with several rows and columns, in a such a way that the number of boxes in each row is less than or equal to the number of boxes in the upper rows. Then, in the S_3 case, we have only three diagrams, corresponding to the three irreducible representations:

$$\square\square\square \qquad \begin{array}{c}\square\\ \square\\ \square\end{array} \qquad \begin{array}{l}\square\square\\ \square\end{array}$$

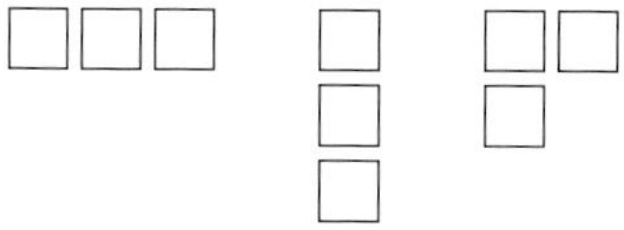

called $[3]$, $[1^3]$ and $[2,1]$, respectively.

As an example of the use of these diagrams, the dimension of the corresponding representation is equal to the number of different ways to place 1,2,3 into the boxes, with the following constraint: The sequence grows up to down in the columns and left to right in the rows (that is, how many tableaux we can construct corresponding to this diagram). In the example:

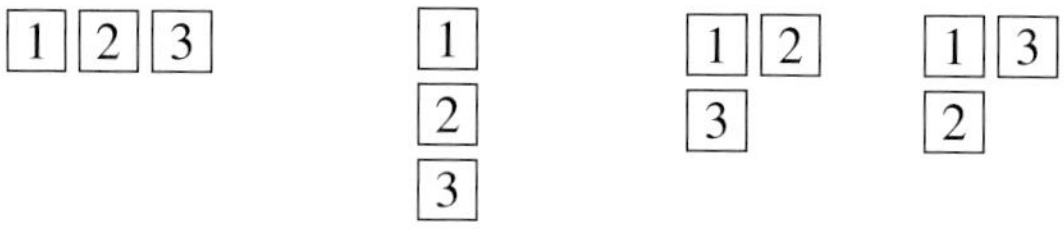

$$\boxed{1}\boxed{2}\boxed{3} \qquad \begin{array}{c}\boxed{1}\\ \boxed{2}\\ \boxed{3}\end{array} \qquad \begin{array}{l}\boxed{1}\boxed{2}\\ \boxed{3}\end{array} \quad \begin{array}{l}\boxed{1}\boxed{3}\\ \boxed{2}\end{array}$$

and the dimension of μ_1 and μ_2 is 1, while the dimension of μ_3 is 2.

It is also an easy task to construct the matrices of the representation μ_3 in the canonical basis of $\mathbb{R}^2$.

$$e \to \begin{pmatrix} 1 & 0 \\ 0 & 1 \end{pmatrix}, \quad (123) \to \begin{pmatrix} -\frac{1}{2} & -\frac{\sqrt{3}}{2} \\ \frac{\sqrt{3}}{2} & -\frac{1}{2} \end{pmatrix}, \quad (132) \to \begin{pmatrix} -\frac{1}{2} & \frac{\sqrt{3}}{2} \\ -\frac{\sqrt{3}}{2} & -\frac{1}{2} \end{pmatrix}$$

$$(12) \to \begin{pmatrix} -1 & 0 \\ 0 & 1 \end{pmatrix}, \quad (13) \to \begin{pmatrix} \frac{1}{2} & -\frac{\sqrt{3}}{2} \\ -\frac{\sqrt{3}}{2} & -\frac{1}{2} \end{pmatrix}, \quad (23) \to \begin{pmatrix} \frac{1}{2} & \frac{\sqrt{3}}{2} \\ \frac{\sqrt{3}}{2} & -\frac{1}{2} \end{pmatrix}$$

since they correspond to the transformations of the dihedral group $\mathcal{D}_3$, they represent the symmetries of the triangle with vertices 1,2,3 (rotations and reflections). This is a faithful representation, usually called the standard representation. The regular representation has dimension 6 and decomposes into the direct sum of μ_1, μ_2 and two copies of μ_3.

The alternating group A_3 is an Abelian group with three elements, $\{e, (123), (132)\}$ (the only group of order 3) and the irreducible representations are those of $\mathbb{Z}_3$, discussed above. To be precise the symmetric and skewsymmetric representations of S_3 reduce to the identity representation of A_3, while the standard representation of S_3 becomes a reducible representation of A_3 which decomposes into the sum of the two other irreducible representations of $\mathbb{Z}_3$ different from the identity representation. After a change of basis (in $\mathbb{C}^2$), the three matrices representing the elements of the group A_3 become:

$$e \to \begin{pmatrix} 1 & 0 \\ 0 & 1 \end{pmatrix}, \quad (123) \to \begin{pmatrix} \omega & 0 \\ 0 & \omega^2 \end{pmatrix}, \quad (132) \to \begin{pmatrix} \omega^2 & 0 \\ 0 & \omega \end{pmatrix}, \quad \omega^3 = 1,$$

explicitly showing the decomposition we have described.

8.8 Canonical decomposition

8.8.1 Canonical decomposition of a representation as sum of invariant spaces

We present in this section a collection of technical results which will be used afterwards in the discussion of the projection operators method, when constructing bases of the linear spaces appearing in the decomposition of reducible representations.

Lemma 8.1

Let D_1 and D_2 be two irreducible matrix representations of the finite group G of order m. For each $g_1 \in G$, we have:

i) If D_1 and D_2 are inequivalent,

$$\sum_{g\in G} [D_2(g^{-1}g_1)]_{ij}[D_1(g)]_{pq} = 0,$$

for any set of indices $1 \leq i, j \leq \dim D_2$, $1 \leq p, q \leq \dim D_1$.

ii) If $D_1 = D_2 = D$, then:

$$\frac{1}{m}\sum_{g\in G} [D(g^{-1}g_1)]_{ij}[D(g)]_{kl} = \frac{1}{n}\,\delta_{il}[D(g_1)]_{kj},$$

where $1 \leq i, j, k, l \leq \dim D = n$.

Proof 8.26 We apply Thm. 8.1 to the linear map: $f = \mu_2(g_1) \circ F$ with $F : V_1 \to V_2$ an arbitrary linear map and μ_1, μ_2 linear representations associated to D_1 and D_2 respectively, then:

i) If μ_1 and μ_2 are not equivalent the average map ϕ_f associated to $f = \mu_2(g_1) \circ F$ vanishes. Thus if A is the matrix associated to F we get:

$$\frac{1}{m} \sum_{g \in G} \sum_{j,p} [D_2(g^{-1} g_1)]_{ij} A_{jp} [D_1(g)]_{pq} = 0 .$$

ii) If $\mu_1 = \mu_2 = \mu$, the linear map ϕ_f is a homothety: $\lambda = \mathrm{Tr}[\rho(g_1)F]/n$, Then:

$$\frac{1}{m} \sum_{j,l} \sum_{g \in G} [D(g^{-1} g_1)]_{ij} A_{jl} [D(g)]_{lk} = \frac{1}{n} \delta_{ik} \sum_{j,l} [D(g_1)]_{lj} A_{jl} .$$

Because A is arbitrary, we get the conclusion.

Corollary 8.8

Let G be a finite group. Then:

i) If D and D' are two disequivalent irreducible matrix representations with characters χ and χ', we have:

$$\sum_{g \in G} \chi'(g^{-1} g_1) \chi(g) = 0 , \qquad \forall g_1 \in G .$$

ii) If D is an irreducible matrix representation and χ its character. Then:

$$\sum_{g \in G} \chi(g^{-1} g_1) \chi(g) = \frac{m}{n} \chi(g_1) , \qquad \forall g_1 \in G,$$

where $n = \dim D$, and $m = |G|$.

Proof 8.27 i) It suffices to take $i = j$, $k = l$ in Lemma 8.1.i and sum over i and k.

ii) It suffices to take $i = j$, $k = l$ in Lemma 8.1.ii and sum over i and k.

Proposition 8.12

Let (μ, V) be a linear representation of the finite group G or order m. For any (equivalence class of) irreducible representation $\mu_\alpha = \mu \mid_{W_\alpha}$ that appears in the decomposition $V = W_1 \oplus \cdots \oplus W_r$ in Thm. 7.1 the linear map defined by

$$P^\alpha = \frac{m_\alpha}{m} \sum_{g \in G} \overline{\chi^\alpha(g)} \mu(g), \tag{8.10}$$

with $m_\alpha = \dim W_\alpha$, is a projector. If μ_α is inequivalent with μ_β, then

$$P^\alpha P^\beta = P^\beta P^\alpha = 0 .$$

Proof 8.28 It is sufficient to check that:

$$P^{\alpha}P^{\alpha} = \frac{m_{\alpha}^2}{m^2}\sum_{g,g'\in G}\overline{\chi^{\alpha}(g)}\,\overline{\chi^{\alpha}(g')}\mu(gg') = \frac{m_{\alpha}^2}{m^2}\sum_{g''\in G}\mu(g'')\sum_{g\in G}\overline{\chi^{\alpha}(g)}\,\overline{\chi^{\alpha}(g^{-1}g'')},$$

from which we obtain the desired result.

Moreover, if $\mu_{\alpha} \not\sim \mu_{\beta}$, then

$$\begin{aligned} P^{\alpha}P^{\beta} &= \frac{m_{\alpha}m_{\beta}}{m^2}\sum_{g,g'\in G}\overline{\chi^{\alpha}(g)}\,\overline{\chi^{\beta}(g')}\mu(gg') \\ &= \frac{m_{\alpha}m_{\beta}}{m^2}\sum_{g''\in G}\mu(g'')\sum_{g\in G}\overline{\chi^{\alpha}(g)}\,\overline{\chi^{\beta}(g^{-1}g'')}, \end{aligned}$$

we obtain the stated result.

Proposition 8.13
With the notation of the previous proposition, if the representation (ρ,V) is unitary, the projector P^{α} is orthogonal.

Proof 8.29 Suppose that the representation μ is unitary. Then, $\mu^{\dagger}(g) = \mu(g^{-1})$, and $\overline{\chi(g)} = \chi(g^{-1})$. Let us check that P^{α} is self-adjoint:

$$P^{\alpha\dagger} = \frac{m_{\alpha}}{m}\sum_{g\in G}\chi^{\alpha}(g)\,\mu^{\dagger}(g) = \frac{m_{\alpha}}{m}\sum_{g\in G}\overline{\chi^{\alpha}(g^{-1})}\,\mu(g^{-1}),$$

then $P^{\alpha\dagger} = P^{\alpha}$.

Theorem 8.12
Let (ρ,V) be a linear representation of the finite group G. There exists a unique decomposition of V as a direct sum:

$$V = \bigoplus_{\alpha\in\widehat{G}} W_{\alpha},$$

of invariant subspaces $W_{\alpha} = P^{\alpha}(V)$ under μ, such that $\mu\,|_{W_{\alpha}} = d_{\alpha}\mu_{\alpha}$.

Proof 8.30 Let $W_{\alpha} = P^{\alpha}(V)$. Because for all $g \in G$,

$$\mu(g)P^{\alpha} = \frac{m_{\alpha}}{m}\sum_{g'\in G}\overline{\chi^{\alpha}(g')}\,\mu(gg') = \frac{m_{\alpha}}{m}\sum_{g''\in G}\overline{\chi^{\alpha}(g^{-1}g'')}\,\mu(g''),$$

and

$$P^{\alpha}\mu(g) = \frac{m_{\alpha}}{m}\sum_{g'\in G}\overline{\chi^{\alpha}(g')}\,\mu(g'g) = \frac{m_{\alpha}}{m}\sum_{g''\in G}\overline{\chi^{\alpha}(g''g^{-1})}\,\mu(g''),$$

and $\chi(g^{-1}g'') = \chi(g''g^{-1})$, we observe that $\mu(g)P^{\alpha} = P^{\alpha}\mu(g)$, and, consequently, W_{α} is invariant under $\mu(g)$, $\forall g \in G$.

Besides, as $\chi(g)$ is a class function, the restriction of $P^{(\alpha)}$ to any subspace that will not be supporting an irreducible representation of G is a homothety (obviously, the ratio would be 1 or 0).

Let $V = W_1 \oplus \ldots \oplus W_r$ any of the decompositions of V corresponding to the previous decomposition of μ. We will denote $\mu_i = \mu\mid_{W_i}$ and χ_i its character; then,

$$P^\alpha\mid_{W_i} = \frac{1}{m}\langle \chi^\alpha, \chi_i \rangle \mathrm{id}_{W_i} = \begin{cases} \mathrm{id}_{W_i} & \text{if } \mu_\alpha \sim \mu_i \\ 0 & \text{if } \mu_\alpha \not\sim \mu_i \end{cases}$$

Because each μ_i is equivalent to a μ_α with $\alpha \in \widehat{G}$, we find that $\sum_{\alpha\in\widehat{G}} P^\alpha = \mathrm{id}_V$, as well as $\mu\mid_{W_\alpha} = d_\alpha \mu_\alpha$.

Corollary 8.9
Let (ρ, V) be a unitary representation of the finite group G. With the notation of the previous theorem, two vectors $v, w \in V$, such that $v \in W^\alpha$, $w \in W^\beta$ with $\alpha \neq \beta$, are orthogonal.

8.8.2 The projection operators method

In the previous section, we have shown that if (μ, V) is a unitary representation of a finite group G, there exists a canonical decomposition of V in invariant subspaces $V = \bigoplus_{\alpha\in\widehat{G}} W_\alpha$, in such a way that the restriction of μ to W_α is a sum of d_α equivalent representation to the irreducible representation μ_α. However, the decomposition of each factor $W_\alpha = \bigoplus_{k=1}^{d_\alpha} V_\alpha$ as a sum of irreducible representations is not unique.

Choose a matrix unitary representation D^α for each equivalence class of representations that appear in the decomposition of μ. For each triple (i, j, α), $1 \leq i, j \leq n_\alpha = \dim V_\alpha$, define the endomorphism of V, P^α_{ij}:

$$P^\alpha_{ij} = \frac{n_\alpha}{m} \sum_{g\in G} \overline{[D^\alpha(g)]_{ij}} \mu(g) .$$

We will denote by P^α_i the endomorphism P^α_{ii} for short.

Theorem 8.13
The endomorphisms P^α_{ij}, $1 \leq i, j \leq n_\alpha = \dim V_\alpha$, satisfy:

$$(P^\alpha_{ij})^\dagger = P^\alpha_{ji}, \quad P^\alpha_{ij} P^\beta_{kl} = \delta_{\alpha\beta} \delta_{jk} P^\alpha_{il} .$$

Proof 8.31 It suffices to see that:

$$(P^\alpha_{ij})^\dagger = \frac{n_\alpha}{m} \sum_{g\in G} [D^\alpha(g)]_{ij} \mu^\dagger(g) = \frac{n_\alpha}{m} \sum_{g\in G} \overline{[D^\alpha(g^{-1})]_{ji}} \mu(g^{-1}) = P^\alpha_{ji} . \tag{8.11}$$

Now, notice that:

$$
\begin{aligned}
P_{ij}^{\alpha} P_{kl}^{\beta} &= \frac{n_\alpha n_\beta}{m^2} \sum_{g,g' \in G} \overline{[D^\alpha(g)]_{ij}}\, \overline{[D^\beta(g')]_{kl}} \mu(gg'). \\
&= \frac{n_\alpha n_\beta}{m^2} \sum_{g \in G} \overline{[D^\alpha(g)]_{ij}} \sum_{g'' \in G} \overline{[D^\beta(g^{-1}g'')]_{kl}} \mu(g'') \\
&= \frac{n_\alpha}{m} \delta_{\alpha\beta} \delta_{jk} \sum_{g'' \in G} \overline{[D^\alpha(g'')]_{il}} \mu(g'') = \delta_{\alpha\beta} \delta_{jk} P_{il}^{\alpha}\,. \qquad (8.12)
\end{aligned}
$$

Corollary 8.10
The operators P_i^α, $i = 1, \ldots, n_\alpha$, are orthogonal projectors (see Sect. 7.3.3, Def. 7.8).

Proof 8.32 From (8.12), identifying $i = j = k = l$ and $\alpha = \beta$, we get: $P_i^\alpha P_i^\alpha = P_{ii}^\alpha$ and $(P_{ii}^\alpha)^\dagger = P_{ii}^\alpha$.

Corollary 8.11
If $i \neq j$, then the operator P_{ij}^α is nilpotent of order two, that is $(P_{ij}^\alpha)^2 = 0$.

Proof 8.33 Because of Thm. 8.13 we get $P_{ij}^\alpha P_{ij}^\alpha = 0$ whenever $i \neq j$.

Definition 8.5 We will say that a vector $v_j \in V$ belongs to the jth row of the irreducible representation μ_α of G ($1 \leq j \leq n_\alpha$), either if it is zero or if there exist $n_\alpha - 1$ vectors $\{v_1 \ldots, v_{j-1}, v_{j+1}, \ldots, v_{n_\alpha}\}$, such that:

$$
\mu(g) v_j = \sum_{k=1}^{n_\alpha} [D^\alpha(g)]_{kj} v_k\,.
$$

Proposition 8.14
The family of endomorphisms P_{ij}^α previously defined satisfy:

$$
\mu(g) P_{ij}^\alpha = \sum_{k=1}^{n_\alpha} [D^\alpha(g)]_{ki} P_{kj}^\alpha\,.
$$

In particular,

$$
\mu(g) P_i^\alpha = \sum_{k=1}^{n_\alpha} [D^\alpha(g)]_{ki} P_{ki}^\alpha\,.
$$

Proof 8.34 First, we realise that:

$$
\mu(g) P_{ij}^\alpha = \frac{n_\alpha}{m} \sum_{g' \in G} \overline{[D^\alpha(g')]_{ij}} \mu(gg') = \frac{n_\alpha}{m} \sum_{g'' \in G} \overline{[D^\alpha(g^{-1}g'')]_{ij}} \mu(g''),
$$

and because D^α is a matrix representation,

$$\mu(g)P_{ij}^\alpha = \frac{n_\alpha}{m}\sum_{g''\in G}\sum_k \overline{[D^\alpha(g^{-1})]_{ik}}\,\overline{[D^\alpha(g'')]_{kj}}\mu(g''),$$

and because D^α can be chosen to be unitary,

$$\mu(g)P_{ij}^\alpha = \sum_{k=1}^{n_\alpha}[D^\alpha(g)]_{ki}P_{kj}^\alpha .$$

Theorem 8.14

For each $v \in V$ and any index j with $1 \le j \le n_\alpha$ such that some $P_{ij}^\alpha v \neq 0$, the vectors $v_i = P_{ij}^\alpha v$ form an orthogonal basis for the irreducible representation μ_α. Moreover, $||v_i|| = ||v_k||$ for all $1 \le i \le k \le n_\alpha$.

Proof 8.35 According to the previous proposition,

$$\mu(g)v_i = \sum_{k=1}^{n_\alpha}[D^\alpha(g)]_{ki}v_k .$$

On the other hand,

$$\langle v_i, v_k\rangle = \langle P_{ij}^\alpha v, P_{kj}^\alpha v\rangle = \langle v, P_{ji}^\alpha P_{kj}^\alpha v\rangle = \delta_{ik}\langle v, P_j^\alpha v\rangle ,$$

then $\{v_i\}_{i=1}^{n_\alpha}$ is an orthogonal basis and $||v_i|| = ||v_k||$, for all $1 \le i \le k \le n_\alpha$.

Corollary 8.12

The orthogonal projector P_i^α projects onto the subspace of vectors in the ith row of the irreducible representation μ_α.

Proof 8.36 In fact, for each non-zero $v \in V$, $P_i^\alpha v = v_i$ is a vector of the ith row of the irreducible representation μ^α, because:

$$\mu(g)P_{ii}^\alpha v = \sum_{k=1}^{n_\alpha}[D^\alpha(g)]_{ki}P_{ki}^\alpha v .$$

Since $\sum_{i=1}^{n_\alpha} P_i^\alpha = P^\alpha$, because of Prop. 8.14, and $\sum_{\alpha\in\widehat{G}} P^\alpha = \mathrm{id}_V$, the algorithm for the decomposition of V as $V = \bigoplus_{\alpha\in\widehat{G}'}\bigoplus_{i=1}^{m_\alpha} V_\alpha^i$, is as follows:

For each index α consider the projection operator P_{11}^α. If $W_1^\alpha = P_{11}^\alpha V$, we choose an orthonormal basis in W_1^α (that will consists of n_α vectors). For instance, $\mathcal{B}_1 = \{w_{l1}^\alpha \mid 1 \le l \le n_\alpha\}$. We define $w_{lj}^\alpha = P_{j1}^\alpha w_{l1}^\alpha$. For any of the indexes l, the vectors $\{w_{lj}^\alpha \mid 1 \le j \le n_\alpha\}$ form an orthonormal basis of the subspace V_α^l because $\mathcal{B}_1$ is orthonormal and,

$$\langle w_{lj}^\alpha, w_{l'j'}^\alpha\rangle = \langle P_{j1}^\alpha w_{l1}^\alpha, P_{j'1}^\alpha w_{l'1}^\alpha\rangle = \langle w_{l1}^\alpha, P_{1j}^\alpha P_{j'1}^\alpha w_{l'1}^\alpha\rangle = \delta_{jj'}\langle w_{l1}, w_{l'1}\rangle = \delta_{jj'}\delta_{ll'} .$$

Moreover,

$$\mu(g)w^{\alpha}_{lj} = \mu(g)P^{\alpha}_{j1}w^{\alpha}_{l1} = \sum_{k=1}^{n_\alpha}[D^{\alpha}(g)]_{kj}P^{\alpha}_{k1}w^{\alpha}_{l1} = \sum_{k=1}^{n_\alpha}[D^{\alpha}(g)]_{kj}w^{\alpha}_{lk}.$$

Notice that there is an ambiguity: The choice of the basis $\mathcal{B}_1 = \{w^{\alpha}_{l1} \mid 1 \leq l \leq n_\alpha\}$ in any subspace.

8.9 An application in quantum mechanics: Spectrum degeneracy

Even if the notions of symmetry and invariance have played an instrumental role in the history of Physics since the times of Galileo and Newton, the relevance of such concepts has heightened with the development of Quantum Mechanics. The reason for this is that, quite early in the development of Quantum Mechanics, the fundamental role played by groups of unitary transformations was revealed. The mathematical framework of the theory based on Hilbert spaces and linear operators, that we will call Dirac's framework, was established quite early (von Neumann). Associated to the given system under study, there is a Hilbert space $\mathcal{H}$ and the pure states of the system are normalised vectors $|\psi\rangle$ (more precisely rays of vectors). The physical observables of the theory are associated to self-adjoint operators (not necessarily bounded) A on $\mathcal{H}$. The fundamental notion that relates the experimental results with the abstract mathematical objects is the statement that the probability of transition or of interference between states (rays) is given by:

$$I(|\bar{\psi}\rangle, |\bar{\phi}\rangle) = \frac{|\langle\psi|\phi\rangle|^2}{\langle\psi|\psi\rangle\langle\phi|\phi\rangle}.$$

The transformations of the theory that will preserve the linear structure of the Hilbert space will be linear transformations and, only the unitary ones among them will preserve the interference between states and can implement the symmetries of the theory. A more detailed and careful analysis, that takes into account the character of rays possessed by the states of the quantum system, would show that antiunitary transformations could also be accepted as symmetries of quantum theories. This is the content of Wigner's theorem.

Thus, if we consider a unitary transformation U of $\mathcal{H}$, the quantum descriptions obtained by means of the vectors $|\psi'\rangle = U|\psi\rangle$ and the operators $A' = UAU^\dagger$, or the vectors $|\psi\rangle$ and the operators A respectively, are equivalent.

On the other hand, after a transformation U on the system, the expected values of the observable A will change:

$$\langle\psi'|A|\psi'\rangle = \langle\psi|U^\dagger AU|\psi\rangle.$$

Then we observe that the observable A is invariant under the transformation U if $A = U^\dagger AU$, or equivalently:

$$UA = AU. \tag{8.13}$$

In such a case we will say that the unitary transformation U is a symmetry of the observable A.

The set of all symmetries of a given observable A is a subgroup of the group $\mathcal{U}(\mathcal{H})$ of all unitary transformations of $\mathcal{H}$, that will be called the group of symmetries of A.

Among the many applications of the theory of representations of groups to Quantum Mechanics, we would like to show here how the knowledge of the symmetry group of a given observable helps us in solving the spectral problem of the associated operator.

Thus, we will consider a group G of symmetry transformations of the system, that is, any element g of G will be realized in the associated Hilbert space $\mathcal{H}$ by an unitary operator $U(g)$ and the fact that performing the transformation $g'g$ is equivalent to perform first the transformation g and then g' translates in:

$$U(g'g) = U(g')U(g) \,. \tag{8.14}$$

We must point out that if the proper ray character of states were taken in consideration rather than just unit vectors, there could appear a complex number $\omega(g',g)$ of modulus one in the right hand side of Eq. (8.14). These factors can often be removed (but not always) and absorbed by redefining the operator $U(g)$ associated to g, so we will assume that Eq. (8.14) holds rather than the more general situation.

Thus, Eq. (8.14), together with $U(e) = I$, shows that the group G is realized in the Hilbert space of the theory as a unitary representation (perhaps up to a factor). What can be said then if the unitary operators $U(g)$ are, in addition, symmetries of the observable A? If this were the case, because (8.13):

$$U(g)A = AU(g) \,,$$

and, in consequence, the proper subspaces of A are invariant under $U(g)$ $\forall g \in G$, hence, they will support a, in general reducible, unitary representation of G. This implies that the multiplicities of the eigenvalues of A cannot be arbitrary, they must correspond to the dimensions of the irreducible representations contained in the corresponding representation U. In addition, this can also be handy in determining the proper subspaces of A because if $|\psi\rangle$ is a vector in a proper subspace, then $U(g)|\psi\rangle$, $\forall g \in G$ will also belong to the same subspace.

Let us assume that the irreducible representations of the group G are known. The representation U will have a canonical decomposition as a direct sum of multiples of irreducibles, and this will allow us to establish "selection rules" between the matrix elements of the operator A.

Actually, we will choose, as in Sect. 8.8.2, a linear basis on each one of the invariant subspaces $P^\alpha\mathcal{H}$, that we will denote as $\{|\psi_{n\alpha k}\rangle\}$. The index k varies from 1 to the dimension of the representation U^α while the index n distinguish the different irreducible representations (all equivalent) within $P^\alpha\mathcal{H}$. If G is a finite group, the relation

$$\langle\psi_{n\alpha k}|U^\dagger(g)AU(g)|\psi_{m\beta l}\rangle = \langle\psi_{n\alpha k}|A|\psi_{m\beta l}\rangle \,,$$

leads to:

$$\sum_{k',l'} \langle\psi_{n\alpha k'}|A|\psi_{m\beta l'}\rangle \overline{[D^\alpha(g)]_{k'k}}[D^\beta(g)]_{ll'} = \langle\psi_{n\alpha k}|A|\psi_{m\beta l}\rangle \,,$$

where D^α and D^β are the matrix representations associated to U^α and U^β in the selected bases, respectively. Adding these expressions for each $g \in G$ and using the orthogonality relations (8.2), we get:

$$\langle \psi_{n\alpha k} | A | \psi_{m\beta l} \rangle = \frac{1}{n_\alpha} \delta_{\alpha\beta} \, \delta_{kl} \sum_{k'} \langle \psi_{n\alpha k'} | A | \psi_{m\alpha k'} \rangle ,$$

that tell us that if $\alpha \neq \beta$ or $k \neq l$, the matrix element vanishes, whereas, if $\alpha = \beta$ and $k = l$, the corresponding matrix element does not depend on k. In other words, the matrix associated to the operator A in this basis, which is adapted to the symmetry group, is quite simple and simplifies the spectral problem of the operator A.

We have already indicated that if we know an eigenvector $|\psi\rangle$ of the operator A, the vectors $U(g)|\psi\rangle$, $\forall g \in G$ will also be eigenvectors with the same eigenvalue whenever G is a symmetry group of A. This means that, if A has non-commutative groups of symmetry, its eigenvalues would be degenerate because, if this were not the case, $U(g)|\psi\rangle$ would have to be proportional to $|\psi\rangle$ $\forall g \in G$, and we would just have a one-dimensional representation of G.

Recall that there exist groups that do not possess non-trivial one-dimensional representations. For instance, the derived group G' (which is normal in G), generated by the commutators of elements in G, always has trivial one-dimensional representations. Hence, the one-dimensional representations of a group G are not faithful (except if G is Abelian, in which case $G' = \{e\}$), and they do, in fact, define representations in the Abelianized group $\mathrm{Ab}(G) = G/G'$. To conclude these comments if, for instance, G is simple and non-Abelian, $\mathrm{Ab}(G)$ reduces to the identity and does not have non-trivial one-dimensional representations.

Among all observables of a quantum system, the Hamiltonian H of the system plays a fundamental role as the generator of the dynamical evolution of the system. The dynamical evolution of the system is determined by Schrödinger's equation:

$$i\frac{d}{dt}|\psi\rangle = H|\psi\rangle ,$$

thus, if the vector $|\psi_0\rangle$ determines an initial state, it will evolve as $|\psi(t)\rangle = U_t|\psi_0\rangle$ with U_t the one-parameter group of unitary transformations determined by H and given as $U_t = \exp(-itH)$. Then, a more detailed study of the symmetry group of H is justified because the symmetry transformations of H will leave invariant the dynamics of the system.

If $U^{(\alpha)}$ is an irreducible representation contained in the representation U of the symmetry group of G of H, the restriction of H to the subspace supporting $U^{(\alpha)}$ must be a multiple of the identity by Schur's Lemma, and the corresponding energy eigenvalue is degenerate. However, not all degeneracies of the spectrum of H can be explained in this way. The use of the theory of representations in the determination of stationary states is based on the following assumption (not always acknowledged explicitly):

Irreducibility postulate. If G is the group of all symmetries of the observable H, then each proper subspace of H supports an irreducible representation of G.

Stated in this way, this statement is not very useful because we do not know which is the group of all symmetries of H. It must be applied judiciously. In general, once we know the Hamiltonian of the system, we can find a group G that we expect would be the group of symmetries of H. Whenever we find a degenerate eigenvalue of the Hamiltonian that cannot be accounted for the invariance under the group G, the irreducibility postulate would indicate that we have overlooked some symmetries and we should look for these hidden symmetries. In other words, an unexpected degeneracy of the spectrum of the Hamiltonian, also known as accidental degeneracy, must be interpreted with the idea that G is a group of symmetries of the Hamiltonian but it is not the largest group of symmetry, and because we are just considering a subgroup G of the largest symmetry group of H, an irreducible representation of such largest symmetry group, when restricted to G would be decomposed as a direct sum of irreducible representations of G.

We will end this discussion by considering the following situtation. Let H_0 be a Hamiltonian such that G is a group of symmetries of H_0. We introduce a perturbation V. In general, only a proper subgroup $G_V \subset G$ of the symmetry group of H_0 will be the symmetry group of the full Hamiltonian. Thus, in general, the degeneracy of the eigenvalues will break down because an irreducible representation of G restricted to G_V will decompose as a direct sum of irreducible representations of G_V. If $G_V = G$, the degeneracy will not change if it were not accidental because G would continue being the group of symmetry of the new Hamiltonian H. However, if the degeneracy were accidental, then even if we introduce a perturbation V with the same group of symmetry G, the new largest group of symmetry of the Hamiltonian H could be smaller than that of H_0 alone and the accidental degeneracy can be broken, that is, the accidental degeneracy can be made apparent.

Chapter 9

Linear Representations of Categories and Groupoids

This chapter will be devoted to introducing the notions of linear representations of categories and groupoids.

Following the path laid in Chapter 1, linear representations of categories will be introduced using quivers, that is, we will discuss first what a linear representation of a quiver is and we will extend it naturally to define the notion of a linear representation of a category. The basic notions of the theory will be introduced: Equivalence, operations with representations and, finally, the fundamental notions of irreduciblity and indecomposability of representations.

Using explicit examples of representations of quivers, it will be shown that irreduciblity and indecomposability are different. The Krull-Schmidt theorem, stating the decomposability of finite-dimensional representations of categories, will be proved, and Gabriel's theorem, characterizing quivers possessing a finite number of indecomposable representations, will be stated.

Linear representation of groupoids are a particular instance of linear representation of categories and, taking advantage of the previous work, the basics of the theory will be discussed again, this time, with more emphasis on the set theoretical aspects. The relation between linear representations of the isotropy groups and the representations of the groupoid itself will be analyzed.

The chapter will be concluded by proving the Jordan-Hölder theorem for representations of groupoids and showing that any finite-dimensional linear representation of a finite groupoid is completely reducible (or semisimple).

9.1 Linear representations of categories

As it was argued before—recall the discussion at the beginning of Chapter 7, Sect. 7.1.1—one of the main reasons to develop a theory of linear representations of abstract mathematical structures is that, in doing so, we are able to provide a realisation of the abstract relations among the objects of the category by means of objects from Linear Algebra: Vector spaces, linear maps, etc., that we understand better and for which there are extremely powerful numerical algorithms.

Groups and groupoids are instances of categories, both naturally associated to arbitrary categories (the groupoid of isomorphisms of a category is the subcategory of all isomorphisms of the given category). Thus, before discussing the notion of linear representation of groupoids, it is natural to consider the notion of a linear representation of a category. In doing so, we will be inspired by the results obtained in the previous chapter on the theory of representations for groups but, at the same time, we will emphasize the differences and how the abstract thinking provided by categorical notions will help us in understanding the true scope of the previous theory.

9.1.1 Linear representations of quivers

Perhaps the best way to start this discussion would be to turn our attention to the original examples provided at the beginning of this work, Sect. 1.1. There, we were using quivers to generate categories. Then, it is natural to first consider the notion of a linear representation of a quiver.

Let $Q = (\Omega, L)$ be a quiver with vertices the elements of the set Ω and links the elements of the set L. We denote, as usual, the source and target maps $s, t \colon L \to \Omega$, and the link l as $l \colon x \to y$, if $s(l) = x$ and $t(l) = y$.

Definition 9.1 A (linear) representation R of the quiver $Q = (\Omega, L)$ is an assignment of a linear space space $R(x) = V_x$ to each vertex $x \in \Omega$ and a linear map $R(l) \colon V_x \to V_y$, to each link $l \colon x \to y$. We will also denote the representation R as (R, V) with V the direct sum of the linear spaces V_x that is $V = \bigoplus_{x \in \Omega} V_x$.

Example 9.1 *We now exhibit some simple examples of representations of quivers. We start with the examples on Sect. 1.1.2. In what follows, we will assume that all our representations are finite dimensional, that is* $\dim V_x = d_x$ *is finite for all* $x \in \Omega$.

1. *The singleton: With the notations in 1.1.2, the singleton is defined as the quiver* $Q_{2,1}$ *with two vertices* v_1, v_2 *and one link* $l \colon v_1 \to v_2$. *We consider the representation:* $R(v_1) = R(v_2) = \mathbb{C}$, $R(1_{v_1}) = R(1_{v_2}) = 1$, $R(l) = \lambda \in \mathbb{C}$.

   More generally, let V_1 *be a linear space attached to* v_1 *and* V_2 *a linear space attached to* v_2. *Any linear map* $L \colon V_1 \to V_2$ *defines a linear representation of the singleton,* $R(l) = L$.

   Conversely, the singleton can also be seen as an abstract diagrammatic description of a linear map between two linear spaces.

For later purposes, we will denote the representation on the one-dimensional linear space $\mathbb{C}$ given by $R(l) = 1$, as $R_{1,1}$ or $1 \to 1$. The representation defined by the trivial linear map $\mathbf{0} \to \mathbb{C}$, will be denoted by $R_{0,1}$ (or $0 \to 1$) and the representation defined by the trivial linear map $\mathbb{C} \to \mathbf{0}$, will be denoted by $R_{1,0}$ (or $1 \to 0$).

2. *The single loop category (the monoid $\mathbb{N}$): Now the quiver $Q_{1,1}$ has only one vertex v and one link $l\colon v \to v$. We may consider the representation: $R(v) = \mathbb{C}$, $R(1_v) = 1$, $R(l) = e^{i\theta} = z$, where z is a complex number of modulus 1. Then, $R(l_n) = R(l^n) = e^{in\theta} = z^n$. The same applies for the extended loop, Example 1.1, the group $\mathbb{Z}$. Notice that the set of all unitary representations of $\mathbb{Z}$ of dimension 1 is the set of complex numbers of modulus 1, denoted $U(1)$, that is, the circle S^1. Because the group $\mathbb{Z}$ is Abelian, irreducible representations are one-dimensional, Prop. 7.7, and they coincide with its characters[1], hence, $\widehat{\mathbb{Z}} = U(1)$.*

   However, a representation of the loop quiver $Q_{1,1}$ does not have to be defined by an invertible linear map, any endomorphism $L\colon V \to V$ will define a representation of it.

3. *The three-dot quiver Q_3. Now we have three objects (recall Fig. 1.6), v_1, v_2 and v_3 and three arrows $l_1\colon v_1 \to v_2$, $l_2\colon v_2 \to v_3$ and $l_3 = l_2 \circ l_1\colon v_1 \to v_3$. Thus, a linear representation of Q_3 consists of three linear spaces V_1, V_2 and V_3 associated, respectively, with v_1, v_2 and v_3, two linear maps $R_1\colon V_1 \to V_2$ and $R_2\colon V_2 \to V_3$ and the composition $R_3 = R_2 \circ R_1$. Thus, the composition of arrows becomes the composition of linear maps (or the multiplication of matrices). Notice that the linear maps do not have to be invertible.*

4. *The star quiver. Consider the quiver of Fig. 9.1. Then a representation of the star quiver with six points is a collection of seven vector spaces $V_1, \ldots, V_7$, the first six attached to the points $v_1, \ldots, v_6$ and V_7 attached to the center v_7, and maps $R(x_k) = L_k\colon V_k \to V_7 = V$, $k = 1, \ldots, 6$. In case all maps L_k are injective, then a representation of the quiver will be family of subspaces of V.*

Definition 9.2 A morphism φ between two representations (R,V), (R',V') of the quiver $Q(\Omega,L)$ is a family of linear maps $\varphi_x\colon V_x \to V'_x$ such that $\varphi_y \circ R(l) = R'(l) \circ \varphi_x$, for any link $l\colon x \to y$. We will denote it by $\varphi\colon (R,V) \to (R',V')$ or just $\varphi\colon V \to V'$ for short.

As in the situation of groupoid homomorphisms, we will say that a morphism of representations $\varphi\colon V \to V'$ is an isomorphism if it is invertible, which amounts to $\varphi_x\colon V_x \to V'_x$ be invertible for every $x \in \Omega$.

It is clear that the collection of all representations (R,V) of a quiver Q are the objects of a category whose morphisms are the morphisms $\varphi\colon V \to V'$. They define

[1]This is a beautiful example, although with a non-finite group, of Pontryagin's duality discussed summarily in Sect. 8.7.2, Thm. 8.11.

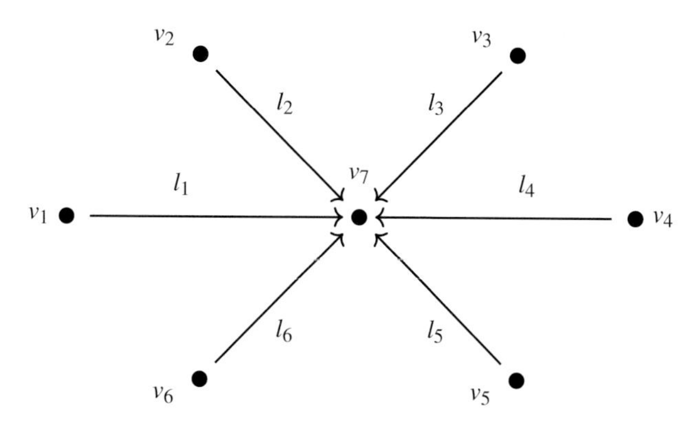

Figure 9.1: The six points star quiver.

a category, called the category of representations of the quiver Q, that will be denoted as $\mathbf{Rep}(Q)$. As in the case of group and groupoids, there is a natural notion of equivalence of representations.

Definition 9.3 Two representations (R,V), (R',V') of the quiver Q are equivalent if there exists an isomorphism $\varphi\colon (R,V) \to (R',V')$. We will indicate it by $R \sim R'$, this means that for every $x \in \Omega$, there exists a linear isomorphism $\varphi_x\colon V_x \to V'_x$, such that $\varphi_y \circ R(l) = R'(l) \circ \varphi_x$, for all links $l\colon x \to y$.

We would like to classify all representations of a given quiver Q up to isomorphisms, what we called the skeleton of the category $\mathbf{Rep}(Q)$.

Example 9.2

1. *In the singleton example, recall Example 9.1.1, note that, given a linear map $R(l) = L\colon V_1 \to V_2$, it is always possible to choose bases in V_1 and V_2, such that the matrix associated to L has the form:*

$$\begin{bmatrix} I_r & 0 \\ 0 & 0 \end{bmatrix} \tag{9.1}$$

   where $r = \operatorname{ran} L$ and I_r is the $r \times r$ identity matrix. Thus, two representations $L\colon V_1 \to V_2$ and $L'\colon V'_1 \to V'_2$ are equivalent if and only if $\dim V_1 = \dim V'_1 = n_1$, $\dim V_2 = \dim V'_2 = n_2$ and $\operatorname{ran} L = \operatorname{ran} L' = r$. The space of equivalence classes of representations of $Q_{2,1}$ is the set of triples of non-negative integers (n_1, n_2, r), with $r \leq n_1, n_2$.

2. *Consider now the single loop, Example 9.1.2. Now, we have to understand the equivalence classes of linear maps $L\colon V \to V$. The equivalence relation is given by $L \sim L'$ if there exists a linear isomorphism $\varphi\colon V \to V$, such that $\varphi \circ L = L \circ \varphi$. The Jordan normal form theorem (see for instance [42, Sect. 1.9] for a detailed proof) establishes that there always exists a basis such that*

the matrix associated with L has the form:

$$\begin{bmatrix} J_{\lambda_1,n_1} & 0 & \cdots & 0 \\ 0 & J_{\lambda_2,n_2} & \cdots & 0 \\ \vdots & & \ddots & \vdots \\ 0 & 0 & \cdots & J_{\lambda_r,n_r} \end{bmatrix} \tag{9.2}$$

where $J_{\lambda,n}$ denotes the Jordan block of order n with eigenvalue λ which is a matrix of the form:

$$J_{\lambda,n} = \begin{bmatrix} \lambda & 1 & 0 & \cdots & 0 \\ 0 & \lambda & 1 & \cdots & 0 \\ & \ddots & \ddots & \ddots & \vdots \\ & & 0 & \lambda & 1 \\ & & & 0 & \lambda \end{bmatrix}. \tag{9.3}$$

We conclude that two representations L and L′ of the loop quiver are equivalent if and only if L and L′ have the same Jordan normal form.

Remark 9.1 It is wonderful that the classification of representations of such simple quivers as $Q_{1,1}$ and $Q_{1,2}$ correspond to basic problems in linear algebra. It is even more important to realize that the classification of other quivers, not so complicated, leads to hard problems. For instance, the classification of all representations of the two-loops quiver $Q_{1,2}$ is notoriously difficult. An equally hard problem is the classification of representations of the star quiver, Example 9.1.4. ∎

When discussing quivers, we moved immediately to the category generated by them and to the significant class of categories generated by quivers and relations. We can do the same now and wonder on the relation between the representations of quivers and the representations of categories generated by them.

Thus, if $\mathbf{C}(Q)$ is the category generated by the quiver $Q(\Omega, L)$, recall the notions introduced at the end of Sect. 1.1.1, then it is clear that any representation (R, V) of Q induces a representation, denoted with the same symbol, R of $\mathbf{C}(Q)$. We only have to define $R(w) = R(l_s) \circ R(l_{s-1}) \circ \cdots \circ R(l_1)$ for all paths $w = l_s l_{s-1} \cdots l_1$ on Q.

More interesting is the situation that arises when we consider relations among the generators L of the category $\mathbf{C}(Q)$. This situation led us to numerous interesting examples, recall Sect. 1.1.3, Exercises 1.2, 1.3, for instance.

Now it is clear that if $\mathcal{R} = \{p_1 = q_1, \ldots, p_s = q_s\}$ is the set of words that are identified and that defines the relations that are satisfied by the generating links L of the quiver Q, not all linear representations of the quiver Q will induce a representation of the category $\mathbf{C}(Q, \mathcal{R}) = \mathbf{C}(Q)/\sim_{\mathcal{R}}$. Only those representations R, such that $R(p_k) = R(q_k)$ for any expression $p_k = q_k \in \mathcal{R}$, $k = 1, \ldots, s$, will define representations of the category $\mathbf{C}(Q, \mathcal{R})$.

Exercise 9.3 *Prove that there is a one-to-one correspondence between representations of the category* $\mathbf{C}(Q,\mathcal{R})$ *and linear representations* (R,V) *of the quiver* Q *such that* $R(p_k) = R(q_k)$ *for all* $p_k = q_k \in \mathcal{R}$, $k = 1,\ldots,s$.

It is also clear that all notions introduced so far: Morphisms, equivalence of representations, etc., extend to the new setting. This invites us to introduce the notion of representations of categories that will be the subject of the coming section.

9.1.2 Linear representations of finite categories

As discussed before, a 'linear representation' of a category is nothing but the realisation of the 'abstract relations' among the objects of the category, expressed by their morphisms in terms of linear algebraic objects, like linear spaces, linear maps, etc.

As a category is essentially a collection of relations or 'arrows', among abstract objects that satisfy an associative composition law, transporting such relations into another category, in this case the category of linear spaces and linear maps, is what we have called a functor, Def. 5.1 (recall the examples introduced in Sect. 5.1.2). Then, it should be obvious that a linear representation of a category is a functor from the given category into the category of linear spaces. More precisely:

Definition 9.4 Let $\mathbf{C}$ be a category. A finite-dimensional linear representation of $\mathbf{C}$ is a functor R from the category $\mathbf{C}$ to the category **FinVect** of finite-dimensional linear spaces and will be denoted as $R\colon \mathbf{C} \to \mathbf{FinVect}$. In other words, the functor R assigns to any object x in the category $\mathbf{C}$ a finite-dimensional linear space V_x, and to any arrow $\alpha\colon x \to y$ a linear map $R(\alpha)\colon V_x \to V_y$ in such a way that $R(\alpha\circ\beta) = R(\alpha)\circ R(\beta)$ and $R(1_x) = \mathrm{id}_{V_x}$.

The first thing we must notice about this definition is that, contrary to the situation with linear representations of groups (see Eq. (7.2)), the linear maps $R(\alpha)$ must not be invertible unless α is an isomorphism.

Proposition 9.1
If $\alpha\colon x \to y$ *is an isomorphism, then* $R(\alpha)\colon V_x \to V_y$ *is an isomorphism of linear spaces and* $R(\alpha^{-1}) = R(\alpha)^{-1}$.

Proof 9.1 Notice that $\alpha\circ\alpha^{-1} = 1_y$, then

$$R(\alpha)R(\alpha^{-1}) - R(1_y) = \mathrm{id}_{V_y} \tag{9.4}$$

and $\alpha^{-1}\circ\alpha = 1_x$, then $R(\alpha^{-1})R(\alpha) = R(1_x) = \mathrm{id}_{V_x}$. Let $\dim V_x = n_x$ and $\dim V_y = n_y$. Assume that $n_x \neq n_y$ and, for instance, $n_x > n_y$, this implies that both $\operatorname{rank} R(\alpha)$, and $\operatorname{rank} R(\alpha^{-1}) \leq n_y$, hence, $\operatorname{rank}(R(\alpha)R(\alpha^{-1})) \leq n_y < n_x$, but, on the other hand, because Eq. (9.4), we have that $\operatorname{rank}(R(\alpha)R(\alpha^{-1})) = n_x$ which is contradictory. Thus, $n_x = n_y$ and then we get that the maps $R(\alpha)$ and $R(\alpha^{-1})$ are the inverse of each other.

Example 9.4 *We exhibit some simple examples of representations of categories. We continue with the examples of quivers discussed before and with the examples in Sect. 1.1.2.*

1. *The extended singleton* $\mathbf{A}_2$. *This category has two objects* v_1, v_2, *two arrows,* $l\colon v_1 \to v_2$ *and* $l^{-1}\colon v_2 \to v_1$, *such that they are inverse of each other. This category is a groupoid that can be identified naturally with the groupoid of pairs of two elements. A representation* R *of* $\mathbf{A}_2$ *will consist of two linear spaces* V_1, V_2 *attached to* v_1 *and* v_2, *respectively, and a linear isomorphism* $R\colon V_1 \to V_2$ *(then* V_1 *and* V_2 *must have the same dimension) corresponding to the link* l. *Clearly* R^{-1} *will be associated to* l^{-1}. *Thus, any linear isomorphism between linear spaces can be considered to be a linear representation of the pair groupoid. Conversely, the pair groupoid can be considered as an abstract diagram describing an invertible map between two linear spaces.*

2. *The three-dot quiver* Q_3. *Now we have three objects (recall Fig. 1.6)* v_1, v_2 *and* v_3 *and three arrows* $l_1 : v_1 \to v_2$, $l_2\colon v_2 \to v_3$ *and* $l_3 = l_2 \circ l_1 : v_1 \to v_3$. *Thus, a linear representation of* Q_3 *consists of three linear spaces* V_1, V_2 *and* V_3 *associated to* v_1, v_2 *and* v_3, *respectively, two linear maps* $R_1 : V_1 \to V_2$ *and* $R_2\colon V_2 \to V_3$ *and the composition* $R_3 = R_2 \circ R_1$. *Thus, the composition of arrows becomes the composition of linear maps (or the multiplication of matrices). Notice that now, the linear maps do not have to be invertible.*

9.1.3 The category of representations: Equivalence of representations

In Sect. 5.4, the notion of natural transformation between functors was introduced using, as a motivation, the notion of equivalence between linear representations of groups. In the previous section, the notion of morphism between two quiver representations was introduced. That is precisely the notion that can be extended to representations of categories and which we discuss below.

Let us recall that a natural transformation $\varphi\colon F \Rightarrow G$ between two functors F and G is a way of 'comparing' them, Def. 5.10, that is, if the functors F, G assign morphisms $F(\alpha)\colon F(x) \to F(y)$ and $G(\alpha)\colon G(x) \to G(y)$ in the category $\mathbf{C}'$ to any morphism $\alpha\colon x \to y$ in the category $\mathbf{C}$, then in order to compare them, we must provide for each object x a morphism $\varphi(x)$, such that $\varphi(y) \circ F(\alpha) = G(\alpha) \circ \varphi(x)$, see Fig. 9.2.

If the morphisms $\varphi(x)$ are isomorphisms, then we may think of the two functors F and G as equivalent, that is, we may use one or the other, as we can pass from one to the other at will by means of the isomorphisms $\varphi(x)$. Observe that if $\varphi(x)$ are isomorphisms, we may define another natural transformation, this time denoted by $\varphi^{-1}\colon G \Rightarrow F$ as $\varphi^{-1}(x) = \varphi(x)^{-1}$, and then the two functors F, G are related among them as $G = \varphi \circ F \circ \varphi^{-1}$.

For instance, let us consider two representations of the singleton category (see Example 9.1.1), that is, two linear isomorphisms $R\colon V_1 \to V_2$ and $R'\colon V_1' \to V_2'$. Then a natural transformation between the functors R and R' consists of linear maps

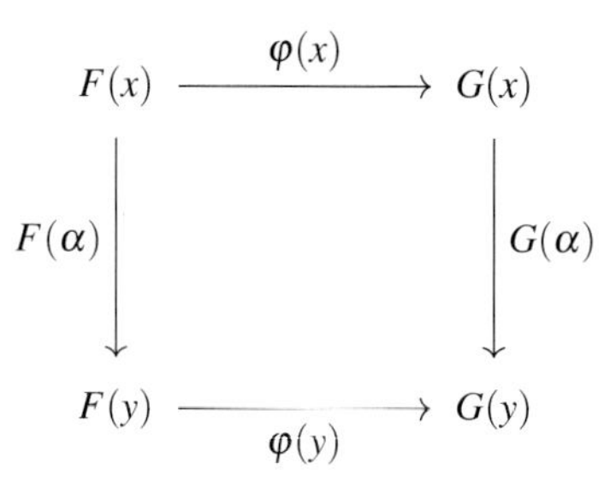

Figure 9.2: A natural transformation $\varphi\colon F \Rightarrow G$.

$T_1\colon V_1 \to V_1'$ and $T_2\colon V_2 \to V_2'$ such that $T_2 \circ R = R' \circ T_1$. If the linear maps T_1, T_2 are isomorphisms, then we say that the functors R and R' are equivalent. In terms of linear maps, this means that $R' = T_2 \circ R \circ T_1^{-1}$, that coincides with the 'natural' idea that two linear maps, such that they are the same up to linear isomorphisms, are equivalent.

Exercise 9.5 *Prove that the equivalence classes of representations of the groupoid* $\mathbf{A}_2$ *are labelled by a natural number n.*

Example 9.6 *Consider now a linear representation of the loop category (see Example 9.1.2), that is, of the Abelian group* $\mathbb{Z}$*. A finite-dimensional linear representation of* $\mathbb{Z}$ *is just an invertible linear map* $R\colon V \to V$*, two representations* R, R' *being equivalent if there exists an isomorphism* $T\colon V \to V$*, such that* $R' = T \circ R \circ T^{-1}$*. It is a fundamental result in Linear Algebra that such equivalence classes of linear maps are characterized by the Jordan block structure of the matrix, that is, by a list* (λ_i, n_i)*,* $i = 1, \ldots, r$*,* $n = \dim V = n_1 + \cdots + n_r$*, where* λ_i *are the eigenvalues of* R *and* n_i *is the size of the corresponding Jordan block (see Example 9.2.2).*

Exercise 9.7 *Compute the equivalence classes of linear representations for the three-dot quiver.*

Notice that given a category $\mathbf{C}$, the family of all its finite-dimensional linear representations $R\colon \mathbf{C} \to \mathbf{FinVect}$ form a category themselves: Its objects are the functors R and its morphisms are natural transformations $\varphi\colon R \Rightarrow R'$, the composition law of two natural transformations being naturally defined as $(\psi \circ \varphi)(x) = \psi(x) \circ \varphi(x)$, for $\varphi\colon R \Rightarrow R'$ and $\psi\colon R' \Rightarrow R''$ two natural transformations.

Such category will be denoted as $\mathbf{Rep}(\mathbf{C})$ and it will be called the category of representations of the category $\mathbf{C}$. Its groupoid of isomorphisms consists of those natural transformations defining equivalences between representations. Hence, the space of equivalence classes of representations of the category $\mathbf{C}$ is the quotient of $\mathbf{Rep}(\mathbf{C})$ by its canonical groupoid. Ultimately, given a category $\mathbf{C}$, we would like to describe its family of equivalence classes of linear representations. Note that the category $\mathbf{Rep}(\mathbf{C})$ of representations of the category $\mathbf{C}$ is not finite (we will see in the coming section that we can always construct the direct sum and tensor product of representations), however, in many instances of interest, the space of its equivalence classes of representations could be described in terms of a finite number of them.

For instance, in the singleton example, the equivalence classes of representations of the category $\mathbf{C}(Q_{2,1})$ are given by the family of three non-negative integers (n,m,r) described above. We will see that, in general, describing them is much harder.

Another example is provided by the representations of groups. The space of equivalence classes of representations is described as direct sums of a finite family of them, the irreducible representations.

In what follows, we will refer to equivalence classes of linear representations of categories just as linear representations, in other words, we will always consider linear representations up to invertible natural transformations, and we will say that R is a linear representation of a given category, understanding that it specifies a representative of the corresponding equivalence class.

9.2 Properties of representations of categories

9.2.1 Operations with representations of categories

There are three basic operations we can perform on linear spaces that can be extended naturally to the category of linear representations: Direct sums, tensor products and duals (see App. A.3 for background material).

1. Direct sums. Given two linear spaces V, W, we can define another linear space $V \oplus W$ in the standard way: Vectors in $V \oplus W$ are pairs of vectors (v,w) (commonly denoted as $v \oplus w$) with the standard component-wise addition law. Notice that $\dim V \oplus W = \dim V + \dim W$. The direct sum defines a functor $\oplus\colon \mathbf{FinVect} \times \mathbf{FinVect} \to \mathbf{FinVect}$ from the Cartesian product category $\mathbf{FinVect} \times \mathbf{FinVect}$ (whose objects are pairs of linear spaces (V,W) and morphims are pairs (L,M) of linear maps $L\colon V \to V'$, $M\colon W \to W'$) into the category $\mathbf{FinVect}$ defined as: $\oplus(V,W) = V \oplus W$ and $\oplus(L,M) = L \oplus M$ with $(L \oplus M)(v,w) = L(v) \oplus M(w)$, for all $v \in V$, $w \in W$.

2. Tensor products. Given two linear spaces V,W we can define the tensor product $V \otimes W$ of V and W as the linear space generated by elements of the form $v \otimes w$, $v \in V$, $w \in W$, with relations $(\lambda v) \otimes w = \lambda(v \otimes w) = v \otimes (\lambda w)$, $(\lambda v + \lambda' v') \otimes w = \lambda v \otimes w + \lambda' v' \otimes w$ and $v \otimes (\lambda w + \lambda' w') = \lambda v \otimes w + \lambda' v \otimes w'$. Notice that $\dim V \otimes W = \dim V \times \dim W$.

   The tensor product defines a functor $\otimes\colon \mathbf{FinVect} \times \mathbf{FinVect} \to \mathbf{FinVect}$ as: $\otimes(V,W) = V \otimes W$ and $\otimes(L,M) = L \otimes M$ with $(L \otimes M)(v \otimes w) = L(v) \otimes M(w)$, for all $v \in V$, $w \in W$.

3. Duals. Given a finite-dimensional linear space V, its dual space V^* is the linear space of all linear maps $\beta\colon V \to \mathbb{C}$. For any linear map $L\colon V \to W$ there is

a canonically defined adjoint map[2] $L^*: W^* \to V^*$: $L^*(\beta)v = \beta(Lv)$, $v \in V$, $\beta \in W^*$.

We may consider the operation * as a functor from the category **FinVect** into the opposite category[3] $\mathbf{FinVect}^{\mathrm{op}}$ (see Def. 9.5.3).

Thus, the category of finite-dimensional linear spaces has a rich structure. In addition to the previous remarks, the operations $\oplus$, $\otimes$ satisfy a family of axioms resembling those of a $*$-algebra (see later on, Sect. 10.3.1, Exer. 10.26):

For all finite-dimensional linear spaces U, V, W we have:

1. Associativity: $(U \oplus V) \oplus W \cong U \oplus (V \oplus W)$, $(U \otimes V) \otimes W \cong U \otimes (V \otimes W)$.
2. Neutral elements: $U \oplus \mathbf{0} \cong U$, $U \otimes \mathbb{C} \cong U$.
3. Commutativity: $U \oplus V \cong V \oplus U$, $U \otimes V \cong V \otimes U$.
4. Distributive law: $U \otimes (V \oplus W) \cong (U \otimes V) \oplus (U \otimes W)$.

Notice that all equivalences above are canonical, that is, there are natural isomorphisms between the given spaces. Thus, for instance the isomorphism $U \otimes V \to V \otimes U$ is defined by means of the natural map $u \otimes v \mapsto v \otimes u$. It is not hard to check the remaining cases.

It is also true that the finite-dimensional linear spaces V and V^* are isomorphic, however they are not canonically isomorphic. In this sense, we will refrain from writing $V \cong V^*$ and we will use the symbol $\cong$ only when the corresponding spaces are naturally identified (i.e., when there is a canonical isomorphism among them). It is, however, true that (see Sect. A.3.2):

$$\mathrm{Hom}\,(V, W) \cong V^* \otimes W\,.$$

The canonical identification is provided by the linear map $V^* \otimes W \to \mathrm{Hom}(V, W)$ induced from the identification of the element $\alpha \otimes w$ in $V^* \otimes W$ with the linear map $V \to W$, such that $v \mapsto \alpha(v)w$ (clearly such map is injective and, by counting dimensions, we conclude that it is a linear isomorphism).

Then, we can add a few more equivalences to the previous list:

1. $(V^*)^* \cong V$.
2. $(U \oplus V)^* \cong U^* \oplus V^*$.
3. $(U \otimes V)^* \cong U^* \otimes V^*$.

Notice that, when we consider maps $f: A \to \mathbb{C}$ from any set A into the complex numbers (for instance), any operations defined on the target space naturally induces a similar operation in the space of functions, by simply performing it point-wise: $(f+g)(a) = f(a) + g(a)$, any $a \in A$.

[2]Do not confound it with the adjoint operator $L^\dagger$ defined on a Hilbert space, Sect. 7.3.3.

[3]The opposite category of a given category is the category with the same objects but all arrows reversed.

In a similar way, thinking on linear representations as 'maps' from $\mathbf{C}$ into the category of finite-dimensional linear spaces **FinVect**, we can induce in them the same operations defined on linear spaces, i.e., duals, direct sums and tensor products.

Definition 9.5

1. Direct sum of representations. Let R, R' be two linear representations of the category $\mathbf{C}$. The direct sum $R \oplus R'$ of the linear representations R and R', is the linear representation defined as $(R \oplus R')(x) = R(x) \oplus R'(x)$ and $(R \oplus R')(\alpha) = R(\alpha) \oplus R'(\alpha)$, for any object x and morphism $\alpha \colon x \to y$.

2. Tensor product of representations. Let R, R' be two linear representations of the category $\mathbf{C}$. The tensor product $R \otimes R'$ of the linear representations R and R' is the linear representation defined as $(R \otimes R')(x) = R(x) \otimes R'(x)$ and $(R \otimes R')(\alpha) = R(\alpha) \otimes R'(\alpha)$ for any object x and morphism $\alpha \colon x \to y$.

3. Dual of a representation. Let R be a linear representation of the category $\mathbf{C}$. The dual representation of R is the linear representation R^* of the opposite category $\mathbf{C}^{\mathrm{op}}$ defined as $R^*(x) = R(x)^*$ for any object x in $\mathbf{C}$ and $R^*(\alpha) \colon V_y^* \to V_x^*$ is the adjoint linear map to $R(\alpha) \colon V_x \to V_y$, i.e., $R^*(\alpha) = R(\alpha)^*$, with $\alpha \colon x \to y$, $V_x = R(x)$ and $V_y = R(y)$.

It is in this sense, we may say that the category of representations $\mathbf{Rep}(\mathbf{C})$ is equipped with the operations $\oplus$, $\otimes$ and $*$.

Example 9.8 *1. The singleton representations. Consider two singleton representations, that is, linear maps $R \colon V_1 \to V_2$ and $R' \colon V_1' \to V_2'$. Then, the direct sum of the representations $R \oplus R'$ is the linear map $R \oplus R' \colon V_1 \oplus V_1' \to V_2 \oplus V_2'$, given by $(R \oplus R')(v \oplus v') = R(v) \oplus R'(v')$, $v \in V_1$, $v' \in V_1'$.*

2. *Continuing with the previous example, we can consider, for instance, the rather trivial representation $R_{0,1}$ of type $0 \to 1$ (Example 9.1.1), i.e., the trivial map $\mathbf{0} \to \mathbb{C}$ and a representation $R \colon V_1 \to V_2$, then $R \oplus R_{0,1} \colon V_1 \to V_2 \oplus \mathbb{C}$, $(R \oplus R_{0,1})(v) = R(v) \oplus \mathbf{0}$.*

   In the same vein, consider the representation $R_{1,0} \colon \mathbb{C} \to \mathbf{0}$ of type $1 \to 0$. Then $R \oplus R_{1,0} \colon V_1 \oplus \mathbb{C} \to V_2$, $R \oplus R_{1,0}(v \oplus \lambda) = R(v)$.

3. *We may finish the considerations raised in the two previous examples by considering now the representation $R_{1,1} \oplus R_{1,0} \oplus R_{0,1} \colon \mathbb{C} \oplus \mathbb{C} \to \mathbb{C} \oplus \mathbb{C}$ given by $R_{1,1} \oplus R_{1,0} \oplus R_{0,1}(\lambda, \lambda') = (\lambda, 0)$, which has associated matrix:*

$$\begin{bmatrix} 1 & 0 \\ 0 & 0 \end{bmatrix}. \tag{9.5}$$

4. *Recall that equivalence classes of singleton representations R (just linear maps $R \colon V_1 \to V_2$) are described by three non-negative integers (n, m, r),*

where $n = \dim V_1$, $m = \dim V_2$ and $r = \operatorname{ran} R$. Then, the representation R has a representative with an associated matrix of the form (9.1), recall Example 9.2. Hence, we can write the representation R as the direct sum of r copies of the representation $R_{1,1}$ of type $1 \to 1$ (see Example 9.1.1), $n-r$ copies of the representation $R_{1,0}$ of type $1 \to 0$ and $m-r$ copies of the representation $R_{0,1}$ of type $0 \to 1$. That is:

$$R = R_{1,1}^{\oplus r} \oplus R_{1,0}^{\oplus n-r} \oplus R_{0,1}^{\oplus m-r} . \tag{9.6}$$

9.2.2 Irreducible and indecomposable representations of categories

There is a natural extension of the notion of irreduciblity and complete reducibility of representations of groups (recall Def. 7.4 and the complete reducibility theorem, Thm. 7.1) to representations of categories. There is, however, as will be shown below, a fundamental difference regarding the situation we are facing here and the discussion of representations of groups, and, is that, in general, representations of categories are not completely reducible. In other words, they cannot be always constructed by adding 'irreducible blocks'. We will call this property decomposability and we will introduce it formally in the following definition.

Given a representation $R\colon \mathbf{C} \to \mathbf{FinVect}$ of the category $\mathbf{C}$, we will say that the representation $R'\colon \mathbf{C} \to \mathbf{FinVect}$ is a subrepresentation of R, if $R'(x) = V'_x$ is a subspace of V_x and $R'(\alpha)\colon V'_x \to V'_y$ is the restriction of $R(\alpha)$ to V'_x, that is, $R(\alpha) \mid_{V'_x} = R'(\alpha)$, for any $\alpha\colon x \to y$. There are always two obvious subrepresentations of any representation R, the representation that assigns the space $\mathbf{0}$ to any object x, and the representation R itself. We will call them the trivial subrepresentations of R. We will say that a subrepresentation is proper if it is not trivial.

Definition 9.6 Given a representation $R\colon \mathbf{C} \to \mathbf{FinVect}$ of the category $\mathbf{C}$, we will say that R is irreducible if it has no proper subrepresentations. If a representation is not irreducible we will say that it is reducible.

We will say that R is indecomposable if there are no representations R', R'' of $\mathbf{C}$ such that $R \cong R' \oplus R''$. If a representation is not indecomposable we will say that it is decomposable.

We will say that a representation R is completely reducible or semisimple if there exist $R_1, \ldots, R_r$ irreducible representations, such that $R \cong R_1 \oplus \cdots \oplus R_r$.

Example 9.9 *If a representation is irreducible it must be indecomposable, however, the converse is not necessarily true. Consider, for instance, the representation of the single loop category $\mathbf{C}(Q_{1,1})$ provided by a linear map $L\colon V \to V$, V a finite dimensional linear space, Example 9.2.2. A subrepresentation will just be a subspace W invariant under L, $L(W) \subset W$. Any eigenvector v of L, $L(v) = \lambda v$, will define a one-dimensional invariant subspace. Thus, irreducible representations of the single-loop category must be one-dimensional.*

We may consider, however, a linear map L whose Jordan canonical form is a block $J_{\lambda,n}$, $n > 1$. This representation is not irreducible (there is a one-dimensional invariant subspace defined by the eigenvector v of $J_{\lambda,n}$), however, it is indecomposable. Notice that, if there were a decomposition $L = L' \oplus L''$, then it would be possible to write the matrix associated to L in block diagonal form, which is impossible because the Jordan normal form of L is $J_{\lambda,n}$.

Example 9.10 *1. Consider a group G. As a category, it has only one object. The notion of linear representation as a group and as a category are the same, and the same are the notions of subrepresentation, irreducibility and complete reducibility. In the case of finite groups, we did not introduce the notion of indecomposability because being irreducible and indecomposable are the same in such situation.*

Exercise 9.11 *Let (μ, V) be a finite dimensional representation of the finite group G. Prove that μ is irreducible if and only if it is indecomposable.*

2. *Consider the category $\mathbf{C}(Q)$ generated by a quiver $Q(\Omega, L)$. Because there is a one-to-one correspondence between representation of the quiver and representations of the category, we may extend the previous definitions to representations of quivers. Thus, for instance, we will say that the representation R of the quiver is irreducible if it has no proper subrepresentations, etc.*

Obviously, all previous notions can be used immediately when our category is a groupoid $\mathbf{G}$ and we can start analyzing the properties of representations of groupoids which will be the main objective of the remaining of this work but, before doing that, we will like to finish the discussion on the theory of representations of quivers by summarizing a few remarkable results collected under the name of Gabriel's theorem.

9.2.3 Representations of quivers and categories

In this section we will come back to the discussion on representations of quivers started in Sect. 9.1.1. Gabriel's theorem, which describes the quivers that only have a finite number of indecomposable representations, will be stated.

Decomposition in indecomposable representations

We have already observed that, for some simple quivers, there are indecomposable representations that are not irreducible, recall, for instance, the discussion in Example 9.9 about the representation of the single loop quiver with a non-trivial Jordan block.

Actually, the Jordan normal form theorem [42, Sect. 1.7] can be seen as an explicit description of the decomposition of a linear representation of the single loop into indecomposable representations. It also shows that there are infinitely many indecomposable representations parametrized by pairs (λ, n). These blocks show the road to obtaining a decomposition theorem (see later Thm. 9.1) similar to the complete reducibility of group representations, Thm. 7.1, for representations of quivers (hence, categories).

On the other hand, the equivalence classes of representations of the singleton, Example 9.2, are parametrized by three non-negative integers (n,m,r) and they can be written as diret sums of indecomposable representations (9.6): $R = R_{1,1}^{\oplus r} \oplus R_{0,1}^{\oplus m-r} \oplus R_{1,0}^{\oplus n-r}$.

Thus, we can change the focus and we may ask if any finite-dimensional representation of a category can be written as a direct sum of indecomposable (instead of irreducible) representations. The answer is positive and this is the content of the so called Krull-Schmidt Theorem discussed below.

Before proceeding to prove it, we will discuss the analogue of Schur's Lemmas for indecomposable representations. Obviously, the statement must be weaker than in ordinary Schur's Lemmas for irreducible group representations, but they still have interest on their own.

Lemma 9.1

(Schur's Lemma for indecomposable representations) Let (R,V) be a finite-dimensional indecomposable representation of the category **C**, *then if $\varphi\colon R \Rightarrow R$ is a natural transformation (or intertwiner), the linear map $\varphi\colon \bigoplus_{x\in\Omega} \varphi_x\colon V \to V$ is either an isomorphism or is nilpotent.*

Proof 9.2 Consider the generalized eigenspaces of φ: $V_\lambda = \{v \in V \mid \exists k, (\varphi - \lambda)^k v = 0\}$. Then, the subspace V_λ defines a subrepresentation of (R,V). Moreover, because of the Jordan normal form theorem, the representation V is a direct sum of them $V = \bigoplus_{i=1}^{s} V_{\lambda_i}$. However, if (R,V) is indecomposable, there can only be one factor in the previous decomposition corresponding to some eigenvalue λ. Then, if $\lambda \neq 0$, φ is an isomorphism. On the contrary, if $\lambda = 0$, φ is nilpotent.

As a consequence of Schur's lemma we get:

Proposition 9.2

Let (R,V) be finite-dimensional indecomposable representation of the category **C**, *and $\varphi_k\colon R \Rightarrow R$, $k = 1,\ldots,r$, are nilpotent intertwiners, then $\varphi = \varphi_1 + \ldots + \varphi_r$ is nilpotent too.*

Proof 9.3 We will prove it by induction on r. The case $r = 1$ is trivial. Assume that the statement is true up to r and consider $r+1$. Let $\varphi = \varphi_1 + \ldots + \varphi_{r+1}$ and assume that it is not nilpotent, then, because of Schur's lemma, Lem. 9.1, φ must be an isomorphism. Then there exists φ^{-1}. The maps $\varphi^{-1} \circ \varphi_k$ are not isomorphisms. In fact, if φ_k is nilpotent of order n_k, then $\varphi_k^{n_k} = 0$, with $\varphi_k^{n_k - 1} \neq 0$. Then $(\varphi^{-1} \circ \varphi_k) \circ \varphi_k^{n_k - 1} = 0$, which is impossible if $\varphi^{-1} \circ \varphi_k$ where isomorphisms (multiplying by its inverse on the left, we will get that $\varphi_k^{n_k - 1} = 0$).

Then, again applying Schur's lemma, we get that $\varphi^{-1} \circ \varphi_k$ are nilpotent for all k, and using the induction hypothesis we get that $\varphi^{-1} \circ \varphi_1 + \cdots + \varphi^{-1} \circ \varphi_r$ is nilpotent.

On the other hand, we get $\varphi^{-1}\circ\varphi=\sum_{k=1}^{r+1}\varphi^{-1}\circ\varphi_k$. From which we obtain that $\mathrm{id}_V-\varphi^{-1}\circ\varphi_{r+1}=\varphi^{-1}\circ\varphi_1+\cdots+\varphi^{-1}\circ\varphi_r$, which is nilpotent. But this is impossible because:

$$
\begin{aligned}
(\mathrm{id}_V-\varphi^{-1}\circ\varphi_{r+1})\circ(\mathrm{id}_V &+ \varphi^{-1}\circ\varphi_{r+1}+(\varphi^{-1}\circ\varphi_{r+1})^2+\cdots \\
\cdots &+ (\varphi^{-1}\circ\varphi_{r+1})^{n_{r+1}-1})=\mathrm{id}_V\,,
\end{aligned}
$$

where n_{r+1} is the order of nilpotency of $\varphi^{-1}\circ\varphi_{r+1}$, that shows that $\mathrm{id}_V-\varphi^{-1}\circ\varphi_{r+1}$ is invertible with inverse:

$$
(\mathrm{id}_V-\varphi^{-1}\circ\varphi_{r+1})^{-1}=\mathrm{id}_V+\varphi^{-1}\circ\varphi_{r+1}+(\varphi^{-1}\circ\varphi_{r+1})^2+\cdots+(\varphi^{-1}\circ\varphi_{r+1})^{n_{r+1}-1}.
$$

We are ready to prove the decomposition theorem for finite-dimensional representations of categories into indecomposable ones.

Theorem 9.1

(Krull-Schmidt Theorem) Any finite-dimensional representation (R,V) of a category $\mathbf{C}$ can be written as a direct sum of indecomposable representations (R_k,V_k), $k=1,\ldots,r$, unique up to reordering of the factors: $R=R_1\oplus\cdots\oplus R_r$.

Proof 9.4 Because the representation (R,V) is finite-dimensional, the existence of the decomposition is obvious. If R is indecomposable, we are done. If R is decomposable, that means that there exists two subrepresentations $(R^{(1)},V_1)$ and $(R^{(2)},V_2)$, such that $R=R^{(1)}\oplus R^{(2)}$. Notice that because R is finite-dimensional and $(R^{(1)},V_1)$, $(R^{(2)},V_2)$ are subrepresentations, they are finite-dimensional too and their dimensions are smaller than $\dim V$. Now, if $R^{(1)}$ and $R^{(2)}$ are indecomposable, we are done. If not, we repeat the argument for each factor. After a finite number of steps, we will get a finite family of indecomposable representations (R_k,V_k), such that $R=\bigoplus_{k=1}^{r}R_k$.

To prove the uniqueness of the previous decomposition, we will proceed by induction on $\dim V$. If $\dim V=1$, the statement is obvious. Now let $\dim V=n$ and suppose that we have two decompositions $R=R_1\oplus\cdots\oplus R_r$ and $R=R_1'\oplus\cdots\oplus R_{r'}'$. We will denote by $i_k\colon V_k\to V$, $i_l'\colon V_l'\to V$ and $p_k\colon V\to V_k$, $p_l'\colon V\to V_l'$ the natural inclusions and projections associated with the previous decompositions. Consider the maps $\varphi_l=p_1\circ i_l'\circ p_l'\circ i_1\colon V_1\to V_1$.

Then, we have $\sum_{l=1}^{r'}\varphi_l=\mathrm{id}_V$. Because of Prop. 9.2, some of the maps φ_l must be isomorphisms (if not, the sum would be nilpotent, which is not the case). So, assume that $\varphi_1\colon V_1\to V_1$ is an isomorphism. However, $\varphi_1=(p_1\circ i_1')\circ(p_1'\circ i_1)$, then $V_1'=\ker p_1\circ i_1'\oplus\operatorname{ran}p_1'\circ i_1$, and because V_1' is indecomposable, we get that both $p_1\circ i_1'\colon V_1'\to V_1$ and $p_1'\circ i_1\colon V_1\to V_1'$ are isomorphisms. Then, we have that $V=V_1\oplus W$, $W=\bigoplus_{i>1}V_i$, and $V=V_1'\oplus W'$, with $W'=\bigoplus_{i>1}V_i'$.

Now, consider the map $\psi\colon W\to W'$, $\psi=p\circ i$, defined as the composition of the natural inclusion and projection maps $i\colon W\to V$ and $p\colon V\to W'$, defined by the previous decompositions, $V=V_1\oplus W$ and $V=V_1'\oplus W'$. To prove that ψ is an isomorphism, all we have to do is to check that $\ker\psi=0$. Let $v\in\ker\psi$, then $v\in V_1'$

as $\ker p = V_1'$. On the other hand, the projection of v to V_1 is zero, that is, $p_1(v) = 0$, then $p_1 \circ i_1'(v) = 0$, but we proved before that the map $p_1 \circ i_1' : V_1' \to V_1$ was an isomorphism, then $v = 0$.

Hence, we conclude that $W \cong W'$ and have dimension smaller than $\dim V$, then, by the induction hypothesis, their factors are the same up to a permutation.

Gabriel's theorem

We may wonder when a representation of a quiver, or equivalently of a finitely generated free category, has only finitely many indecomposable linear representations.

Definition 9.7 Let $Q = (\Omega, L)$ be a finite quiver. We will say that Q is of finite type if it has finitely many indecomposable linear representations.

A linear representation of the quiver Q determines a linear representation of its freely generated category (which is a monoid) $\mathbf{C}(Q)$. We cannot expect to have a "nice" theory of representations of free objects. The free group generated by one element F_1 is Abelian, but the free groups F_n, $n \geq 2$ are non-Abelian (see Appendix B.1) and its behaviour from the point of view of their representation theory is 'wild'[4].

However, we have found some examples of representations of categories that are not so bad. Recall the singleton (two-dots) and three-dots quivers whose categories where finite and their theory of representations quite simple. They only have a finite number of indecomposable representations. Something similar happens for the n-dots quiver. Thus, the previous definition, Def. 9.7, is not empty, the n-dots quiver is of finite type for all n.

Are there other instances of finite-type quivers, or is there a classification of them? P. Gabriel [27] proved a remarkable theorem giving a positive answer to all these questions. It so happens that the answer provided by Gabriel brings us back to the theory of Dynkin diagrams, in fact, simply-laced diagrams displayed in Fig. 9.3.

We will state Gabriel's theorem without proving it (we will refer the reader to [26, Sects. 6.2–6.8] for a detailed proof of the theorem).

Theorem 9.2

(Gabriel's theorem) A connected quiver Q is of finite type if and only if the corresponding unoriented graph is a Dynkin diagram of type A_n, D_n, E_6, E_7 or E_8, i.e., any of the graphs displayed in Fig. 9.3.

[4] Technically, we say that these groups are not amenable.

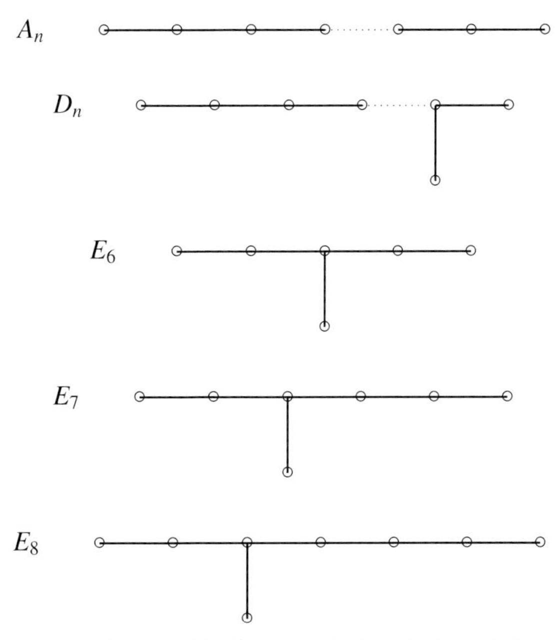

Figure 9.3: Dynkin diagrams: A_n, D_n, E_6, E_7 and E_8.

9.3 Linear representations of groupoids

9.3.1 *Basic definitions and properties*

After the previous discussions, that can be considered as a preparation for the coming sections, we can address the main problem considered in this book, the study of representations of finite groupoids. For the sake of completeness and to fix the notations, we will repeat part of the definitions that we have encountered so far by adapting them, when necessary, to the case of groupoids.

Groupoids are particular instances of categories, hence, everything that was said for linear representations of categories also applies to groupoids. Thus, a linear representation of a groupoid $\mathbf{G}$ will be a functor R from the category $\mathbf{G}$ to the category of linear spaces **Vect**.

Definition 9.8 A linear representation of a groupoid $\mathbf{G} \rightrightarrows \Omega$ is a functor R from the category $\mathbf{G}$ to the category **Vect** of linear spaces. That is, a linear representation is an assignment of a finite dimensional linear space $V_x := R(x)$ to any object $x \in \Omega$, and of a linear map $R(\alpha)\colon V_x \to V_y$, to any morphism $\alpha\colon x \to y$ in the groupoid, in such a way that $R(\beta \circ \alpha) = R(\beta)R(\alpha)$ for every composable pair $\alpha\colon x \to y$, $\beta\colon y \to z$. Moreover, $R(1_x) = \mathrm{id}_{V_x}$.

A finite-dimensional representation of a groupoid $\mathbf{G} \rightrightarrows \Omega$ is a functor R from the category $\mathbf{G}$ to the category **FinVect** of finite dimensional linear spaces. A linear representation will be denoted as $R\colon \mathbf{G} \to \mathbf{Vect}$.

We shall prove in the next Proposition that if the groupoid is connected then all spaces $V_x = R(x)$ corresponding to the linear representation R are isomorphic, hence, the linear spaces defined by any representation of a finite groupoid are isomorphic on each orbit of the groupoid and they possess the same dimension n if the representation is finite dimensional.

Proposition 9.3
Let R be a linear representation of the groupoid $\mathbf{G}$. Then, on each connected component of the groupoid, the linear spaces $V_x = R(x)$ are all isomorphic. In particular, if the groupoid is transitive, all linear spaces V_x are isomorphic.

Proof 9.5 It suffices to observe that the linear maps $R(\alpha)\colon V_x \to V_y$ are linear isomorphisms because $\mathrm{id}_{V_x} = R(1_x) = R(\alpha^{-1} \circ \alpha) = R(\alpha^{-1})R(\alpha)$, and $\mathrm{id}_{V_y} = R(1_y) = R(\alpha \circ \alpha^{-1}) = R(\alpha)R(\alpha^{-1})$.

As in the case of groups, often we will refer to the linear representation $R\colon \mathbf{G} \to \mathbf{Vect}$ of the groupoid $\mathbf{G} \rightrightarrows \Omega$ as a pair (R,V), where V, called the total space of the representation (or the support of the representation) is the direct sum of the linear spaces V_x, that is:

$$V = \bigoplus_{x \in \Omega} V_x \,. \tag{9.7}$$

Notice that, if Ω is finite with elements numbered as $x_1, \ldots, x_n$, for instance, then $V = V_{x_1} \oplus \cdots \oplus V_{x_n}$ is a finite-dimensional linear space.

Definition 9.9 Let $\mathbf{G} \rightrightarrows \Omega$ be a groupoid and (R,V) a linear representation. Let $\mathcal{O}$ be a connected component of $\mathbf{G}$, then we call the dimension n of the linear space V_x, $x \in \mathcal{O}$, the local dimension of the representation R on the connected component $\mathcal{O}$. Notice that the dimension could vary from connected component to connected component.

Because of the previous results, if $\mathbf{G}$ is connected and R is finite dimensional, $\dim V = n|\Omega|$, with n the local dimension of the representation. Sometimes we will just call n the dimension of the representation, even though it is not the dimension of the total space V of the representation.

In particular, if the groupoid $\mathbf{G}$ is a group, i.e., it possesses just one object (that may be identified with the neutral element of the group), a linear representation R of the groupoid becomes a standard linear representation of the group, that is $R(g)\colon V \to V$, V is a linear space (the one associated to the neutral element), $R(g)$

is a linear map, such that $R(g^{-1}) = R(g)^{-1}$ for any element g in the group, and $R(gg') = R(g)R(g')$ for any g, g' elements in the group $\mathbf{G}$.

One important property in the theory of linear representations of groupoids that needs to be highlighted, is that given a linear representation R of the groupoid $\mathbf{G}$, by restricting it to the isotropy groups G_x, it defines a family μ_x of linear representations of them. More precisely:

Definition 9.10 Let R be a linear representation of the groupoid $\mathbf{G}$, the restriction of R to the isotropy group G_x will be denoted by $\mu_x \colon G_x \to \mathrm{End}(V_x)$, $\mu_x(g) = R(g) \colon V_x \to V_x$ for all $g \in G_x$, and it will be called the isotropy representation of R at x.

How different are the isotropy representations of a given representation? Not much, let us see why.

Proposition 9.4
Let (R, V) be a linear representation of a groupoid $\mathbf{G}$ and x, y two objects in the same connected component. Then the isotropy representations (μ_x, V_x) and (μ_y, V_y) of the isotropy groups G_x and G_y are equivalent. In particular, if $\mathbf{G}$ is transitive (or connected), then all isotropy representations are equivalent.

Proof 9.6 Let x, y be objects in the same connected component, then there is a morphism $\alpha \colon x \to y$. This means that $\Phi = R(\alpha) \colon V_x \to V_y$ is a linear isomorphism and the map $\varphi \colon G_y \to G_x$ defined as $\varphi(g) = \alpha^{-1} \circ g \circ \alpha$, $g \in G_y$, is a group isomorphism. We will use it to identify the two groups G_x and G_y. Then, a trivial computation shows that $\Phi \circ \tilde{\mu}_x(g) = \mu_y(g) \circ \Phi$, where $\tilde{\mu}_x(g) = \mu_x(\varphi(g))$, $g \in G_x$, namely:

$$\Phi \circ \mu_x(\varphi(g)) = R(\alpha)R(\alpha^{-1} \circ g \circ \alpha) = R(g \circ \alpha) = R(g)R(\alpha) = \mu_y(g) \circ \Phi. \quad (9.8)$$

9.3.2 Properties of representations of groupoids

The natural notion of equivalence between linear representations of a groupoid is given, as in the case of categories, by the notion of natural transformations among functors, Sect. 9.1.3.

Definition 9.11 Given two linear representations $R, S \colon \mathbf{G} \to \mathbf{Vect}$, we will say that R and S are equivalent if there exists an invertible natural transformation $\varphi \colon R \Rightarrow S$ between the functors R and S (see Fig. 9.4).

More explicitly, the representations (R, V), (S, W), of the groupoid $\mathbf{G} \rightrightarrows \Omega$ are equivalent if there exists a family of linear isomorphisms $\varphi(x) \colon V_x \to W_x$, $x \in \Omega$ such that:

$$\varphi(y) \circ S(\alpha) = R(\alpha) \circ \varphi(x), \qquad \forall \alpha \colon x \to y \in \mathbf{G}. \quad (9.9)$$

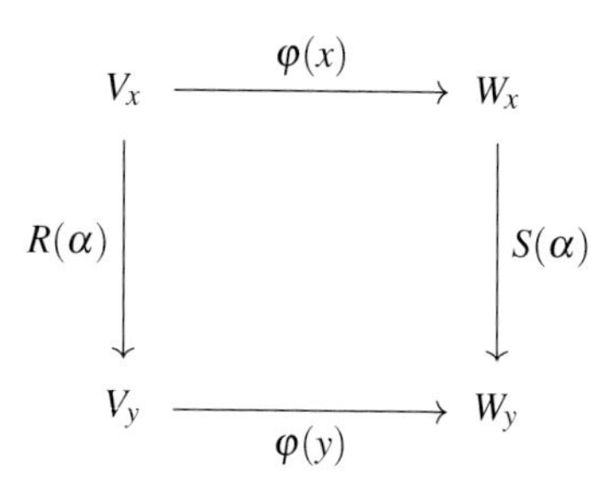

Figure 9.4: A natural transformation $\varphi\colon R \Rightarrow S$.

Again, the restriction of the notion of equivalence of linear representations to the class of groups provides the standard notion of equivalence of linear representations of groups.

The family of linear representations of groupoids is a category itself, recall Sect. 9.1.3, whose objects are the representations (R,V) and its morphisms are natural transformations $\varphi\colon R \Rightarrow S$. We may consider the category $\mathbf{Rep}(\mathbf{G})$ whose objects are equivalence classes of representations of the groupoid $\mathbf{G}$. This category is the structure we would like to study for a given groupoid $\mathbf{G}$ as it provides all possible linear 'snapshots' of the given abstract groupoid. We will find out that, similar to the theory of linear representations of finite groups, a full description of this category will be provided and it will be shown that it is constructed from a finite number of 'elementary' blocks, the irreducible representations of the groupoid.

Given two representations (R,V), (S,W), a natural transformation $\varphi\colon R \Rightarrow S$ will also be called, as in the case of groups, Def. 7.10, an intertwiner among them. The set of all intertwiners, i.e., families of linear maps $\varphi(x)\colon V_x \to W_x$ satisfying (9.9), will be denoted as $\mathrm{Hom}_{\mathbf{G}}(V,W)$.

9.3.3 Operations with representations

Definition 9.12 A subrepresentation R' of a linear representation R of the groupoid $\mathbf{G}$ is a functor $R'\colon \mathbf{G} \to \mathbf{Vect}$ such that, for each object x, $V'_x = R'(x) \subset R(x) = V_x$, and the restriction of the linear map $R(\alpha)$ to the subspace V'_x coincides with $R'(\alpha)$, in such case, we will write $R' \subset R$.

Notice that, if $R' \subset R$ is a subrepresentation of R, then the subspaces $V'_x = R'(x) \subset V_x = R(x)$ are invariant under $\mathbf{G}$ in the sense that $R(\alpha)v_x \in V'_y$ for any vector $v_x \in V'_x$ and $\alpha\colon x \to y$.

Direct sum and disjoint unions

Similarly, as it was done with groups, Sect. 7.2.2, and categories, Sect. 9.2.1, given two linear representations R, S of the groupoid $\mathbf{G}$, we may define its direct sum $R \oplus S$ and tensor product $R \otimes S$.

More specifically, we define the direct sum $R \oplus S$ of two representations (R,V), (S,W) as the linear representation of the groupoid $\mathbf{G}$ determined by the functor that assigns the linear space $R(x) \oplus S(x) = V_x \oplus W_x$ ($R(x) = V_x$ and $S(x) = W_x$) to any object x, and a linear map $R(\alpha) \oplus S(\alpha) \colon V_x \oplus W_x \to V_y \oplus W_y$ to any $\alpha \colon x \to y$. The direct sum of (R,V) and (S,W) will be denoted as $(R \oplus S, V \oplus W)$. We proceed similarly with the tensor product (see below).

In addition to this, given a subrepresentation $R' \subset R$, we can also define the quotient representation R/R':

Definition 9.13 If $R' \subset R$ is a subrepresentation of R, then we may define a representation T of $\mathbf{G}$ on the quotient spaces V_x/V'_x, by means of $T(\alpha)(v_x + V'_x) = R(\alpha)(v_x) + V'_x$, $v_x \in V_x$. We will call this representation the quotient representation of R by R' and we denote it by $T = R/R'$.

An important difference with the case of groups is that the direct sum of linear spaces allows to define a new representation on the disjoint union of groupoids.

Proposition 9.5
Let $\mathbf{G}_1 \rightrightarrows \Omega_1$ and $\mathbf{G}_2 \rightrightarrows \Omega_2$ be two groupoids and (R_1,V_1), (R_2,V_2) linear representations of $\mathbf{G}_1$ and $\mathbf{G}_2$, respectively. Then, there is a natural representation (R,V) of the disjoint union $\mathbf{G}_1 \bigsqcup \mathbf{G}_2$ defined by the direct sum of (R_1,V_1) and (R_2,V_2), as $R(x_a) = R_a(x_a) = (V_a)_{x_a}$, $R(\alpha_a) = R_a(\alpha_a)$, $a = 1,2$ and called, consequently, the direct sum of (R_1,V_1) and (R_2,V_2). The total space of the representations R is $V_1 \oplus V_2$ and it will be denoted by $R_1 \oplus R_2$ as well.

Exercise 9.12 *Prove the previous proposition.*

Example 9.13 *Let G_i, (μ_i, V_i), $i = 1,\ldots,n$, be a family of groups and linear representations of them, then, the groupoid $\mathbf{G} = \bigsqcup_{i=1}^n G_i$ inherits a linear representation $R = \bigoplus_{i=1}^n \mu_i$ defined on the total space $V = \bigoplus_{i=1}^n V_i$ given by the direct sum of the representations μ_i. Of course, all this in the sense described in Prop. 9.5, because $V = \bigoplus_{i=1}^n V_i$ does not make sense as a direct sum of group representations.*

Note that, given a group G and a family of representations (μ_i, V_i) of the group G, we can form the direct sum of the given representations $\mu = \bigoplus_i \mu_i$ on the linear space $V = \bigoplus_i V_i$ as $\mu(g)(v_1 \oplus \cdots \oplus v_n) = (\mu_1(g)v_1) \oplus \cdots \oplus (\mu_n(g)v_n)$, $v_i \in V_i$, i.e., the group G acts at the same time on all spaces V_i, on each one with the representation μ_i, however, the action of the groupoid $\mathbf{G} = \bigsqcup_{i=1}^n G_i$ on the direct sum $V = \bigoplus_{i=1}^n V_i$ happens on each space V_i with the representation μ_i of the group G_i (that could be all different).

Example 9.14 *Given a groupoid* $\mathbf{G} \rightrightarrows \Omega$*, any family of representations* (μ_x, V_x)*,* $x \in \Omega$ *will define a linear representation of the fundamental groupoid* $\mathbf{G}_0$ *given by the direct sum* $(\bigoplus_x \mu_x, \bigoplus_x V_x)$*.*

Proposition 9.6
Given a groupoid $\mathbf{G} \rightrightarrows \Omega$*, any representation* (R, V) *of* $\mathbf{G}$ *decomposes as the direct sum of representations* $(R_{\mathcal{O}}, V_{\mathcal{O}})$ *of any of its connected components* $\mathcal{O}$*:* $R = \bigoplus_{\mathcal{O}} R_{\mathcal{O}}$*,* $V = \bigoplus_{\mathcal{O}} V_{\mathcal{O}}$*.*

Proof 9.7 Clearly, any representation (R, V) of $\mathbf{G}$ induces by restriction a representation on each of its subgroupoids. The subgroupoids $\mathbf{G}_{\mathcal{O}}$, obtained by restricting the groupoid to its connected components, are maximal normal subgroupoids of $\mathbf{G}$. Then, we denote by $(R_{\mathcal{O}}, V_{\mathcal{O}})$ the corresponding restriction of the representation R to any of them.

On the other hand, because of the first structure theorem, Thm. 3.1, $\mathbf{G} = \bigsqcup_{\mathcal{O}} \mathbf{G}_{\mathcal{O}}$, the restrictions $R_{\mathcal{O}}$ of the representation R define the representation $\bigoplus_{\mathcal{O}} R_{\mathcal{O}}$. Checking that $R = \bigoplus_{\mathcal{O}} R_{\mathcal{O}}$ is straighforward.

Because of the canonical decomposition of any representation R of a groupoid $\mathbf{G}$ as a direct sum of representations defined on its connected components stated in the previous theorem, in what follows we will restrict to this situation.

Tensor product

Similarly, as in the situation with groups, we define the tensor product $(R \otimes S, V \otimes W)$ of the representations (R, V) of the groupoid $\mathbf{G} \rightrightarrows \Omega$ and (S, W) of the groupoid $\mathbf{G}' \rightrightarrows \Omega'$ as the linear representation of the direct product groupoid $\mathbf{G} \times \mathbf{G}'$ determined by the functor that assigns the linear space $(R \otimes S)(x, x') = R(x) \otimes S(x') = V_x \otimes W_{x'}$ to any object $(x, x') \in \Omega \times \Omega'$, and the linear map $(R \otimes S)(\alpha, \alpha') = R(\alpha) \otimes S(\alpha') \colon V_x \otimes W_{x'} \to V_y \otimes W_{y'}$, to any $\alpha \colon x \to y \in \mathbf{G}$, $\alpha' \colon x' \to y' \in \mathbf{G}'$.

In the particular instance that the groupoids $\mathbf{G}$ and $\mathbf{G}'$ are the same, there is a natural restriction of the tensor product representation to a representation of $\mathbf{G}$, also called, the tensor product representation: $(R \otimes S)(x) = R(x) \otimes S(x) = V_x \otimes W_x$ and $(R \otimes S)(\alpha) = R(\alpha) \otimes S(\alpha) \colon V_x \otimes W_x \to V_y \otimes W_y$.

Remark 9.2 With respect to the natural operations induced in the category $\mathbf{Rep}(\mathbf{G})$ by the operations $\oplus$ and $\otimes$, the category of equivalence classes of representations of a groupoid becomes a monoidal category. We will not use this structure in what follows as we are not intending to address the extension of the Tannaka-Krein duality (see for instance [39] and references therein) to the category of groupoids, a problem that will be dealt with elsewhere. ■

9.3.4 Irreducible and indecomposable representations

As in the general case of categories, any representation (R,V) of a groupoid $\mathbf{G}$ has two trivial subrepresentations: (R,V) itself, and the zero representation, that is, the functor $\mathbf{0}$ that assigns the zero vector space and the zero map to any object. Any subrepresentation of a linear representation R of the groupoid $\mathbf{G}$, different from the two trivial subrepresentations R and $\mathbf{0}$, will be called a proper subrepresentation. Again, as in the general situation, we will be interested in representations that do not possess proper subrepresentations. Then, we repeat for the sake of the current discussion, the definition of irreducible and indecomposable representations (see Def. 9.6).

Definition 9.14 A representation (R,V) of a groupoid $\mathbf{G}$ will be said to be irreducible it it has no proper subrepresentations.

We get some immediate consequences of the definitions.

Proposition 9.7
Any irreducible representation of a finite groupoid is finite-dimensional.

Proof 9.8 In fact, if $R\colon \mathbf{G} \to \mathbf{Vect}$ is irreducible, we can choose a non-zero vector v_x in each space V_x for each $x \in \Omega$. Then, define the space $W_x = \mathrm{span}\{R(\alpha)v_y \mid \alpha \in \mathbf{G}_-(x)\}$ consisting of the linear span of all vectors obtained from vectors v_y and morphisms $\alpha\colon y \to x$. The space W_x is finite dimensional because $\mathbf{G}$ is finite and it clearly constitutes a finite-dimensional subrepresentation of R.

Moreover, a simple induction argument shows that any finite-dimensional representation contains an irreducible representation.

The following observation will establish the deep relation existing between irreducible representations of a groupoid and those of its isotropy groups. Given a representation (R,V) of a groupoid $\mathbf{G}$ it is immediate that it restricts to a representation (μ_x, V_x) of any isotropy group G_x of $\mathbf{G}$. It is not hard to prove that if (R,V) is irreducible, so it is (μ_x, V_x). What could come as a little suprise is that the converse is also true.

Theorem 9.3
Let $\mathbf{G} \rightrightarrows \Omega$ be a finite connected groupoid and (R,V) a linear representation. Then, (R,V) is irreducible if and only if any of the representations (μ_x, V_x) of the isotropy groups defined by restriction of R are irreducible.

Proof 9.9 The main observation to prove the previous assertions is that the linear isomorphisms $R(\alpha)\colon V_x \to V_y$, $\alpha\colon x \to y$, act as intertwiners of the representations μ_x, μ_y. In fact, the computation done in the proof of Prop. 9.4 shows, Eq. (9.8):

$$R(\alpha) \circ \mu_x(\varphi(g)) = \mu_y(g) \circ R(\alpha)\,. \tag{9.10}$$

Then, suppose that the restricted representations (μ_x, V_x) are irreducible (they are all equivalent on the other hand). Because Eq. (9.10), and Schur's Lemma $R(\alpha)$ is a multiple of the identity and the representation (R,V) is irreducible.

Consider, on the other hand, that (R,V) is irreducible. If for some $x_0 \in \Omega$, (μ_{x_0}, V_{x_0}) were not irreducible there would be a proper invariant subspace $W_{x_0} \subset V_{x_0}$. Because **G** is connected, choose a family of morphisms $\alpha_x \colon x_0 \to x$, one for any $x \in \Omega$. Then, define the family of subspaces $W_x = R(\alpha_x) W_{x_0}$.

It is not hard to check that the subspaces W_x define a proper subrepresentation of (R,V). Recall that given a morphism $\alpha_x \colon x_0 \to x$, it induces a group isomorphism $\varphi_x \colon G_{x_0} \to G_x$ and any $\alpha \colon x_0 \to x$ can be written as $\alpha_x \circ \gamma$ for some $\gamma \in G_{x_0}$. Then any linear isomorphism $R(\alpha) \colon V_{x_0} \to V_x$, $\alpha \colon x_0 \to x$ can be written as $R(\alpha) = R(\alpha_x) \circ R(\gamma)$, but then:

$$R(\alpha) v_{x_0} = (R(\alpha_x) \circ R(\gamma)) v_{x_0} = R(\alpha_x)(R(\gamma) v_{x_0}) .$$

However, W_{x_0} is invariant, so $R(\gamma) v_{x_0} \in W_{x_0}$ and, by definition of W_x, we get $R(\alpha) v_{x_0} \in W_x$ Thus, the spaces W_x are invariant under R and they are proper subspaces because the linear maps $R(\alpha_x)$ are all isomorphisms. This leads to a contradiction, then the representations (μ_x, V_x) must be irreducible.

The proof of the previous theorem shows something else that we state explicitly as it will be used later on.

Lemma 9.2

Given an invariant subspace W_x with respect to the restriction (μ_x, V_x) at some $x \in \Omega$ of a finite-dimensional representation (R,V) of the connected groupoid **G**, *there is a subrepresentation (R', W) of (R,V), such that $R'(x) = W_x$.*

Proof 9.10 This is the essence of the 'if' part of the proof of Thm. 9.3.

We may think of the previous result as a property of 'spreading' on connected components of groupoids. In other words, a property referred to an isotropy group G_x spreads out over its connected component. This, together with the relation between irreducible representations of groupoids and the irreducibility of their restrictions to the isotropy groups, determines the structure of the space of intertwiners of the groupoid. We realize that the space of intertwiners $\mathrm{Hom}_{\mathbf{G}}(V,W)$ is determined by the space of intertwiners of the representations (μ_x, V_x) obtained by restriction of the representation (R,V) to the isotropy groups G_x, that is:

Proposition 9.8

Let $\mathbf{G} \rightrightarrows \Omega$ *be a finite connected groupoid and (R,V), (R,V), are two finite-dimensional representations of* **G**. *Then, the space of intertwiners* $\mathrm{Hom}_{\mathbf{G}}(V,W)$ *is isomorphic, albeit non-canonically, to the space of intertwiners* $\mathrm{Hom}_{G_x}(V_{1,x}, V_{2,x})$ *of* *the corresponding isotropy representations at an arbitrary point $x \in \Omega$:*

$$\mathrm{Hom}_{\mathbf{G}}(V,W) \cong \mathrm{Hom}_{G_x}(V_x,W_x)\,. \tag{9.11}$$

Proof 9.11 Clearly, any intertwiner $\varphi\colon R \Rightarrow S \in \mathrm{Hom}_{\mathbf{G}}(V,W)$ defines a family $\varphi_x\colon V_x \to W_x$ of intertwiners, each one in $\mathrm{Hom}_{G_x}(V_x,W_x)$, actually we get that if $S\circ\varphi = \varphi\circ R$, then, for any element $\gamma_x \in G_x$, we have for all $v \in V_x$:

$$\varphi_x(\mu_x(\gamma_x)(v)) = (\varphi_x\circ R(\gamma_x))(v) = (S(\gamma_x)\circ\varphi_x)(v) = \nu_x(\gamma_x)(\varphi_x(v),$$

with $\mu_x(\gamma_x) = R(\gamma_x)\mid_{V_x}$ and $\nu_x(\gamma_x) = S(\gamma_x)\mid_{W_x}$, the restrictions of the representations R and S to V_x and W_x, respectively. The previous relation shows that the restriction of φ to V_x is in $\mathrm{Hom}_{G_x}(V_x,W_x)$. This defines a linear map $\varphi \mapsto \varphi\mid_{V_x}$ from $\mathrm{Hom}_{\mathbf{G}}(V,W)$ to $\mathrm{Hom}_{G_x}(V_x,W_x)$.

Conversely, let $\varphi_x \in \mathrm{Hom}_{G_x}(V_x,W_x)$ be an intertwiner between the isotropy representations μ_x, ν_x, obtained by restricting the representations R and S to G_x. We should now construct an intertwiner $\varphi\colon R \Rightarrow S$. For that, we will use an idea exploited exhaustively in the first part of this work (recall Prop. 5.7) and choose an identification between the isotropy groups. As the groupoid $\mathbf{G}$ is connected, we proceed to choose an identification between all the isotropy groups G_y with G_x by using a section σ of the canonical projection $\pi\colon \mathbf{G} \to \mathbf{G}(\Omega)$.

Then, we define the linear map $\varphi_y\colon V_y \to W_y$, for all $y \in \Omega$, as $\varphi_y = S(\sigma(y,x))\circ \varphi_x\circ R(\sigma(x,y))$, i.e.:

$$\varphi_y(v_y) = S(\sigma(y,x))(\varphi_x(R(\sigma(x,y))(v_y)))\,,$$

for $v_y \in V_y$. Note that with the standard conventions for sections, recall Sect. 6.5.1: $\sigma(y,x)\colon x \to y$, $R(\sigma(y,x))\colon V_x \to V_y$, and $\sigma(x,y) = \sigma(y,x)^{-1}$. Note that the definition is consistent because when applied to x, we obtain φ_x again (and, in checking this, we use that φ_x is an intertwiner).

Then, finally, we check that $S(\alpha)\circ\varphi_x = \varphi_y\circ R(\alpha)$. For that, we observe that, given the section σ, any morphism $\alpha\colon x\to y$, can be written uniquely as $\alpha = \sigma(y,x)\circ\gamma_x$, $\gamma_x \in G_x$ and $\alpha = \gamma_y\circ\sigma(y,x)$, $\gamma_y \in G_y$. Then we get:

$$\begin{aligned}
\varphi_y\circ R(\alpha) &= S(\sigma(y,x))\circ\varphi_x\circ R(\sigma(x,y))\circ R(\alpha)\\
&= S(\sigma(y,x))\circ\varphi_x\circ R(\sigma(x,y))\circ R(\gamma_y\circ\sigma(y,x))\\
&= S(\sigma(y,x))\circ\varphi_x\circ R(\sigma(x,y)\circ\gamma_y)\circ R(\sigma(y,x))\\
&= S(\sigma(y,x))\circ\varphi_x\circ R(\gamma_x\circ\sigma(x,y))\circ R(\sigma(y,x))\\
&= S(\sigma(y,x))\circ\varphi_x\circ R(\gamma_x)\circ R(\sigma(x,y)\circ\sigma(y,x))\\
&= S(\sigma(y,x))\circ S(\gamma_x)\circ\varphi_x = S(\sigma(y,x)\circ\gamma_x)\circ\varphi_x = S(\alpha)\circ\varphi_x\,.
\end{aligned}$$

where we have used the definition of φ_y (first equality); the decomposition of α as $\gamma_y\circ\sigma(y,x)$ and the fact the R is a representation (second and third equalities); the identity $\sigma(x,y)\circ\gamma_y = \gamma_x\circ\sigma(x,y)$ (fourth equality); the representation property of R again (in the fifth equality); the intertwiner property of φ_x (sixth equatlity) and the representation property of S and the factorization of α again in the last two equalities.

Thus, we obtain immediately a version of Schur's lemma for irreducible representations of groupoids.

Theorem 9.4

Let $\mathbf{G} \rightrightarrows \Omega$ *be a finite connected groupoid and* (R,V) *a linear representation. Then* (R,V) *is irreducible if and only if* $\mathrm{Hom}_{\mathbf{G}}(V,V) \cong \mathbb{C}$*. Moreover, if* (R,V)*,* (S,W) *are irreducible representations, then* $\mathrm{Hom}_{\mathbf{G}}(V,W) = \mathbb{C}$*, if they are equivalent, and* $\mathrm{Hom}_{\mathbf{G}}(V,W) = \mathbf{0}$ *if they are not.*

Proof 9.12 Note that, because of Prop. 9.8, $\mathrm{Hom}_{\mathbf{G}}(V,V) \cong \mathrm{Hom}_{G_x}(V_x,V_x)$, but because of Thm. 9.3 (R,V) is irreducible iff (μ_x,V_x) is irreducible for all x and because of Thm. 7.5, (μ_x,V_x) is irreducible iff $\mathrm{Hom}_{G_x}(V_x,V_x) \cong \mathbb{C}$, then $\mathrm{Hom}_{\mathbf{G}}(V,V) \cong \mathbb{C}$.

Regarding the second statement, the proof follows again immediately from Thm. 7.5. If R, S are equivalent, then the isotropy representations μ_x, ν_x are equivalent and, consequently, $\mathrm{Hom}_{G_x}(V_x,W_x) \cong \mathbb{C}$, however, if they are disequivalent, $\mathrm{Hom}_{G_x}(V_x,W_x) = \mathbf{0}$.

Before turning our attention to elucidating the structure of a given representation, we will also repeat the notion of indecomposability in the current context.

Definition 9.15 A linear representation (R,V) of the groupoid $\mathbf{G}$ will be said to be decomposable if, given a subrepresentation (R',V') of R, there exists another subrepresentation (R'',V'') of R, such that $R = R' \oplus R''$. We will say that the representation R is indecomposable if it is not decomposable.

Notice that any irreducible representation is indecomposable, however the converse is not necessarily true as it was shown in Example 9.9. However in the case of groupoids, as in the case of groups, Thm. 7.1, the converse holds true. We will end up this chapter by proving this important result.

Definition 9.16 A linear representation $R\colon \mathbf{G} \to \mathbf{Vect}$ of a groupoid $\mathbf{G}$ will be said to be completely reducible (or semisimple) if it is equivalent to a direct sum of irreducible representations. In particular a finite dimensional linear representation $R\colon \mathbf{G} \to \mathbf{FinVect}$ will be completely reducible if it is equivalent to a finite direct sum $R_1 \oplus \cdots \oplus R_r$ of irreducible representations of $\mathbf{G}$.

The remainder of this section will be devoted to proving the following theorem.

Theorem 9.5

Any finite-dimensional linear representation R *of a finite groupoid* $\mathbf{G}$ *is completely reducible and is equivalent to the direct sum of irreducible representations whose factors and multiplicities are uniquely determined. Moreover, any indecomposable finite-dimensional linear representation of a finite groupoid is irreducible.*

Hence, if $\mathbf{G}$ is a finite groupoid, as in the case of groups, and as a consequence of the previous theorem, we can concentrate on the study of its finite-dimensional linear representations and its decomposition in irreducible ones.

Proof 9.13 The existence of a decomposition in irreducibles is an easy consequence of Thm. 9.3 and the observation after it. In fact, let (R,V) be a finite-dimensional representation of the finite connected groupoid $\mathbf{G}$. Consider the restriction μ_x of the representation R to the isotropy group G_x. Such representation is completely reducible, Thm. 8.6, then consider its decomposition as sum of irreducibles $\mu_x = \bigoplus_{\alpha \in \widehat{G}_x} n_\alpha \mu_x^\alpha$ and $V_x = \bigoplus_{\alpha \in \widehat{G}_x} n_\alpha W_x^\alpha$, with W_x^α the support of the irreducible representation μ_x^α.

Because of Lemma 9.2, we get that fixed a point $x_0 \in \Omega$, we can extend the corresponding invariant subspaces $W_{x_0}^\alpha$ at x_0 all over the groupoid defining subrepresentations of (R,V) denoted as (R_α, W_α). Moreover, as $W_{x_0}^\alpha \cap W_{x_0}^\beta = \mathbf{0}$, at x_0, the same holds at any other point and because their sum is the full space V_x, the same happens at any other point. Thus, $\bigoplus n_\alpha W_\alpha = V$. Finally, because the restriction of any one of the representations (R_α, W_α) to G_{x_0} is the irreducible representation $\mu_{x_0}^\alpha$ (by construction), and, because of Thm. 9.3, (R_α, W_α) is irreducible.

This proves that any finite-dimensional representation is completely reducible. To prove the uniqueness, we can either wait until we develop similar techniques to those used in the case of groups, i.e., the theory of characters in Sect. 12.1, or we can use more abstract reasoning under the name of Jordan-Hölder Theorem. We will choose to do this in this case, as it will serve us as a preparation for a slightly more general version of it that will be used later on (see Sect. 11.3).

Thus, the conclusion of the proof, that is, the uniqueness of the decomposition, will be an immediate consequence of the following results.

Definition 9.17 Given a linear representation $R\colon \mathbf{G} \to \mathbf{Vect}$ of the groupoid $\mathbf{G}$, a finite filtration of R is a sequence of subrepresentations R_k, $k = 0, 1, \ldots, n$, such that $\mathbf{0} = R_0 \subset R_1 \subset R_2 \subset \cdots \subset R_{n-1} \subset R_n = R$.

Lemma 9.3

Any finite dimensional representation R of a finite groupoid $\mathbf{G}$ admits a finite filtration $\mathbf{0} = R_0 \subset R_1 \subset \cdots \subset R_n = R$, such that the successive quotients R_i/R_{i-1} are irreducible. Such filtration will be called a composition series for the representation R.

Proof 9.14 We will assume that the groupoid is connected and the proof is by induction in n, the dimension of the representation. The statement is obviously true for $n = 1$.

Let R be a finite dimensional representation. If R is not irreducible, let $R' \subset R$ be an irreducible subrepresentation. Then, consider the representation R/R' and apply the induction hypothesis.

Theorem 9.6

(Jordan-Hölder theorem for groupoids) Let R be a finite-dimensional representation of the finite groupoid $\mathbf{G}$. Let $\mathbf{0} = R_0 \subset R_1 \subset R_2 \subset \cdots \subset R_{n-1} \subset R_n = R$ and $\mathbf{0} = R'_0 \subset R'_1 \subset R'_2 \subset \cdots \subset R'_{m-1} \subset R'_m = R$ be filtrations of R, such that the representations R_i/R_{i-1} and R'_k/R'_{k-1} are irreducible for all i and k (in particular R_1 and R'_1 are irreducible). Then, $n = m$ and there exists a permutation σ, such that $R_{\sigma(i)}$ is equivalent to R'_i.

Proof 9.15 Again, assuming that $\mathbf{G}$ is connected, the proof is by induction on the dimension of the representation n. The case $n = 1$ is trivial.

Suppose that $n > 1$ and R_1 is not equivalent to R'_1 (if R_1 and R'_1 were equivalent we apply the induction hypothesis). In such case, the intersection of the subspaces corresponding to R_1 and R'_1 are zero, that is, $R_1(x) \cap R'_1(x) = \mathbf{0}$ (if not, the intersection will define a proper subrepresentation of both R_1 and R'_1 in contradiction with the assumption that they are irreducible). Then, we can consider the subrepresentation $R_1 \oplus R'_1 \subset R$ and the quotient $R/(R_1 \oplus R'_1)$. Consider a filtration $\mathbf{0} = T_0 \subset T_1 \subset \cdots \subset T_p = R/(R_1 \oplus R'_1)$, such that the quotients $S_i = T_i/T_{i-1}$ are irreducible (that exists because of Lemma 9.3). Then, R/R_1 has a filtration with successive quotients $R'_1, T_1, \ldots, T_p$ and another with successive quotients $R_2/R_1, \ldots, R_n/R_{n-1}$ and R/R'_1 has two filtrations with succesive quotients $R_1, T_1, \ldots, T_p$ and $R'_2/R'_1, \ldots, R'_m/R'_{m-1}$. Then, by the induction hypothesis, the collection of irreducible representations on both filtrations coincide.

Exercise 9.15 *Prove the uniqueness of the decomposition of a finite-dimensional representation (R,V) of the groupoid $\mathbf{G}$ by using the composition series given by $\widetilde{R}_i = \bigoplus_{k=1}^{i} R_k$, $i = 1, \ldots, r$, where $R = \bigoplus_{i=1}^{r} R_i$ is any decomposition in irreducibles of R.*

Chapter 10

Algebras and Representations: The Algebra of a Groupoid

The notion of the algebra of a groupoid, a notion that would be extremely helpful in uncovering the properties of groupoids and their representations, will be introduced. First, the basic definitions and properties of associative algebras will be discussed and a wealth of examples will be shown.

The notion of linear representation of algebras, that leads to the important notion of left-modules over the given algebra, will be introduced. The basic properties of representations of algebras (or left-modules) will be studied and the relation between linear representations of algebras and linear representations of groupoids will be explored. The basic results regarding this relation will be established.

10.1 Algebras

Dealing with representations of groupoids (and categories) from the point of view of functors, as discussed in the previous chapters, is getting more and more cumbersome. The minutiae of the description becomes heavier when exploring the deeper structure of them. In this chapter, we will introduce a notion that will prove to be extremely helpful in this regard.

In the same way that the introduction of the categorical way of thinking proved to be valuable to understand better the basic structures of groupoids (even the notion of representation itself!), the notion of the algebra of a groupoid will prove to be the right way to address deeper structural questions in the theory of representations, mainly because linear representations become A-modules, with A the algebra of the given groupoid.

Thus, we will have a new way of thinking about representations and their classification, that of understanding the structure of modules. In this sense, the main notion that will be raised is that of simple and semisimple modules, notions that will be discussed in the next chapter and that will provide the key to obtaining the structure theorems for the theory of representations of groupoids we are looking for.

10.1.1 Algebras: The basic examples

The notion of algebra is an abstraction of the algebraic properties of the space of functions on a set (which constitutes the basic example of commutative algebras) and of the set of square matrices of a given order (which constitutes the basic example of non-commutative algebras).

We will begin with the commutative case. Thus, if Ω is a set (finite or not), we will denote as usual by $\mathcal{F}(\Omega)$ the set of complex-valued functions on Ω, i.e., $\mathcal{F}(\Omega) = \{f\colon \Omega \to \mathbb{C}\}$. On one side, $\mathcal{F}(\Omega)$ is a complex linear space with the pointwise addition of functions:

$$(f+g)(x) = f(x)+g(x)\,, \qquad f,g \in \mathcal{F}(X)\,, \quad x \in \Omega\,,$$

and multiplication by scalars:

$$(\lambda f)(x) = \lambda(f(x))\,, \qquad \lambda \in \mathbb{C}\,.$$

In addition to this, the pointwise product of functions:

$$(f \cdot g)(x) = f(x)g(x)\,,$$

defines a commutative ring structure on $\mathcal{F}(\Omega)$. That is, the product "$\cdot$" is associative and commutative and the operations $+$ and $\cdot$ satisfy the obvious compatibility axioms inherited from the algebraic structure of complex numbers that come under the name of distributive laws.

In this particular instance, there is even a unit element for the product given by the constant function that takes the value 1 for any $x \in X$, and that will be denoted as $\mathbf{1}$. The set $\mathcal{F}(\Omega)$ with the structures above is called the algebra of functions on the set Ω and is the model for commutative algebras.

In the particular instance of Ω being a finite set, containing n elements x_k, $k = 1,\ldots,n$, a function f in $\mathcal{F}(\Omega)$ is described by the values $f_k = f(x_k)$ that takes on each one of the elements of Ω, that is, f is described by the list of numbers $(f_1,\ldots,f_n) \in \mathbb{C}^n$. Hence, $\mathcal{F}(\Omega)$ is isomorphic as a linear space to $\mathbb{C}^n$. We will identify the function f with the numerical vector with components f_k. Notice that the pointwise product $f \cdot g$ is obtained by multiplying the corresponding components[1] of both vectors $f_k g_k$.

Let us consider now the space $M_n(\mathbb{C})$ of square matrices of dimension n with complex entries. The space of matrices of any size is a linear space with respect

[1]This product is also known in matrix analysis as the Hadamard product.

to the standard addition of matrices and multiplication by scalars. In particular, the space of square matrices of dimension n is a linear space of dimension n^2 (a basis is provided by the matrices E_{jk}, $j,k=1,\ldots,n$, whose entries are all zero, except the (j,k) entry which is one).

In addition to the sum of matrices, there is also the product of matrices that corresponds to the composition of linear maps (see Appendix A.4 for a glossary of Matrix Algebra). The set $M_n(\mathbb{C})$ with the sum and product of matrices operations is a non-commutative ring with unit the identity matrix I_n. As usual, the product of two matrices A,B will be denoted as AB. Moreover, the linear space and ring structures are compatible in the sense that

$$\lambda(A+B)=\lambda A+\lambda B\,,\quad (\lambda+\mu)A=\lambda A+\mu A\,,$$

and,

$$(\lambda\mu)A=\lambda(\mu A)\,,\quad (\lambda A)B=A(\lambda B)\,,$$

for all matrices A,B and all scalars λ,μ. Thus, the space of square matrices of a given dimension n will be called the algebra of matrices of dimension n.

Notice that we may consider the Abelian algebra over the finite set $\{1,2,\ldots,n\}$ discussed above as a subset of $M_n(\mathbb{C})$ by mapping any element $f=(f_1,\ldots,f_n)$ into the diagonal matrix $D(f)=\mathrm{diag}(f_1,\ldots,f_n)$. Moreover, $D(f\cdot g)=D(f)D(g)$ and the standard product of matrices restricted to diagonal ones becomes the Hadamard product. Thus, the Abelian algebra $\mathcal{F}(n)=\mathbb{C}^n$ is a subalgebra[2] of $M_n(\mathbb{C})$.

With the previous examples in mind we may define algebras formally.

Definition 10.1 An algebra A is a complex linear space together with a product $\star$, such that $(A,+,\star)$ is a ring and, in addition, satisfies the following compatibility properties with respect to the product by scalars:

$$(\lambda x)\star y=x\star(\lambda y)=\lambda(x\star y)\,,\quad \forall x,y\in A\,,\quad \lambda\in\mathbb{C}\,.$$

Notice that $1x=x$ and $0\star x=0x=0$. If the ring $(A,+,\cdot)$ has a unit element $\mathbf{1}$, i.e., $\mathbf{1}\star x=x\star\mathbf{1}$ for all $x\in A$, the algebra is called unital. An algebra such that the product is commutative, $x\star y=y\star x$, $x,y\in A$, will be called commutative or Abelian.

Notice that if the algebra A has a unit $\mathbf{1}$, it is unique. Any algebra A can be turned into a unital algebra. If A does not have a unit element, we may consider the extended algebra $\tilde{A}=A\oplus\mathbb{C}$, with the product $(x,\lambda)\star(y,\mu)=(x\star y+\lambda y+\mu x,\lambda\mu)$.

Exercise 10.1 *Prove that the linear space $\tilde{A}=A\oplus\mathbb{C}$ with the product $\star$ defined above is an algebra and the element $\mathbf{1}=(0,1)$ is a unit.*

[2]However, as it will be discussed later on, this is not the proper substructure to look at when dealing with algebras, but that of ideals. In this case, it can be proved that $M_n(\mathbb{C})$ has no proper ideals, see later on Prop. 10.1 and the algebra is simple. Actually, all simple finite algebras are exactly of this form, Thm. 11.6).

In what follows, we will assume that all abstract algebras we use will be unital and finite-dimensional.

The algebra of functions $\mathcal{F}(\Omega)$ is an Abelian algebra, as the algebra of complex polynomials $\mathbb{C}[x]$ in the variable x, while the algebra of matrices $M_n(\mathbb{C})$ is not Abelian.

Example 10.2 *For any algebra A, we can define another algebra denoted by* A^{op} *which is defined over the same linear space as A, but whose product is given by* $x \star_{\mathrm{op}} y = y \star x$. *The algebra* A^{op} *will be called the opposite algebra of A. It is obvious that* $(A^{\mathrm{op}})^{\mathrm{op}} = A$.

Exercise 10.3 *Prove that* $(A, +, \star_{\mathrm{op}})$ *is an algebra with unit the unit element of A.*

10.1.2 Ideals

Definition 10.2 Given an algebra A, a subalgebra B of A is a subset $B \subset A$ such that B is a linear subspace of A, $B \star B \subset B$, and the linear space B with the operation $\star$ is an algebra itself.

Exercise 10.4 *Prove that* $B \subset A$ *is a subalgebra if* $\lambda b \in B$ *for any* $b \in B$, $\lambda \in \mathbb{C}$, $b - b' \in B$ *and* $b \star b' \in B$, *for all* $b, b' \in B$.

Example 10.5

1. *Let* Ω *be a set and* $\sim$ *an equivalence relation on* Ω. *Consider the set of functions* $\mathcal{F}(\Omega/\sim)$ *of functions in* Ω *which are constant on equivalence classes, i.e.,* $f \in \mathcal{F}(\Omega/\sim)$ *if* $x \sim y$, *then* $f(x) = f(y)$. *Clearly* $\mathcal{F}(\Omega/\sim)$ *is a subalgebra of* $\mathcal{F}(\Omega)$. *Note that* $\mathcal{F}(\Omega/\sim)$ *is unital as* $\mathbf{1} \in \mathcal{F}(\Omega/\sim)$.

   However, note that if $\Omega' \subset \Omega$, *the algebra* $\mathcal{F}(\Omega')$ *is not a subalgebra of* $\mathcal{F}(\Omega)$ *(a way to extend the functions in* Ω' *to* Ω *must be given).*

2. *Consider the algebra* $M_n(\mathbb{C})$ *of matrices, then the set of upper triangular matrices is a subalgebra (clearly, it is a linear subspace and the product of two upper triangular matrices is upper triangular). The same happens with diagonal matrices.*

Exercise 10.6 *Find two subalgebras* M, N *of the algebra of matrices* $M_n(\mathbb{C})$ *such that* $M_n(\mathbb{C}) = MN$, *that is, such that any matrix A can be written as the product of a matrix in M and another one in N and such that* $M \cap N = 0$.

As in the case of groups and, more generally, groupoids, where subsets generate subgroups or subgroupoids, respectively, any subset $S \subset A$ determines a subalgebra $A(S)$ of A called the subalgebra generated by it and defined as the smallest subalgebra of A containing S. Its explicit description is given by all elements of the form $x = \lambda_0 \mathbf{1} + \lambda_i s_i + \lambda_{ij} s_i \star s_j + \cdots + \lambda_{i_1 \ldots i_r} s_{i_1} \star s_{i_2} \star \cdots \star s_{i_r}$, with $r \in \mathbb{N}$, $\lambda_0, \lambda_i, \lambda_{ij}, \ldots, \lambda_{i_1 \ldots i_r} \in \mathbb{C}$ and $s_i, \ldots, s_{i_k} \in S$. An element x of the form before will be called a polynomial on the elements of S.

Note that if the elements $s_i \in S$ commute with each other, i.e., $s_i \star s_j = s_j \star s_i$, for all $s_i, s_j \in S$, then the polynomials x on the elements $s_i \in S$ are the same as the standard polynomials on the variables s_i and the algebra generated by S is just the algebra of standard polynomials on S, i.e., $A(S) = \mathbb{C}[S]$.

Exercise 10.7 *Show that the intersection of subalgebras is a subalgebra again. Then prove that, given a set $S \subset A$, the subalgebra generated by S is the intersection of all subalgebras containing S.*

Interesting as it is, the notion of subalgebra is not the key to unlocking the structure of algebras. As in the case of groups and groupoids, the natural equivalence relation induced by a subgroup is not, in general, compatible with the group structure. Normal subgroups (or normal subgroupoids) are those such that the set of equivalence classes inherit a group (or groupoid) structure from the corresponding group or groupoid. Only in the case of Abelian groups, any subgroup is (obviously) normal. Indeed, any subspace of a linear space is a normal subgroup of the total linear space with respect to the Abelian group structure provided by vector addition. In this sense, any subalgebra is a linear subspace of the total algebra, so it is the product operation that causes trouble.

If we are given a linear subspace V of an algebra A, the natural equivalence relation induced by it is that two elements $x, y \in A$ are equivalent if $y - x \in V$. Thus, equivalence classes have the form $x + V$. Then, if we want to multiply two equivalence classes, say $x + V$ and $y + V$, the natural choice would be $(x+V) \star (y+V) = x \star y + V$. However, this will not work unless $x \star V \subset V$ (and $V \star y \subset V$ too), in which case, the product operation among classes is well defined and satisfies the axioms of an algebra themselves. This leads to the notion of ideal of a given algebra.

Definition 10.3 Given an algebra A, a left ideal I is a subspace $I \subset A$ such that $x \star a \in I$ for all $x \in A$, $a \in I$. In an analogous way, we define a right ideal as a subspace $I \subset A$, such that $a \star x \in I$ for all $x \in A$, $a \in I$. An ideal $I \subset A$ will be called bilateral, or just an ideal if there is no risk of confusion, if it is both a left and a right ideal. Sometimes bilateral ideals will be denoted as $I \triangleleft A$.

Notice that (left, right and bilateral) ideals are subalgebras themselves. However, not all subalgebras are ideals. For instance, consider the subspace of diagonal matrices $D(n)$ of the algebra $M_n(\mathbb{C})$. Such subspace is a subalgebra (the product of two diagonal matrices is again diagonal) but it is not an ideal (the product with any matrix with off diagonal terms turns the diagonal matrix into a non-diagonal one).

Of course, the total algebra and the zero subspace are bilateral ideals of any algebra. A nonzero bilateral ideal different from the total algebra will be called proper. Note that a proper ideal cannot contain the unit **1** as it would be the total algebra.

Definition 10.4 Let A be an algebra, then an element $x \in A$ will be called invertible if there exists x', such that $x \star x' = x' \star x = \mathbf{1}$. In such case, the element x' will be denoted by x^{-1} and is unique. Note that $\mathbf{1}^{-1} = \mathbf{1}$.

Exercise 10.8 *Prove that the set of invertible elements on the algebra A form a group. Such group will be called the group of invertible elements of the algebra A.*

Exercise 10.9 *Prove that if the ideal I contains an invertible element then $I = A$.*

Proposition 10.1
The algebra $M_n(\mathbb{C})$ has no proper bilateral ideals.

Proof 10.1 We prove it by induction on n. For $n = 1$, the assertion is trivial since the algebra $M_1(\mathbb{C}) = \mathbb{C}$ does not have proper ideals.

Let us assume that the statement is true for n. The algebra $M_n(\mathbb{C})$ can be considered as a subalgebra of $M_{n+1}(\mathbb{C})$ in a natural way identifying any $n \times n$ matrix A with the $(n+1) \times (n+1)$ matrix:

$$\tilde{A} = \begin{bmatrix} A & 0 \\ 0 & 0 \end{bmatrix}.$$

Let J be a proper ideal of $M_{n+1}(\mathbb{C})$. Consider the intersection $J' = J \cap M_n(\mathbb{C})$. Clearly J' is an ideal of $M_n(\mathbb{C})$, hence, because of the induction hypothesis, it cannot be proper.

If $J' = \{\mathbf{0}\}$, then any matrix in J must be zero except for entries in the last row or the last column. However, it is trivial to check that the product of any such matrix by the appropriate matrices in $M_{n+1}(\mathbb{C})$ will create nonzero entries in $M_n(\mathbb{C})$, which is impossible.

If, alternatively, $J' = J \cap M_n(\mathbb{C}) = M_n(\mathbb{C})$, take the unit matrix $I_n \in M_n(\mathbb{C}) \in J$. Then, multiplying it by the matrix in $M_{n+1}(\mathbb{C})$ that exchanges the last two rows, we get that the matrix:

$$\tilde{I}_n = \begin{bmatrix} 1 & 0 & \cdots & & \\ 0 & 1 & \cdots & & \\ & & \ddots & & \\ & & \cdots & 0 & 0 \\ & & \cdots & 0 & 1 \end{bmatrix}.$$

is in J, but then the invertible matrix $I_n + \tilde{I}_n$ is in J and $J = M_{n+1}(\mathbb{C})$, which again contradicts our hypothesis.

On the contrary, the algebra $\mathcal{F}(\Omega)$ of functions on a given set X has many ideals, for instance, given a point $x \in X$, the set $\mathcal{J}_x$ of functions vanishing at x is a bilateral ideal. Not only that, such ideal is maximal in the sense that, if J is another ideal, such that $\mathcal{J}_x \subset J$, then either $J = \mathcal{J}_x$ or $J = \mathcal{F}(\Omega)$.

Example 10.10 *1. Consider the Abelian algebra of polynomials $\mathbb{C}[x]$ on the variable x, then any ideal $I \triangleleft \mathbb{C}[x]$ is of the form $I = p(x)\mathbb{C}[x]$ for some polynomial $p(x)$. Ideals of this form are said to be principal and the algebra $\mathbb{C}[x]$ is said to be a domain of principal ideals. The proof of this fact follows the same pattern as the proof that any ideal of the ring $\mathbb{Z}$ is of the form $n\mathbb{Z}$, sometimes denoted (n).*

Exercise 10.11 *Prove that any ideal in* $\mathbb{Z}$ *is principal and use it as a model to prove that any ideal in* $\mathbb{C}[x]$ *is principal.*

2. *Consider the algebra* $M_{n_1,\ldots,n_r}(\mathbb{C})$ *of block diagonal matrices of sizes* $n_1,\ldots,n_r$, $n_1+\cdots+n_r=n$. *That is, a matrix* $A\in M_{n_1,\ldots,n_r}(\mathbb{C})$ *is such that its non-zero elements* A_{ij} *are those such that* $0<i,j\leq n_1,\ldots,n_1+\cdots+n_k<i,j\leq n_1+\cdots+n_{k+1}$, $k=1,\ldots,r-1$. *We will show later that* $M_{n_1,\ldots,n_r}(\mathbb{C})=M_{n_1}(\mathbb{C})\oplus\cdots\oplus M_{n_r}(\mathbb{C})$. *The set of matrices corresponding to a single block* $M_{n_k}(\mathbb{C})$, *form a subalgebra and they define an ideal.*

Remark 10.1 The set of integer numbers (or the set of rational numbers) do not form an algebra, they are just a ring, as the product by scalars (complex numbers) is not defined. In spite of this the notion of ideal applies in exactly the same way for rings and is, as in the case of algebras, the main conceptual tool to discuss their structure. Thus, for instance, the subgroups $n\mathbb{Z}$ are ideals of the ring $\mathbb{Z}$. ▮

As in the case of subalgebras, given a subset $S\subset A$, there is a natural ideal associated to S (we will consider bilateral ideals only). We will call it the ideal generated by S and it will be denoted as $I(S)$ or just (S). It is defined as the smallest ideal in A containing S. Clearly the elements of the ideal $I(S)$ are all of the form $a_1\star s_1\star a_1'+\cdots a_r\star s_r\star a_r'$, for all $r\in\mathbb{N}$, $s_k\in S$, $a_k,a_k'\in A$ for all $k=1,\ldots,r$.

Exercise 10.12 *Prove that the intersection of ideals is again an ideal and that, as a consequence, the ideal generated by a set is the intersection of all ideals containing S.*

Proposition 10.2
Let A be an algebra and $I\subset A$ *a bilateral ideal. Then, the space of equivalence classes of elements in A determined by I as a subspace of A, denoted as* A/I, *naturally inherits the structure of an algebra from the operations of A. Such algebra will be called the quotient algebra of A with respect to the ideal I.*

Proof 10.2 As an ideal I is a subspace of the linear space structure of A, it induces an equivalence relation given by $x\sim y$ iff $y-x\in I$ or, in other words, two elements are equivalent if we can get one from the other by adding elements in I. Thus, the class determined by I containing the element $x\in A$ is the set $\{x+a\mid a\in I\}$, commonly denoted by $x+I$.

There is a natural addition operation induced by the addition in the algebra, given by:

$$(x+I)+(y+I)=(x+y)+I,$$

and the space of classes A/I with the operation $+$ is a linear space (see Appendix A, Prop. A.3).

We would like to define the product of two classes as:

$$(x+I)\star(y+I)=x\star y+I,$$

(i.e., the product of the two classes is determined by the product of any two representatives of them). That this product is well defined is a direct consequence of the defining property of ideals, that is, if x and x' define the same class, $x+I=x'+I$, i.e., $x'=x+a$ for some $a\in I$, then $(x'+I)\star(y+I)=(x+a)\star y+I=x\star y+a\star y+I=x\star y+I=(x+I)\star(y+I)$ (notice $a\star y\in I$ because I is a right-ideal). Similarly, if $y'+I=y+I$, then there will be an element $a'\in I$ such that $y'=y+a'$, hence we get $(x+I)\star(y'+I)=(x+I)\star(y+I)$ because $x\star a'\in I$ as I is a left-ideal.

The rest of the proof is a routine checking of the algebraic properties of the natural operations $+$ and $\star$ induced in the space of classes.

Remark 10.2 It is relevant to point out that if A is an algebra with unit element $\mathbf{1}$, then A/I has a unit element too, given by the class $\mathbf{1}+I$. ∎

Example 10.13

1. *Consider the ideal $\mathcal{J}_x\subset\mathcal{F}(\Omega)$, then the quotient algebra $\mathcal{F}(\Omega)/\mathcal{J}_x$ can be identified with the trivial one-dimensional algebra $\mathbb{C}$ (the identification being the assignment to any class $f+\mathcal{J}_x$ the value $f(x)$ of the function f at x.*

2. *Consider the ideal $\mathcal{J}_S\subset\mathcal{F}(\Omega)$ of functions vanishing on the subset $S\subset\Omega$. Then, the quotient algebra $\mathcal{F}(\Omega)/\mathcal{J}_S$ can be identified with the algebra $\mathcal{F}(S)$ of functions on S. The assignment given by $f+\mathcal{J}_S\mapsto f\mid_S$.*

3. *Consider the ideal $(x^2)\subset\mathbb{C}[x]$, where (x^2) denotes the ideal generated by the polynomial x^2, in other words, (x^2) is the set of all polynomials of the form $x^2p(x)$. Then, the quotient algebra $\mathbb{C}[x]/(x^2)$ is two-dimensional. Its elements have the form $a\mathbf{1}+b\mathbf{x}$, where $\mathbf{1}$ denotes the class corresponding to the polynomial 1, that is $\mathbf{1}=1+(x^2)$, and $\mathbf{x}$ denotes the class corresponding to the polynomial x, i.e., $\mathbf{x}=x+(x^2)$. Notice that $\mathbf{x}\cdot\mathbf{x}=0$.*

Exercise 10.14 *An interesting, but harder exercise, consists of describing the quotient algebra $\mathbb{C}[x]/(p(x))$ where $p(x)$ is an arbitrary complex polynomial with simple roots. Hints: start describing the quotient algebras $\mathbb{C}[x]/(x-a)$ and $\mathbb{C}[x]/((x-a)(x-b))$.*

10.1.3 Morphisms of algebras

As well as with groups and groupoids, there is a natural notion of maps between algebras, we will call them homomorphisms of algebras and they are the natural morphisms of a category whose objects are the algebras themselves.

Definition 10.5 Given two algebras A, B, a map $\varphi\colon A \to B$ is a homomorphism of algebras if φ is a linear map between the corresponding linear structures, i.e, $\varphi(x+y) = \varphi(x) + \varphi(y)$, $\varphi(\lambda x) = \lambda \varphi(x)$ for all $x, y \in A$, $\lambda \in \mathbb{C}$, and φ is a ring homomorphism between the corresponding ring structures, that is:

$$\varphi(x \star y) = \varphi(x) \star \varphi(y), \qquad \forall x, y \in A.$$

Notice that if A and B have units $\mathbf{1}_A$, $\mathbf{1}_B$, respectively, then, if φ is an algebra homomorphism, $\varphi(\mathbf{1}_A) = \mathbf{1}_B$. Moreover, if x is an invertible element, i.e., there exists $x^{-1} \in A$, such that $x \star x^{-1} = x^{-1} \star x = \mathbf{1}_A$, then $\varphi(x)$ is invertible too and $\varphi(x^{-1}) = (\varphi(x))^{-1}$.

Definition 10.6 A homomorphism of algebras $\varphi\colon A \to B$ which is a bijection, will be called an isomorphism, in which case, we will also write $A \cong_\varphi B$, or just $A \cong B$.

As in the case of homomorphisms of groups, we define $\ker\varphi$ as the set of elements $a \in A$ which are mapped into 0, that is, $\ker\varphi = \{a \in A \mid \varphi(a) = 0\}$. In the same way, we define $\operatorname{ran}\varphi$ as the range $\varphi(A)$ of φ.

Proposition 10.3
(First isomorphism theorem for algebras) Let $\varphi\colon A \to B$ an algebra homomorphism, then $\ker\varphi$ is a bilateral ideal of A; $\varphi(A) \subset B$ is a subalgebra of B, and $A/\ker\varphi \cong \varphi(A)$ are isomorphic as algebras.

Exercise 10.15 *Prove Prop. 10.3. Prove that $\varphi\colon A \to B$ is injective, hence, we will call it a monomorphism, if and only if $\ker\varphi = 0$. Prove that the canonical projection $\pi\colon A \to A/I$, where $I \lhd A$, is a homomorphism of algebras.*

Example 10.16 *Consider the algebra $M_n(\mathbb{C})$ of square matrices. The algebra $M_n(\mathbb{C})^{\mathrm{op}}$ is canonically isomorphic to $M_n(\mathbb{C})$, the isomorphism provided by the map $A \mapsto A^T$, that is, each matrix A is mapped into its transpose. Then, $(A \star B)^T = B^T \star A^T = A^T \star_{\mathrm{op}} B^T$ (where we are denoting by $\star$ the standard product of matrices). Thus, $M_n(\mathbb{C})^{\mathrm{op}} \cong M_n(\mathbb{C})$.*

10.1.4 Operations with algebras

As in the case of linear spaces, Appendix A, Sect. A.3, and linear representations, there are some basic operations with algebras that will be used often in what follows.

Direct sum

Given two algebras A, B, we define the direct sum $A \oplus B$ of them, as the algebra defined on the direct sum of the corresponding linear spaces with product:

$$(a,b) \star (a',b') = (a \star a', b \star b'), \qquad a, a' \in A, \quad b, b' \in B.$$

It is an easy exercise to check that the structure thus defined is actually an algebra. It is common, as in the case of linear spaces, to denote the elements $(a,b) \in A \oplus B$ as $a \oplus b$ (note that $\lambda(a \oplus b) = \lambda a \oplus \lambda b$). If A and B are unital, the unit in $A \oplus B$ is $\mathbf{1}_A \oplus \mathbf{1}_B$.

The subsets $\{(a,0) \mid a \in A\} \subset A \oplus B$ and $\{(0,b) \mid b \in B\} \subset A \oplus B$, are ideals of $A \oplus B$, denoted just as A and B, respectively, and $A \oplus B/A = B$, $A \oplus B/B = A$.

Example 10.17 *Consider again the algebra $M_{n_1,n_2,\ldots,n_r}(\mathbb{C})$ of block diagonal matrices of orders $n_1, n_2, \ldots, n_r$, $n_1 + \cdots + n_r = n$ discussed in Example 10.10.2. It is clear that this algebra is (isomorphic to) the direct sum $M_{n_1}(\mathbb{C}) \oplus \cdots \oplus M_{n_r}(\mathbb{C})$.*

Tensor product

Given two algebras A, B, we define the tensor product $A \otimes B$ of them as the algebra defined on the tensor product of the corresponding linear spaces and product $\star$ defined on monomials as:

$$(a \otimes b) \star (a' \otimes b') = (a \star a') \otimes (b \star b'), \qquad a, a' \in A, \quad b, b' \in B.$$

The product extends by linearity to arbitrary elements $a_1 \otimes b_1 + \cdots + a_r \otimes b_r$ in the tensor product $A \otimes B$.

Again, it is an easy exercise to check that the product thus defined on the tensor product of the linear spaces A and B satisfies the axiom of an algebra. If both A and B are unital, then $A \otimes B$ is unital too, and its unit is given by $\mathbf{1}_A \otimes \mathbf{1}_B$.

It is important to note that $a \otimes 0 = 0$, then we cannot define subalgebras A and B of $A \otimes B$ (much less ideals) as we did in the case of the direct sum. However if the algebras A, B are unital we may define the subalgebras $A \otimes \mathbf{1}_B = \{a \otimes \mathbf{1}_B \mid a \in A\}$ and $\mathbf{1}_A \otimes B$. However, they are not ideals.

Exercise 10.18 *Show that $A \otimes \mathbf{1}_B$ and $\mathbf{1}_A \otimes B$ are actually subalgebras of $A \otimes B$, but that they are not ideals.*

Example 10.19 *It is interesting to check that if Ω and Ω' are finite sets, then $\mathcal{F}(\Omega) \otimes \mathcal{F}(\Omega') \cong \mathcal{F}(\Omega \times \Omega')$, the isomorphism provided by the identification $(f \otimes g)(x,x') = f(x)g(x')$, $f \in \mathcal{F}(\Omega)$, $g \in \mathcal{F}(\Omega')$, $x \in \Omega$, $x' \in \Omega'$.*

Example 10.20 *It is not difficult to see that $M_n(\mathbb{C}) \otimes M_m(\mathbb{C}) \cong M_{nm}(\mathbb{C})$, however, contrary to the previous example, the isomorphism is not canonical.*

Remark 10.3 Contrary to the situation with linear spaces, where, given a linear space V, its covectors define another linear space V^* called the dual of V, the dual of an algebra is not an algebra[3]. ∎

---

[3]It inherits the structure of a coalgebra, but we will not discuss this issue here.

Operations with homomorphisms

If $\varphi\colon A \to B$ and $\varphi'\colon A' \to B'$ are homomorphisms of algebras we can define the corresponding homomorphisms between the algebras $A \oplus B$ and $A' \oplus B'$ and $A \otimes B$ and $A' \otimes B'$, respectively, as follows:

$$(\varphi \oplus \varphi')(a \oplus a') = \varphi(a) \oplus \varphi'(a'),$$

$$(\varphi \otimes \varphi')(a \otimes a') = \varphi(a) \otimes \varphi'(a'), \quad \forall a \in A, a' \in A'.$$

Exercise 10.21 *Prove that $\varphi \oplus \varphi'$ and $\varphi \otimes \varphi'$ are homomorphisms of algebras. Describe* $\ker(\varphi \oplus \varphi')$ *and* $\ker(\varphi \otimes \varphi')$.

Example 10.22 *At the begining of this section, we presented the algebra of matrices $M_n(\mathbb{C})$ as the fundamental model for non-commutative algebras. We may consider it from a slightly more abstract viewpoint. Let V be a linear space. The space of endomorphisms of V, i.e., the space of all linear maps $L\colon V \to V$, form an algebra, denoted by* $\mathrm{End}(V)$ *(Appendix A.4.2).*

Now, if $L\colon V \to W$ is a linear isomorphism, then it induces a natural map $\varphi_L\colon \mathrm{End}(W) \to \mathrm{End}(V)$ *by $\varphi_L(M) = L^{-1} \circ M \circ L$. It is easy to check that φ_L is an isomorphism of algebras.*

Exercise 10.23 *Prove that the only homomorphisms of algebras* $\varphi\colon \mathrm{End}(W) \to \mathrm{End}(V)$ *are isomorphisms of the form described in Example 10.22.*

After all these preparations, it is already time to introduce some specially interesting examples of algebras, the algebras defined by categories and groupoids.

10.2 The algebra of a category

Let $\mathbf{C}$ be a category, finite if you wish, with objects denoted by $x, y, \ldots$, and arrows denoted $\alpha, \beta, \ldots$. Consider (similar to Example 7.1.5 in Sect. 7.1.2) the complex linear space generated by the arrows α, that is, consider the set of all formal finite linear combinations:

$$\boldsymbol{\psi} = \sum_{k=1}^{s} \psi_k \, \boldsymbol{\alpha}_k,$$

with $s \in \mathbb{N}$, and ψ_k complex numbers. Observe that we are denoting the arrows $\boldsymbol{\alpha}$ in boldface to indicate that they are being considered as abstract symbols, but retaining its algebraic nature (contrary to the notation used in Example 7.1.5 where the notation $|k\rangle$ was intended to represent abstract vectors). Even if the sums are finite and the arrows in the category can be labeled as above by the numbers $1, 2, \ldots, s$, $s < \infty$, sometimes we will prefer a more abstract notation for the elements $\boldsymbol{\psi}$, as follows:

$$\boldsymbol{\psi} = \sum_{\alpha \in \mathrm{Mor}(\mathbf{C})} \psi_\alpha \, \boldsymbol{\alpha},$$

where all coefficients ψ_α are null except a finite number of them in case that we were considering an infinite category. In what follows, the reference to Mor(**C**) in the sums will be omitted unless it could create confusions.

The sum and multiplication by scalars are defined as:

$$\psi + \varphi = \sum_\alpha (\psi_\alpha + \varphi_\alpha)\, \alpha \,,$$

and,

$$\lambda \psi = \sum_\alpha (\lambda \psi_\alpha) \alpha \,.$$

We will denote such space as $\mathbb{C}[\mathbf{C}]$. There is a canonical composition law in $\mathbb{C}[\mathbf{C}]$ given by:

$$\psi \cdot \varphi = \sum_{\gamma = \alpha \circ \beta} (\psi_\alpha \varphi_\beta)\, \gamma \,, \tag{10.1}$$

where the sum runs over all pairs of arrows α, β in **C** that can be composed, i.e., $\alpha \colon x \to y$, and $\beta \colon y \to z$. Alternatively, we may have written:

$$\psi \cdot \varphi = \sum_{\alpha,\beta} (\psi_\alpha \varphi_\beta)\, \delta_{\alpha,\beta}\, \alpha \circ \beta \,,$$

where $\delta_{\alpha,\beta}$ is the indicator that takes the value 1 if α and β can be composed and zero otherwise.

Proposition 10.4

*Let **C** be a category. The space $\mathbb{C}[\mathbf{C}]$ with the composition law $\cdot$ given in Eq. (10.1) is an associative algebra, called the algebra of the category **C**. Moreover, if the category **C** has a finite number of objects, the algebra $\mathbb{C}[\mathbf{C}]$ is unital with unit*

$$\mathbf{1} = \sum_{x \in \mathrm{Ob}(\mathbf{C})} 1_x \,.$$

Proof 10.3 The proof is 'mechanical', that is, there are no conceptual difficulties involved, just a routine checking of the axioms defining an algebra.

Thus, for instance, we check the associativity of the product:

$$(\psi \cdot \varphi) \cdot \xi = \sum_{\alpha,\beta} (\psi_\alpha \varphi_\beta)\, \delta_{\alpha,\beta}\, (\alpha \circ \beta) \cdot \xi = \sum_{\alpha,\beta,\gamma} (\psi_\alpha \varphi_\beta \xi_\gamma) \delta_{\alpha,\beta} \delta_{\alpha\circ\beta,\gamma} (\alpha \circ \beta) \circ \gamma \,. \tag{10.2}$$

The composition law $\circ$ is associative, then $(\alpha \circ \beta) \circ \gamma = \alpha \circ (\beta \circ \gamma)$. Moreover, the factor $\delta_{\alpha,\beta} \delta_{\alpha\circ\beta,\gamma}$ will be different from zero only if both $\delta_{\alpha,\beta}$ and $\delta_{\alpha\circ\beta,\gamma}$ are different from zero, but if $\delta_{\alpha\circ\beta,\gamma}$ is different from zero, this means that β can be composed with γ, then the previous factor will be different from zero only if α can be composed with β and β with γ. However, this is clearly the same conclusion we get examining the factor $\delta_{\alpha,\beta\circ\gamma}\delta_{\beta,\gamma}$, then we get that the right hand side of the expression (10.2) can be written as:

$$\sum_{\alpha,\beta,\gamma} \psi_\alpha \varphi_\beta \xi_\gamma \, \delta_{\alpha,\beta\circ\gamma}\delta_{\beta,\gamma} \, \alpha \circ (\beta \circ \gamma) = \psi \cdot (\varphi \cdot \xi) \, .$$

It was claimed that the algebra $\mathbb{C}[\mathbf{C}]$ is unital. Clearly, $\mathbf{1} \cdot \varphi = \sum_{x,\alpha} \varphi_\alpha \, \delta_{1_x,\alpha} \, 1_x \circ \alpha = \sum_\alpha \varphi_\alpha \, \alpha = \varphi$. Similarly, it is proved that $\varphi \cdot \mathbf{1} = \varphi$ for every φ.

Remark 10.4 Note that the unit element $\mathbf{1}$ can always be constructed whenever the collection of objects of the category is finite. As this is not always the case, consider for instance the category of pairs of integer numbers $\mathbf{G}(\mathbb{Z})$, then the algebra $\mathbb{C}[\mathbf{G}(\mathbb{Z})]$ does not have a unit element (what would be in such case the meaning of $\sum_{k\in\mathbb{Z}} 1_k$?). There is, however, a formal way of adding a unit to any algebra, as we did right after Def. 10.1. Thus, we will always assume that our algebras are unital. ∎

As indicated in the begining of the proof of the previous theorem, it seems that there is no conceptual content in the proof that $\mathbb{C}[\mathbf{C}]$ is an associative algebra but only some more or less routine computations. However, the algebra of a category plays a significant role in what we are doing, then the previous proposition could not be just the result of mechanical computations. Actually, the previous proposition is just a particular setting of a general abstract construction of algebras from a set of generators and relations.

Thus, the algebra $\mathbb{C}[\mathbf{C}]$ is just the associative algebra generated by the arrows of the finite category $\mathbf{C}$ and the relations provided by the composition of arrows. In this sense, the notation $\mathbb{C}[\mathbf{C}]$ is consistent with the notation used for the algebra of polynomials $\mathbb{C}[x]$, where now such algebra is just the algebra generated by a single arrow x over a single object, that is, our 'quiver loop', which generates the category denoted by $\mathbf{P_{1,1}}$, recall Sect. 1.1.2, then we have that the algebra of complex polynomials $\mathbb{C}[x]$ is just the algebra of the category $\mathbf{P}_{1,1}$ generated by the loop.

Example 10.24 *The singleton algebra. Let us consider first, the algebra of the singleton, that is, the category generated by the quiver consisting of two dots and a single link,* $\mathbf{P}_{2,1}$, *Sect. 1.1.2. We denote the two objects by the symbols x, y and the arrows as 1_x, 1_y and $\alpha \colon x \to y$. Elements on $\mathbb{C}[\mathbf{P}_{2,1}]$ will have the form:*

$$\varphi = \varphi_x \mathbf{1}_x + \varphi_y \mathbf{1}_y + \varphi_\alpha \, \alpha \, .$$

This algebra has dimension 3 and a basis is provided by $\mathbf{1}_x$, $\mathbf{1}_y$ and α. Given an algebra A, whenever we have a linear basis $\{e_k\}$ for it, we can compute the family of coefficients c^l_{jk} given by the composition of the basis elements,

$$e_j \cdot e_k = \sum_l c^l_{jk} e_l \, ,$$

called the structure constants of the algebra (with respect to the given basis). In our case, we get:

$$\mathbf{1}_x \cdot \mathbf{1}_x = \mathbf{1}_x \quad \mathbf{1}_y \cdot \mathbf{1}_y = \mathbf{1}_y\,, \quad \mathbf{1}_x \cdot \alpha = \alpha\,, \quad \alpha \cdot \mathbf{1}_y = \alpha\,,$$

and all other products vanish. Using a more abstract notation, $e_1 = \mathbf{1}_x$, $e_2 = \mathbf{1}_y$, $e_3 = \alpha$, *we get:*

$$e_1 \cdot e_3 = e_3\,, \quad e_3 \cdot e_2 = e_3\,, \quad e_1 \cdot e_1 = e_1\,, \quad e_2 \cdot e_2 = e_2\,. \tag{10.3}$$

and $e_1 \cdot e_2 = 0$, $e_3 \cdot e_1 = 0$, $e_3 \cdot e_3 = 0$, *etc.*

Just looking at the structure constants of an algebra we can learn a lot about it and its properties. For instance, if we would like to find a linear representation of our algebra then we realise just looking at the formulae in (10.3) that the elements e_1 *and* e_2 *should be represented by projectors, that the element* e_3 *should be represented by a nilpotent operator, etc. Actually, it is simple to check that the linear span of the following* 2×2 *matrices provide a representation of our algebra* $\mathbb{C}[\mathbf{P}_{2,1}]$:

$$E_1 = \begin{bmatrix} 1 & 0 \\ 0 & 0 \end{bmatrix}, \quad E_2 = \begin{bmatrix} 0 & 0 \\ 0 & 1 \end{bmatrix}, \quad E_3 = \begin{bmatrix} 0 & 1 \\ 0 & 0 \end{bmatrix}.$$

Actually, this representation provides a two-dimensional irreducible representation of the singleton category.

Example 10.25 *The loop category. The algebra of the loop category corresponds to the additive monoid of natural numbers* $\mathbb{N}$*, again, recall Sect. 1.1.2. There is just one object, and the arrows are labeled by the natural numbers* n *with composition given by addition. An element* ψ *in* $\mathbb{C}[\mathbf{P}_{1,1}]$ *will be an expression of the form:*

$$\psi = \sum_{n\in\mathbb{N}} \psi_n\, \mathbf{n}\,,$$

where only a finite number of the coefficients ψ_n *are different from zero. Notice that composition is given by:*

$$\psi \cdot \phi = \sum_{n,m} \psi_n \phi_m\, (\mathbf{n}+\mathbf{m})\,.$$

10.3 The algebra of a groupoid

10.3.1 The ⋆-algebra of a finite groupoid

As groupoids are categories whose arrows are all invertible, we simply take the notion of the algebra of a category to define the algebra of a groupoid.

Definition 10.7 Given a groupoid $\mathbf{G} \rightrightarrows \Omega$ we will denote by $\mathbb{C}[\mathbf{G}]$ the algebra associated to it, called the algebra of the groupoid, which is defined as the algebra generated by the elements α in the groupoid and relations provided by the groupoid composition law. That is, elements a of the algebra $\mathbb{C}[\mathbf{G}]$ are formal linear combinations of elements in $\mathbf{G}$ with complex coefficients: $a = \sum_{\alpha \in \mathbf{G}} a_\alpha \, \alpha$, $a_\alpha \in \mathbb{C}$. The composition law is given by:

$$a \cdot b = \sum_{(\beta,\alpha) \in \mathbf{G}_2} (a_\alpha \, b_\beta) \, \alpha \circ \beta \,, \tag{10.4}$$

with $a = \sum_\alpha a_\alpha \, \alpha$, $b = \sum_\beta b_\beta \, \beta$.

If the groupoid $\mathbf{G} \rightrightarrows \Omega$ is finite, the algebra $\mathbb{C}[\mathbf{G}]$ is unital and finite dimensional with dimension $|\mathbf{G}|$.

Notice that the composition of morphisms on the r.h.s. of Eq. (10.4) is only among composable pairs α, β and we have removed the boldface style in the notation representing the abstract arrows α.

There is, in addition to the structure of associative algebra, a natural antilinear involution map $*$ in $\mathbb{C}[\mathbf{G}]$ defined as:

$$a = \sum_\alpha a_\alpha \, \alpha \mapsto a^* = \sum_\alpha \bar{a}_\alpha \, \alpha^{-1} \, .$$

Exercise 10.26 *Prove that the map* *, *satisfies:* $(a^*)^* = a$, $(a+b)^* = a^* + b^*$, $(\lambda a)^\star = \bar{\lambda} a^\star$, $(a \cdot b)^* = b^* \cdot a^*$, $\mathbf{1}^* = \mathbf{1}$.

The unital associative algebra $\mathbb{C}[\mathbf{G}]$ equipped with the involution $*$ satisfying the properties listed in Exercise 10.26 receives the name of a $*$-algebra[4].

Example 10.27

1. *The algebra of groups. Groups are particular instances of groupoids thus, given a group G, we define the algebra of the group G as its groupoid algebra $\mathbb{C}[G]$. Thus, elements in the group algebra $\mathbb{C}[G]$ are just linear combinations $a = \sum_{g \in G} a_g \, g$. The group algebra of a group plays a relevant role in the study of its linear representations, aspects that will be discussed in the coming chapter.*

2. *The algebra $\mathbb{C}[\mathbb{Z}_n]$. As a particular instance of algebras of groups, we may consider the algebra of the Abelian group $\mathbb{Z}_n$. Its elements will have the form $a = \sum_{k=0}^{n-1} a_k \mathbf{k}$ and the composition law will be $a \cdot b = \sum_{k,l=0}^{n-1} a_k b_l \, (\mathbf{k} + \mathbf{l})$.*

   The algebra $\mathbb{C}[\mathbb{Z}_n]$ is Abelian of dimension n and can be identified with polynomials on the variable x such that $x^n = 1$. This algebra was discussed before (when $n = 2$) in Example 10.13.3 and Exercise 10.14, and is the quotient of the algebra of polynomials $\mathbb{C}[x]$ by the ideal generated by the polynomial $x^n - 1$, that is, $\mathbb{C}[\mathbb{Z}_n] = \mathbb{C}[x]/(x^n - 1)$.

[4] Even more, there is a canonical norm $|| \cdot ||$ in $\mathbb{C}[\mathbf{G}]$ that makes it into a C^*-algebra, even if we will not use this structure in the current work.

3. *We will discuss now what is arguably the most relevant example of a groupoid algebra, the algebra of the groupoid of pairs.*

   Consider a finite set Ω and the groupoid of pairs $\mathbf{G}(\Omega)$. If we label the objects in Ω as x_i, $i = 1, \ldots, n$, $n = |\Omega|$, then the elements in the algebra $\mathbb{C}[\mathbf{G}(\Omega)]$ will have the form:

$$a = \sum_{i,j=1}^{n} a_{ij}\,(i,j)\,,$$

   with a_{ij} complex numbers and (i,j) denoting the pair (x_i, x_j), that is the morphism $(i,j)\colon x_j \to x_i$. The composition law is just:

$$a \cdot b = \sum_{i,j=1}^{n} \left(\sum_{k=1}^{n} a_{ik} b_{kj} \right) (i,j)\,.$$

   In other words, the composition of $a \cdot b$ is the element in the algebra with coefficients $\sum_{k=1}^{n} a_{ik} b_{kj}$. However, this is nothing else but the (i,j) entry of the product of the matrices defined by the coefficients a_{ij} and b_{ij} associated with a and b, respectively. Hence, we conclude that, once a choice in the ordering of the elements in the set Ω is made, the algebra of the groupoid of pairs can be indentified with the algebra of matrices $M_n(\mathbb{C})$.

10.3.2 Operations with groupoids and their algebras

We would like to investigate the structure of the algebras of the groupoids obtained by disjoint union and product of groupoids.

In order to discuss that, we would first establish a general result on the relation between isomorphisms of groupoids and homomorphisms of algebras.

Proposition 10.5
Let $\mathbf{G} = \mathbf{G}_1 \sqcup \mathbf{G}_2$ be the disjoint union of the groupoids $\mathbf{G}_a \rightrightarrows \Omega_a$, $a = 1,2$, then the algebra of the disjoint union is the direct sum of the corresponding algebras, that is $\mathbb{C}[\mathbf{G}_1 \sqcup \mathbf{G}_2] \cong \mathbb{C}[\mathbf{G}_1] \oplus \mathbb{C}[\mathbf{G}_2]$. Clearly, if $\mathbf{G}$ is finite, $\dim \mathbb{C}[\mathbf{G}_1 \sqcup \mathbf{G}_2] = \dim \mathbb{C}[\mathbf{G}_1] + \dim \mathbb{C}[\mathbf{G}_2] = |\mathbf{G}_1| + |\mathbf{G}_2|$.

Proof 10.4 Given an element $a = \sum_{\alpha \in \mathbf{G}} a_\alpha \alpha \in \mathbb{C}[\mathbf{G}]$ there is a canonical decomposition as $a = \sum_{\alpha \in \mathbf{G}_1} a_\alpha \alpha + \sum_{\alpha \in \mathbf{G}_2} a_\alpha \alpha = a_1 + a_2$, that determines the isomorphism between $\mathbb{C}[\mathbf{G}]$ and $\mathbb{C}[\mathbf{G}_1] \oplus \mathbb{C}[\mathbf{G}_2]$.

Example 10.28 *Let $\mathbf{G} \rightrightarrows \Omega$ be a groupoid and $\mathbf{G}_0$ its fundamental subgroupoid, then $\mathbb{C}[\mathbf{G}_0] = \bigoplus_{x \in \Omega} \mathbb{C}[G_x]$.*

Example 10.29 *The direct sum of the algebras of matrices $M_n(\mathbb{C}) \oplus M_m(\mathbb{C})$ is the algebra of the disjoint union of the groupoids of pairs $\mathbf{G}(n) \sqcup \mathbf{G}(m)$.*

Proposition 10.6
Let $\mathbf{G} = \mathbf{G}_1 \times \mathbf{G}_2$ *be the direct product of the finite groupoids* $\mathbf{G}_a \rightrightarrows \Omega_a$, $a = 1, 2$, *then the algebra of the product is the tensor product of the corresponding algebras, that is* $\mathbb{C}[\mathbf{G}] \cong \mathbb{C}[\mathbf{G}_1] \otimes \mathbb{C}[\mathbf{G}_2]$. *Moreover* $\dim \mathbb{C}[\mathbf{G}_1 \times \mathbf{G}_2] = \dim \mathbb{C}[\mathbf{G}_1] \times \dim \mathbb{C}[\mathbf{G}_2] = |\mathbf{G}_1| \times |\mathbf{G}_2|$.

Proof 10.5 The isomorphism is provided by the linear extension of the map $(\alpha_1, \alpha_2) \mapsto \alpha_1 \otimes \alpha_2$ from $\mathbf{G}_1 \times \mathbf{G}_2 \to \mathbb{C}[\mathbf{G}_1] \otimes \mathbb{C}[\mathbf{G}_2]$. The order of the direct product $\mathbf{G}_1 \times \mathbf{G}_2$ is $|\mathbf{G}_1| \times |\mathbf{G}_2|$, which is the dimension of $\mathbb{C}[\mathbf{G}_1] \otimes \mathbb{C}[\mathbf{G}_2]$, hence, the induced map is surjective and it provides the natural identification between the algebras.

Example 10.30 *The algebra of cyclic groupoids. Let* $\mathfrak{C}_n(m)$ *the cyclic groupoid of* n *positions with shift of order* m, *Sect. 3.2.3, Def. 3.14. Then, because of Prop. 3.6,* $\mathfrak{C}_n(m) \cong \mathbf{G}(n) \times \mathbb{Z}_s$, *we get that its algebra is given by* $\mathbb{C}[\mathfrak{C}_n(m)] \cong \mathbb{G}[\mathbf{G}(n)] \otimes \mathbb{C}[\mathbb{Z}_s]$. *Moreover, because of examples 10.27.2 and 10.27.3, we have that* $\mathbb{C}[\mathfrak{C}_n(m)] = M_n(\mathbb{C}) \otimes \mathbb{C}[x]/(x^n)$.

10.4 Representations of Algebras

10.4.1 Modules and representations of algebras

It is a common theme in this book to look for the different realizations in terms of linear objects of a given algebraic structure, presently algebras. Generically, we call any such realization a linear representation of the algebraic structure; in the particular instance of algebras we state:

Definition 10.8 Let A be an algebra. A linear representation of the algebra A is a homomorphism of algebras $\rho : A \to \mathrm{End}(V)$, with V a linear space called the support of the representation ρ. Given a vector $v \in V$, the vector $\rho(a)(v)$ will be denoted, if there is no risk of confusion, as $a \cdot v := \rho(a)(v)$.

Then a linear representation of the algebra A satisfies that $(ab) \cdot v = a \cdot (b \cdot v)$ for all $a, b \in A$, $v \in V$. A linear representation ρ of the algebra A on the linear space V will be denoted as (ρ, V).

There is a more concise and well established, albeit exotic, name for a linear representation of an algebra A: A 'left-A-module'. A left-module is the same as a linear representation of an algebra.

On the other hand, modules, and this is a common way of presenting them, can also be thought, as a generalisation of the notion of linear spaces, where scalars are replaced by elements of an algebra (or a ring). This superposition of abstract notions creates a rich interplay between them that we will explore in what follows.

Let us consider a simple example first. Let V be a finite-dimensional linear space over the real numbers. The essential feature of a linear space is that we can make linear combinations of their elements, that is, we can form the element $\lambda_1 v_1 + \ldots + \lambda_s v_s$ out of the collection of real numbers $\lambda_1, \ldots, \lambda_s$ and vectors $v_1, \ldots, v_s$.

Suppose that, instead of real numbers $\lambda_1, \ldots, \lambda_s$, we restrict ourselves to using only integer numbers $n_1, \ldots, n_s$. We can form linear combinations of the form $n_1 v_1 + \ldots n_s v_s$ with vectors $v_1, \ldots, v_s$ in V. It is clear that all axioms regarding the sum and the multiplication by scalars of the linear space V still hold when we replace the field $\mathbb{R}$ with the ring $\mathbb{Z}$. We will call such structure a (left-) module over the ring $\mathbb{Z}$.

We must point out at this moment that, in spite of the obvious similarity, the structure of the space V when we use the integers instead of the reals as scalars becomes completely different. Let us try to understand why.

It is a well-known fact that any linear space V (real or complex) has a basis (see Appendix A.4, Thm. A.3). In the finite-dimensional situation, this means that there exists a family of linearly independent vectors $u_1, \ldots, u_n$ in V, such that any vector $v \in V$ can be written (in a unique way) as $v = \lambda_1 u_1 + \ldots + \lambda_n u_n$. Once a basis is selected, we may identify V with $\mathbb{R}^n$ by using the coordinates $\lambda_1, \ldots, \lambda_n$ associated to any vector. In particular, if V has dimension one, then any non-zero vector u is a basis and any other vector $v \in V$ can be written as $v = \lambda u$.

In particular, we can identify a one-dimensional real linear space with $\mathbb{R}$, assigning to any vector v the coordinate λ. However, as it was suggested before, we can replace the real numbers by integers as scalars, and consider the one-dimensional linear space $\mathbb{R}$ as a module over the integers[5]. The first question we may try to answer is if there is a basis for it, that is, if there exists a collection of numbers $x_1, \ldots, x_r$, such that any other real number x can be written (uniquely) as:

$$x = n_1 x_1 + \ldots + n_r x_r\,, \tag{10.5}$$

for some collection of integers $n_1, \ldots, n_r$. The answer is clearly not. It can be proved in a number of ways. Perhaps the simplest one is counting how many numbers x can be constructed by using linear combinations of the form (10.5).

It may be argued that the reason is that we are just using a finite family of numbers $x_1, \ldots, x_r$, but even if we enlarge the collection of the basis elements by allowing a countable one, that is, we select numbers $\{x_k\}_{k=1}^{\infty}$, we would still be able to construct a countable family of x's (note that $\infty^2 = \infty$). We could construct a generating set by considering *all* real numbers, that is, by considering an uncountable family of elements, but then the decompositions will not be unique and, in any case, this way of proceeding will not be very helpful.

Then, we conclude that even the simplest linear space, when considered as a $\mathbb{Z}$-module becomes quite complicated and our intuition on the structure of linear spaces breaks down. For instance, if we are given a linear space V and within it a linear subspace U, we may always find out a supplementary subspace W such that $V = U \oplus W$ (actually this constitutes a fundamental theorem in the theory of linear spaces, see Thm. A.1 in Appendix A). Think for instance of the line λv, $\lambda \in \mathbb{R}$ defined by any vector $v \in V$ as the subspace U and any supplementary one W defined as a hyperplane passing through the origin. One way to show this is by using linear

[5]Recall that the integers $\mathbb{Z}$ form a ring, not a field.

basis adapted to U, that is, we pick up a linear basis of U and then we enlarge it to get a linear basis of the total space. The vectors we add to complete the linear basis will span a supplementary linear space W. However this intuition fails when considering modules. In example 10.40.4 a simple example of a submodule not possessing a supplementary will be discussed.

Much closer to the situations we are studying is to consider matrices as 'scalars' instead of integers. Let us consider the algebra $\mathcal{A}$ of $n \times n$ complex matrices that we will denote by $A, B, C, \ldots$. The linear space V will be the complex space $\mathbb{C}^n$ whose elements will be denoted by u, v, etc. The most important observation is that we can multiply elements v in V on the left by elements A in the algebra $\mathcal{A}$ by using the standard matrix product. The standard properties of the product matrix-vector, $A(u+v) = Au + Av$, $(A+B)u = Au + Bu$, $A(Bu) = (AB)u$, $A(\lambda u) = (\lambda A)u$ are just the axioms of an $\mathcal{A}$-module. Notice that the field $\mathbb{C}$ is naturally acontained in $\mathcal{A}$ as a subalgebra by associating to any complex number λ the matrix $\lambda \mathbb{I}$. In this sense, the restriction of the $\mathcal{A}$-module structure of V to the subalgebra $\mathbb{C}$ recovers the linear space structure.

Given two vectors $u, v \in \mathbb{C}^n$ there is always a matrix[6] A, such that $v = Au$. This means that there are not subspaces in $\mathbb{C}^n$ which are invariant under multiplication by arbitrary matrices. Then we will say that $\mathbb{C}^n$ is a simple $\mathcal{A}$-module (but it is not as a $\mathbb{C}$-module). Linear spaces which are direct sums of simple modules will be called semisimple, and this is the main idea that will be developed in Chapter 11.

With the previous examples in mind, we may define modules formally.

Definition 10.9 A (left) module M over the algebra A is a linear space M and a homomorphism of algebras $\rho \colon A \to \mathrm{End}(M)$ or, in other words, a linear representation of the algebra A with support the linear space M. Alternatively, a module M over A (sometimes called an A-module) is a linear space and a map $\cdot \colon A \times M \to M$, such that for any $a \in A$, the map $\lambda_a \colon M \to M$, $\lambda_a(m) = a \cdot m$, $m \in M$, is linear and:

$$a \cdot (b \cdot m) = (ab) \cdot m \,, \qquad \forall a, b \in A \,, \quad m \in M \,.$$

The assignment $a \cdot m = \rho(a)(m)$ establishes the relation between both definitions.

Remark 10.5 Similarly we can define right A-modules as linear spaces M with a map $\cdot \colon M \times A \to M$, such that for any $a \in A$, the map $\rho_a \colon M \to M$, $\rho_a(m) = m \cdot a$, $m \in M$, is linear and:

$$(m \cdot a) \cdot b = m \cdot (ab) \,, \qquad \forall a, b \in A \,, \quad m \in M \,.$$

Exercise 10.31 *Let A be an Abelian algebra. Prove that, given a left A-module M, the map $M \times A \to M$, given by $m \cdot a := a \cdot m$, defines a right A-module structure on M.*

[6] For instance, if u and v are not colinear, we may use a reflection (called a Householder reflection) with respect to the axis passing through the midpoint of the segment joining them and orthogonal to the line joining u, v in the plane defined by them (see later on, Sect. 11.1.2, for a non-trivial extension of this simple result).

Due to the previous exercise, there is no distinction between left and right modules over Abelian algebras. Notice, however, that they are different notions in the case of non-commutative algebras. We will just insist that left A-modules are the same as linear representations of algebras. Note that we might as well define a 'contravariant' notion of representation with $\rho(ab) = \rho(b)\rho(a)$ that will correspond to the notion of right A-module. ∎

Example 10.32 *Consider the Abelian algebra $A = \mathbb{C}[x]$ of complex polynomials on the single variable x. Let V be a linear space and $L\colon V \to V$, be a linear map. Consider the representation of the algebra A on V defined by $\rho_L(p)(v) = p(L)v$, where $p(x) = p_0 + p_1 x + \cdots + p_n x^n$ is a polynomial, $v \in V$, and $p(L)$ is the linear map: $p(L) = p_0 + p_1 L + \cdots + p_n L^n$.*

Exercise 10.33 *Prove that $(p+q)(L) = p(L) + q(L)$ and $(pq)(L) = p(L)q(L)$.*

The previous construction shows that any linear map $L\colon V \to V$ defines a linear representation of the Abelian algebra $\mathbb{C}[x]$, that is a $\mathbb{C}[x]$-module, on V.

As in the case of representations of groups, categories and groupoids, there is a natural notion of equivalence of representations of algebras.

Definition 10.10 Let A be an algebra and (ρ_1, V_1), (ρ_2, V_2) two linear representations. We will say that they are equivalent if there exists an isomorphism $\varphi\colon V_1 \to V_2$ such that $\rho_2(a)(\varphi(v_1)) = \varphi(\rho_1(a)(v_1))$, for all $a \in A$, $v_1 \in V_1$.

The relation above is clearly an equivalence relation and, as in the case of groupoids, we will be interested in describing the equivalence classes of equivalence relations of representations of algebras.

Example 10.34 *Consider, as in Example 10.32, the Abelian algebra of polynomials $\mathbb{C}[x]$. Let V be a linear space. Let c be a complex number and ρ_c the representation defined as $\rho_c(p)(v) = p(c)v$, where $p = p(x)$ is a polynomial and $v \in V$. Note that, because $(pq)(c) = p(c)q(c)$, ρ_c is a linear presentation of the algebra of polynomials.*

Exercise 10.35 *Shows that all representations ρ_c, $c \in \mathbb{C}$ are disequivalent.*

From the point of view of modules, the situation is as follows.

Definition 10.11 Let M, N be two A-modules, then a linear map $\varphi\colon M \to N$, such that $\varphi(a \cdot m) = a \cdot \varphi(m)$, for all $a \in A$ and $m \in M$, will be called a homomorphism of A-modules. A homomorphism of modules φ will be called monomorphism, epimorphism, or isomorphism, if it is injective, surjective or bijective, respectively. An isomorphism of A-modules $\varphi\colon M \to N$ is also called an equivalence between the representations M and N of A (recall Def. 10.10).

In the language of representations, a homomorphism of A-modules, $\varphi\colon M \to N$, is also called an intertwiner between the representations ρ_M and ρ_N of A on the linear spaces M and N, respectively, that is, a homomorphism, such that: $\rho_N(a)\circ\varphi = \varphi\circ\rho_M(a)$, for all $a \in A$.

The linear space of all homomorphisms $\varphi\colon M \to N$ between the A-modules M and N will be denoted by $\mathrm{Hom}_A(M,N)$. In the particular instance that $M = N$, we will also write $\mathrm{End}_A(M)$.

Note that $\mathrm{Hom}(M,N) = M^* \otimes N$ (see App. A.3.2). Then, we will also denote $\mathrm{Hom}_A(M,N)$ as $M^* \otimes_A N$. In this sense, we write $\mathrm{End}_A(M) = M^* \otimes_A M$ and we easily find, using the properties of the operations between linear spaces, Prop. A.9, that:

$$\mathrm{End}_A(\oplus_{k=1}^r M_k) \cong \bigoplus_{k,l=1}^{r} M_k^* \otimes_A M_l\,, \tag{10.6}$$

Example 10.36 *Let M be an A-module. Then, the space of endomorphisms of M is itself an A-module. Clearly, it is a linear space and we may define the multiplication by elements a of A as $(a\cdot L)(m) = a\cdot(L(m))$ for all $m \in M$. Thus, the space* $\mathrm{End}(M)$ *defines a representation of A.*

If M is a finite-dimensional linear space of dimension n, then choosing a linear basis $\{v_k \mid k = 1,\dots,n\}$, we may identify $\mathrm{End}(M)$ *with the direct sum $nM = M^{\oplus n} = M\oplus\cdots\oplus M$, by means of the map that assigns to any linear map L the vectors $L(v_1)\oplus\cdots\oplus L(v_n)$. Conversely, given the vector $u_1\oplus\cdots\oplus u_n$, we can define the linear map:*

$$L(\lambda_1 v_1 + \cdots + \lambda_n v_n) = \lambda_1 u_1 + \cdots \lambda_n u_n\,.$$

Hence, the natural A-module structure of $\mathrm{End}(M)$ *becomes the A-module structure on $M^{\oplus n}$, given by $a\cdot(u_1\oplus\cdots\oplus u_n) = a\cdot u_1\oplus\cdots\oplus a\cdot u_n$.*

Example 10.37 *Consider the algebra A as a left A-module, i.e., the left regular representation of A. Then, we may consider homomorphisms of A-modules $\varphi\colon A \to A$, that is, linear maps satisfying $\varphi(ab) = a\varphi(b)$, for all $a,b \in A$. Notice that $\varphi(a) = \varphi(a\mathbf{1}_A) = a\varphi(\mathbf{1}_A)$. Hence, if $\varphi \in \mathrm{End}_A(A)$, it is characterized by $\varphi(\mathbf{1}_A) \in A$. Moreover,* $\mathrm{End}_A(A)$ *is an algebra with the composition of linear maps.*

Proposition 10.7
Let A be an algebra, then $\mathrm{End}_A(A) \cong A^{\mathrm{op}}$.

Proof 10.6 Consider, as in Example 10.37, the algebra $\mathrm{End}_A(A)$ of endomorphisms of the left A-module A, with multiplication given by composition of linear maps. Let $\varphi,\psi\colon A \to A$ be two homomorphisms of A-modules, then their composition $(\varphi\circ\psi)(a) = \varphi(a\psi(\mathbf{1}_A)) = \varphi((a\psi(\mathbf{1}_A))\mathbf{1}_A) = (a\psi(\mathbf{1}_A))\varphi(\mathbf{1}_A) = a(\psi(\mathbf{1}_A)\varphi(\mathbf{1}_A))$, hence the composition of φ and ψ corresponds to the product of the element $\psi(\mathbf{1}_A)$ associated to ψ and the element $\varphi(\mathbf{1}_A)$ associated to φ (in the opposite order!), then we have proved that $\mathrm{End}_A(A) = A^{\mathrm{op}}$.

10.4.2 Submodules and subrepresentations

Definition 10.12 Let $\rho\colon A \to \mathrm{End}(V)$ be a linear representation of the algebra A. A subrepresentation of the representation (ρ, V) is subspace $U \subset V$ invariant under A, i.e., such that $\rho_a(m) \in U$ for any $m \in U$, $a \in A$.

Example 10.38 *Consider the representation ρ_c of the algebra of polynomials on a linear space V, discussed in Example 10.34. Then, any subspace of V defines a subrepresentation. In particular, if we choose a non-zero vector v, the one-dimensional subspace defined by it is invariant under ρ_c (because $\rho_c(p)v = p(c)v$).*

The notion of subrepresentation of a representation of an algebra can be easily restated in the language of modules.

Definition 10.13 Let M be a module over the algebra A. A submodule of M is a linear subspace $N \subset M$, such that $a \cdot m \in N$ for all $a \in A$ and $m \in N$. In other words, a submodule of M is a linear subspace which is invariant with respect to the linear representation of A on M, that is, a subrepresentation of (ρ, V).

Given a linear representation (ρ, V) of the algebra A, both the total space V and the zero subspace $\{\mathbf{0}\}$ provide subrepresentations of (ρ, V). They are trivial and we do not want to consider them. A subrepresentation different from them will be called proper. Correspondingly, given a A-module M, any submodule different from M and $\{\mathbf{0}\}$ will be called proper.

Any (left) ideal $\mathcal{J} \subset A$ provides a representation of A. In other words, any ideal $\mathcal{J}$ of A is a submodule of the module defined by A itself.

Example 10.39

1. *Consider the algebra A of functions on the set Ω and the ideal $\mathcal{J}_S$ of functions vanishing on the subset S. Then $\mathcal{J}_S$ is a submodule of A.*

2. *Consider the algebra of polynomials on the variable x and the ideal (p) generated by a given polynomial $p(x)$, then (p) defines a submodule of $\mathbb{C}[x]$.*

3. *Consider the algebra $A = M_n(\mathbb{C}) \oplus M_m(\mathbb{C})$ and the ideal $M_n(\mathbb{C}) \oplus \{\mathbf{0}\}$. It defines a submodule of A.*

4. *Consider the algebra of polynomials $\mathbb{C}[x]$ again. Let $L\colon V \to V$ be a linear map and ρ_L the representation defined in Example 10.34. Let $W \subset V$ be a subspace of V such that $L(W) \subset W$, then W is a submodule of V with respect to the represention ρ_L, i.e., W determines a subrepresentation of ρ_L.*

5. *Consider, as in Example 10.36, an A-module M which is a finite-dimensional linear space of dimension n, then choosing a linear basis $\{v_k \mid k = 1, \ldots, n\}$, we may identify $\mathrm{End}(M)$ with the direct sum $nM = M^{\oplus n} = M \oplus \cdots \oplus M$. Then the natural representation of the algebra A on the space of endomorphisms $\mathrm{End}(M) \cong M^{\oplus n}$ has natural submodules given by the spaces of homomorphisms of the form $0 \oplus \cdots \oplus u_k \oplus \cdots \oplus 0$.*

10.4.3 Operations with representations

As in the case of categories and groupoids, there are natural operations among representations of algebras: Addition and tensor product.

Definition 10.14 Let (ρ, V), (ρ', V') be two linear representations of the algebra A, then we define the direct sum $\rho \oplus \rho'$ of the two representations as the linear representation of A on the linear space $V \oplus V'$ given by: $(\rho \oplus \rho')(a)(v \oplus v') = \rho(a)v \oplus \rho'(a)v'$ for all $v \in V, v' \in V'$, $a \in A$ or, using the compact notation: $a \cdot (v \oplus v') = a \cdot v \oplus a \cdot v'$.

Sometimes we will just denote the direct sum $(\rho \oplus \rho', V \oplus V')$ of the representations (ρ, V) and (ρ', V') by $\rho \oplus \rho'$ or $V \oplus V'$ for short (if there is no risk of confusion).

Notice that the representations $\rho \oplus \rho'$ and $\rho' \oplus \rho$ are equivalent. We will also denote it as $V \oplus V' \cong V' \oplus V$.

In a completely similar way, we define the tensor product of two representations of A.

Definition 10.15 Let (ρ, V), (ρ', V') be two linear representations of the algebra A, then we define the tensor product $\rho \otimes \rho'$ of the two representations as the linear representation of A on the linear space $V \otimes V'$ given by: $(\rho \otimes \rho')(a)(v \otimes v') = \rho(a)v \otimes \rho'(a)v'$ for all monomials $v \otimes v' \in V \otimes V'$, $a \in A$ or, using the compact notation: $a \cdot (v \otimes v') = a \cdot v \otimes a \cdot v'$.

Similar to the direct sum, we will denote the tensor product of two representations either as $\rho \otimes \rho'$ or $V \otimes V'$ if there is no risk of confusion. Clearly, $V \otimes V' \cong V' \otimes V$.

Again, as in the case of categories and groupoids, there is a natural notion of irreducible and indecomposable representation of algebras that we will repeat for the sake of completness.

Definition 10.16 An irreducible representation of the algebra A is a linear representation (ρ, V) of A, such that it has no proper subrepresentations or, equivalently, that has no proper submodules.

A representation (ρ, V) of the algebra A will be said to be indecomposable if, for any submodule $W \subset V$, there does not exist another submodule U such that $V = W \oplus U$.

A representation which is not irreducible, that is, it has invariant subspaces, will be said to be reducible. In the same way, a representation which is not indecomposable, that is, such that it has an invariant subspace with an invariant supplementary subspace, will be said to be decomposable.

Note that, if a representation is irreducible, is indecomposable but the converse is not necessarily true (see Example 10.40.3 below). Then, if a representation is decomposable, is reducible.

Even if this will be the subject of a full discussion in Chapter 11, we can anticipate the main notions that will guide part of our research in what follows. It is clear that, given a representation (ρ, V) or an algebra, it would be desirable to be able to 'break it down' to small blocks that are easier to understand, these blocks could be either irreducible or indecomposable representations.

Determining whether this is possible or not for the algebras of groupoids is going to be the main task of the rest of the book. We already know that this is true for representations of groups, as stated by the main theorem of the theory, Thm. 7.1. However, we do not know if this is true for groupoids or more general categories. We will provide a positive answer to the first question in the following chapters.

Definition 10.17 A representation (ρ, V) of the algebra A will be said to be completely reducible (or semi-simple[7]) if $V = \bigoplus_{k=1}^{r} V_k$ and $\rho = \bigoplus_{k=1}^{r} \pi_k$, where (π_k, V_k) are irreducible representations of A.

Example 10.40

1. *Let c be a complex number. The linear representation ρ_c of the algebra of polynomials on x on the one-dimensional linear space $\mathbb{C}$, given by multiplication by $p(c)$, i.e., $\rho_c(p)(z) = p(c)z$, $z \in \mathbb{C}$, $p = p(x)$ a polynomial, is irreducible. However, the same representation on a linear space V of dimension $n > 1$ is reducible and completely reducible.*

2. *Let ρ be the natural representation of the algebra of matrices $M_n(\mathbb{C})$ on $\mathbb{C}^n$. The representation ρ is irreducible (recall, however, that the representation of the permutation group S_n on $\mathbb{C}^n$ is reducible). On the other hand, the natural representation of the algebra $M_n(\mathbb{C}) \oplus M_m(\mathbb{C})$ on $\mathbb{C}^n \oplus \mathbb{C}^m$ is reducible (for instance, the subspaces $\mathbb{C}^n \oplus \mathbf{0}$ and $\mathbf{0} \oplus \mathbb{C}^m$ are invariant under $M_n(\mathbb{C})$ on $\mathbb{C}^n$.*

3. *Let ρ_L be the representation of the algebra of polynomials on a linear space V associated to the linear map $L \colon V \to V$. Suppose that L has a Jordan canonical normal form with a single block $J_{\lambda,n}$, $n = \dim V$, λ an eigenvalue of L, that is, there is a basis, such that the matrix associated to L has the form given in (9.3), then, the representation ρ_L is indecomposable but it is not irreducible. Clearly, the subspace generated by the eigenvector of L is invariant, but there is not an invariant supplementary subspace.*

The previous examples display a variety of situations regarding irreducible and indecomposable representations, mainly that there are indecomposable representations which are not irreducible. However, if we want to solve the problem of complete reducibility for algebras, we should know first if, given a representation (ρ, V) of the algebra A, is there always an irreducible representation contained in V? The answer to this question happens to be affirmative.

[7]Compare with Def. 11.2.

Proposition 10.8

Let A be an algebra and (ρ, V) a finite-dimensional representation of it. Then, it has an irreducible subrepresentation $W \subset V$.

Proof 10.7 If V is irreducible, then $W = V$. If V is reducible, then it has a proper subrepresentation $V' \subset V$, $\dim V' < \dim V$. Again if V' is irreducible we are done. If not, we repeat the argument. Then, after a finite number of steps (because the representation V is finite-dimensional), we will find a subrepresentation V' which is irreducible.

Remark 10.6 The previous result is not true if we drop the finiteness assumption on the dimension of the representation. The key is contained in the proof the existence of indecomposable but not irreducible representations in Example 10.17. Notice that there, we used an eigenvector of the Jordan block $J_{\lambda,n}$ to show that the representation is not irreducible, but if we allow an infinite-dimensional representation, there are Jordan blocks (infinite-dimensional of course) without eigenvectors. ▮

10.4.4 Schur's Lemma for algebras

Once we have arrived here, it is clear that we could try to follow the path that we have already walked with groups and groupoids, that is, determining under what conditions a linear representation is comptetely reducible, try to characterize the irreducible ones, etc.

For that, we could try to reproduce the main tools we have so successfully used before: Schur's Lemma and a theory of characters. Schur's Lemma is straightforward and we will state right now. The theory of characters will be developed in Chapter 12 for representations of algebras of groupoids. However, we will devote the coming chapter to discussing the notion of semi-simplicity.

Theorem 10.1

Let A be an algebra and (ρ_M, M), (ρ_N, N) two non-zero finite-dimensional linear representations of A. Let $\varphi : M \to N$ be a non-zero homomorphism of A-modules or, in other words, an intertwiner for the representations M and N.

If M is irreducible, then φ is a monomorphism. If N is irreducible, then φ is an epimorphism and, if both M and N are irreducible, φ is an isomorphism. Moreover, if $M = N$ is finite-dimensional and irreducible, then φ is a homothety.

Proof 10.8 If M is irreducible, because $\ker\varphi$ is a submodule, then, because φ is non-zero, $\ker\varphi = 0$ and φ is a monomorphism.

If N is irreducible, $\operatorname{ran}\varphi$ is a submodule, then, because φ is non-zero, $\operatorname{ran}\varphi = N$ and φ is an epimorphism.

If M and N are both irreducible, then φ is monomorphism and epimorphism.

Finally, if $M = N$ is finite-dimensional, consider an eigenvalue λ of φ, that is $\varphi(m) = \lambda m$ for some $m \neq 0$. Then, $\varphi - \mathrm{id}_M$ is a homomorphism of A-modules. If M is irreducible, then, because $\ker(\varphi - \mathrm{id}_M)$ is non-zero, we get $\ker(\varphi - \mathrm{id}_M) = M$ and for any $m \in M$, $\varphi(m) = \lambda m$. In addition, computing the trace of φ we get: $\mathrm{Tr}\,\varphi = \lambda \dim M$ and $\lambda = \mathrm{Tr}\,\varphi / \dim M$.

10.5 Representations of groupoids and modules

We have already seen both in the case of groups and finite categories that Schur's lemmas, in spite of their simplicity, have deep and subtle consequences, however, before we start exploring them in the context of algebras, a task that we reserve for the coming chapter, we will end this one by discussing the intimate relationship between the theory of representations of groupoids and modules: There is a natural correspondence between linear representations of finite groupoids $\mathbf{G}$ and $\mathbb{C}[\mathbf{G}]$-modules.

More specifically, let $\mathbf{G} \rightrightarrows \Omega$ be a finite groupoid and a finite-dimensional linear representation $R \colon \mathbf{G} \to \mathbf{FinVect}$ of $\mathbf{G}$, there is a $\mathbb{C}[\mathbf{G}]$-module V associated to it. The linear space V is the total space of the representation R which is defined as, recall Eq. (9.7): $V = \bigoplus_{x\in\Omega} V_x$, with $V_x = R(x)$. The action of the algebra $\mathbb{C}[\mathbf{G}]$ on V is given by:

$$a \cdot v = \bigoplus_{x\in\Omega} \left(\sum_{\alpha\in G_+(x)} a_\alpha R(\alpha) v_x \right),$$

where $v = \bigoplus_{x\in\Omega} v_x$ and $a = \sum_\alpha a_\alpha \alpha$, in particular $\alpha \cdot v_x = R(\alpha)(v_x)$ for any $\alpha \colon x \to y \in \mathbf{G}$.

Conversely, if V is a finite-dimensional $\mathbb{C}[\mathbf{G}]$-module, consider any element $a \in \mathbb{C}[\mathbf{G}]$ as a linear map on V. The units 1_x of the groupoid $\mathbf{G}$ define elements in $\mathbb{C}[\mathbf{G}]$ whose associated linear maps define a family of projectors P_x on V, $P_x(v) = 1_x \cdot v$. Notice that, because $1_x^2 = 1_x$, $1_x \cdot 1_y = 0$, $x \neq y$, and $\sum_{x\in\Omega} 1_x = \mathbf{1}$, it follows that:

$$P_x^2 = P_x, \qquad P_x P_y = \delta(x,y) P_x, \qquad \sum_{x\in\Omega} P_x = \mathrm{id}_V. \tag{10.7}$$

The last condition is commonly referred saying that the family of projetors P_x define a resolution of the identity.

If we denote by V_x the range of the projector P_x, i.e., $V_x = P_x(V)$, then because of (10.7), we get $V = \bigoplus_{x\in\Omega} V_x$, and the family of maps $R(\alpha) \colon V_x \to V_y$, $\alpha \colon x \to y$, defined as:

$$R(\alpha) v_x = \alpha \cdot v_x, \qquad v_x \in V_x,$$

defines a linear representation of the groupoid $\mathbf{G}$. Hence we have proved the first part of the following proposition.

Proposition 10.9
Let $\mathbf{G} \rightrightarrows \Omega$ *be a finite groupoid. Then there is a one-to-one correspondence between finite-dimensional linear representation* $R\colon \mathbf{G} \to \mathbf{FinVect}$ *of* $\mathbf{G}$ *and left-*$\mathbb{C}[\mathbf{G}]$*-modules* V*, i.e., homomorphisms* $\rho\colon \mathbb{C}[\mathbf{G}] \to \mathrm{End}(V)$*. Moreover,* R *is irreducible if and only if* ρ *is irreducible.*

Proof 10.9 The correspondence has been established in the previous paragraphs. The relation between both concepts is given by the formula: $R(\alpha)v_x = \rho(\alpha)(v_x)$, with $\alpha\colon x \to y$, and $v_x \in V_x$.

If R is irreducible, then ρ is irreducible too. Note that, if there were a proper submodule $M \subset V$, the subspaces $M_x = P_x(M)$, will define a non-trivial subrepresentation of R. The converse follows the same reasoning, if there were a family $M_x \subset V_x$ of proper R-invariant subspaces, then $M = \bigoplus_{x\in\Omega} M_x$ defines a proper submodule.

In what follows, because of the previous proposition, we will use these notions interchangeably, i.e., linear reprsentations of groupoids and left-$\mathbb{C}[\mathbf{G}]$-modules.

The previous identification of linear representations of the groupoid $\mathbf{G}$ and $\mathbb{C}[\mathbf{G}]$-modules clearly extends to the rest of the definitions, in particular, as stated in the Proposition, to irreducible representations.

There are two natural representations of a groupoid, the regular representation and the fundamental representation. We will finish this chapter by describing the fundamental representation, leaving the discussion of the regular representation to Sect. 12.3 were an extensive use of it will be made.

10.5.1 The fundamental representation of a finite groupoid

The fundamental representation $\pi_0\colon \mathbb{C}[\mathbf{G}] \to \mathrm{End}(\mathcal{H}_\Omega)$ of the finite groupoid $\mathbf{G} \rightrightarrows \Omega$, is supported on the linear space $\mathcal{H}_\Omega$ freely generated by the elements of Ω, that is, vectors in $\mathcal{H}_\Omega$ are formal linear combinations $\psi = \sum_{x\in\Omega} \psi_x x$, $\psi_x \in \mathbb{C}$.

The linear space $\mathcal{H}_\Omega$ has dimension $|\Omega|$ and carries a canonical basis $\{x\}$ given by the elements x of Ω themselves.

As in the case of the defining representation of the group of permutations, Sect. 7.1.5, Eq. (7.6), we can denote the abstract basis vectors $\{x\}$, determined by the elements $x \in \Omega$, as $|x\rangle$ (the famous Dirac's abstract 'kets').

The existence of the canonical basis $\{x\}$ allows us to introduce a natural inner product $\langle\cdot,\cdot\rangle$ on $\mathcal{H}_\Omega$ such that the vectors x form an orthonormal basis[8], that is $\langle x, y\rangle = \delta(x,y)$.

The representation π_0 is defined as follows: Let $\alpha\colon x \to y$ be a groupoid morphism, then define the linear map

$$\pi_0(\alpha)\left(\sum_{z\in\Omega} \psi_z |z\rangle\right) = \sum_{z\in\Omega} \psi_z \delta(x,z)|y\rangle\,,$$

[8]The notation $\mathcal{H}_\Omega$ has been chosen bearing in mind the notation for Hilbert spaces introduced in Sect. 7.3.3.

i.e., the linear map $\pi_0(\alpha)$ transforms the vector $|x\rangle$ into the vector $|y\rangle$, and maps all others to zero.

Exercise 10.41 *Prove that $\pi_0(a^*) = \pi_0(a)^\dagger$, where $(\cdot)^\dagger$ denotes the Hermitean conjugate (or adjoint, recall Sect. 7.3.3) with respect to the inner product $\langle \cdot, \cdot \rangle$, i.e., $\langle \pi_0(a)^\dagger(\psi), \phi \rangle = \langle \psi, \pi_0(a)(\phi) \rangle$, for all $\psi, \phi \in \mathcal{H}_\Omega$.*

Exercise 10.42 *Prove that the fundamental representation π_0 is unitary, that is, for any unitary element a in $\mathbb{C}[\mathbf{G}]$ (i.e., any element such that $a^* \cdot a = a \cdot a^* = \mathbf{1}$), $\pi_0(a)$ is a unitary operator: $\pi_0(a)^{-1} = \pi_0(a)^\dagger$.*

Proposition 10.10

Let $\mathbf{G}$ be a finite connected groupoid, then the fundamental representation π_0 is irreducible.

Proof 10.10 Suppose that $V \subset \mathcal{H}_\Omega$ is a $\mathbb{C}[\mathbf{G}]$-submodule, then $V_x = P_x(V)$, with $P_x = \pi_0(1_x)$ the corresponding projector on $\mathcal{H}_\Omega$, defines a subrepresentation of π_0. However, either V_x is the zero space or $\mathbb{C}x$ (because $P_x(\mathcal{H}_\Omega) = \mathbb{C}x$). If $V_x \neq \{\mathbf{0}\}$, then, given $y \in \Omega$, $V_y \neq \{\mathbf{0}\}$, because there is $\alpha \colon x \to y$, and $y = \alpha \cdot x = \pi_0(\alpha)(x) \in V_y$. Then, we conclude, $\oplus_{x \in \Omega} V_x = V = \mathcal{H}_\Omega$.

Example 10.43

1. *Let G be a group, the fundamental representation of the group G as a groupoid is the trivial representation $\mu_0(g) = 1$, for all $g \in G$.*

2. *Let $\mathbf{G}(\Omega)$ be the groupoid of pairs of a finite set Ω with n elements, then the fundamental representation π_0 of $\mathbf{G}(\Omega)$ is equivalent to the canonical representation of the algebra of matrices $M_n(\mathbb{C})$ on $\mathbb{C}^n$. This representation is obviously irreducible.*

3. *Let Γ be a group acting transitively on the finite set Ω. Consider the action groupoid $\mathbf{G}(\Gamma, \Omega)$. The fundamental representation π_0 of $\mathbf{G}(\Gamma, \Omega)$ is supported in the space of functions on Ω and is irreducible. Compare this result with the representation of the group Γ on the space of function on Ω. This representation is not irreducible! (recall Example 7.6).*

Chapter 11

Semi-simplicity

This chapter, of a slightly more advanced character than the preceding ones, will address the study of the structure of the algebra of a groupoid. As will be shown, the groupoid algebra is semi-simple, hence, it decomposes as direct sum of simple algebras. This is not the case, however, with algebras of categories. Some interesting applications and examples will be discussed.

11.1 Irreducible representations of algebras

Last chapter was ended promising to unveil some of the consequences of Schur's Lemmas, Thm. 10.1. We will start by studying the structure of subrepresentations of direct sums of irreducible representations, Thm. 11.1, and, after this, we will discuss transitivity properties of representations, Thm. 11.3.

In what follows, A will denote a finite-dimensional associative algebra with unit. Let (ρ_k, V_k), $k = 1, \ldots, r$ be a finite family of irreducible representations of A or, in other words, a finite family of left-A-modules V_k. Consider the direct sum of the representations $\rho = \oplus_{k=1}^r n_k \rho_k$, with n_k non-negative integers denoting the multiplicity of the representation ρ_k in ρ. Note that the support space for ρ will be $V = \bigoplus_{k=1}^r n_k V_k$, with $n_k V_k = V_k^{\oplus n} = V_k \oplus \overset{n\,\text{times}}{\cdots} \oplus V_k$.

11.1.1 Subrepresentations of direct sums of irreducible representations

If V_k denotes the support space of the irreducible representation $\rho_k \colon A \to \mathrm{End}\,(V_k)$, then there are no invariant subspaces, i.e., V_k has no A-submodules. Such A-modules are called simple, while the direct sum of a finite family of simple modules will be called semi-simple, recall Sect. 10.4.3, Def. 10.17. The study of semi-simplicity, as the immediate generalization of 'simple' or 'irreducible', will constitute the subject of this chapter. Our first task will be to describe the subrepresentations of a direct sum of irreducible representations of A.

Remark 11.1 We should warn the reader that the form of the direct sum $V = \bigoplus_{k=1}^{r} n_k V_k$ could create a misleading impression. There could be invariant subspaces, and, in fact, there are plenty, whose intersections with the simple factors V_k will be zero. We were not stressing this fact when discussing the structure of representations of groups because we obtained a decomposition theorem right away using the theory of characters, but it is obvious that we may construct invariant subspaces W in $V_1 \oplus V_2$, V_1, V_2 irreducible representations of a group G, with $W \cap V_1 = W \cap V_2 = \mathbf{0}$. Then, what can be said about W?

The reader may consider, for instance, as a simple example the group $\mathbb{Z}_2$, and $V_1 = V_2 = \mathbb{C}$ the support of the trivial representation μ_0. Then, the subspace $W \subset V_1 \oplus V_2 = \mathbb{C}^2$ spanned by the vector $[1,1]$ is, obviously, invariant and intersects V_1 and V_2 at $\mathbf{0}$. Less simple examples are easy to construct explicitly. ∎

Exercise 11.1 *Consider the Abelian group $\mathbb{Z}_m$ and the representation $n_0V_0 \oplus \cdots \oplus n_{m-1}V_{m-1}$, where $V_l = \mathbb{C}$ is the one-dimensional linear space support of the lth irreducible representation of $\mathbb{Z}_m$, recall Sect. 8.7.1, whose characters are given in the Table 8.2. Construct an invariant subspace W of dimension $k < n_0 + \cdots + n_{m-1}$, containing the representations with characters the mth–roots of unity w_l, $l = 0, 1, \ldots, m-1$, with multiplicities $k_0, \ldots, k_{m-1} \geq 0$ and such that the intersection with each one of the subspaces V_l is zero.*

The following theorem provides a general answer to these questions.

Theorem 11.1
Let (ρ_k, V_k), $k = 1, \ldots, r$ be a finite family of pairwise non-equivalent finite-dimensional irreducible representations of the algebra A. Let (ν, W) be a subrepresentation of the direct sum representation $\rho = \oplus_{k=1}^{r} n_k \rho_k$ (that has support space $V = \bigoplus_{k=1}^{r} n_k V_k$). Then, (ν, W) is isomorphic to a direct sum of the form $\oplus_{k=1}^{r} m_k \rho_k$, in other words, containing the same factors as ρ but with possibly smaller multiplicities $m_k \leq n_k$. The inclusion $i \colon W \to V$ is a direct sum of inclusions $i_k \colon m_k V_k \to n_k V_k$ of the form $i_k(v_1 \oplus \ldots \oplus v_{m_k}) = \sum_{j_1, \ldots, j_{n_k}} A_{1j_1} v_{j_1} \oplus \ldots \oplus A_{m_k j_{n_k}} v_{j_{n_k}}$.

Proof 11.1 The proof is by induction. Consider $n = \sum_k n_k$. Then, if $n = 1$, the statement is obvious.

Let $n > 1$, $V = \oplus_k n_k V_k$ as in the statement of the theorem and $W \subset V$ a subrepresentation. Denoting by $i \colon W \to V$, the canonical inclusion, we may consider the projection $p_k = P_k \circ i$, where $P_k \colon V \to n_k V_k$ is the projector onto $n_k V_k$ defined by the decomposition of V.

Let V_l be an irreducible representation contained in W (that always exists because of Prop. 10.8). Then, we can define the map $p_l \circ i_l \colon V_l \to n_l V_l$ and because of Schur's Lemma, it must be of the form $v_l \mapsto c_1 v_1 \oplus \cdots \oplus c_{n_l} v_{n_l}$ with not all coefficients $c_{l'}$ zero (the map must be injective). Then, we map W to $W' \oplus V_l$ by using a linear map $L(v_1^{(1)} \oplus \cdots \oplus v_l^{(l)} \oplus \cdots \oplus v_r^{(r)}) = v_1^{(1)} \oplus \cdots \oplus L_l(v_l^{(l)}) \oplus \cdots \oplus v_r^{(r)}$, where $v_k^{(k)} \in n_k V_k$, and $L_l \colon n_l V_l \to n_l V_l$ is any invertible linear map, such that $L_l(c_1 v_1 \oplus \cdots \oplus c_{n_l} v_{n_l}) = u_1 \in V_l$ with u_1 a fixed vector in V_l. We conclude by observing that $W' \subset n_1 V_1 \oplus \cdots \oplus (n_l - 1)V_l \oplus \cdots \oplus n_r V_r$ and using the induction hypothesis.

Remark 11.2 The previous theorem can be understood by observing that any subrepresentation of a direct sum of irreducible representations has the same structure (with smaller multiplicities) but it could be 'twisted' or 'mixed' inside each subspace $n_l V_l$ by a family of matrix transformations (the coefficients $A_{l\,j_l}$ that appear in the theorem). ▮

11.1.2 *Irreducibility and transitivy*

At the beginning of the previous chapter, we were discussing the properties of the algebra of endomorphisms $\mathrm{End}(V)$, in particular, it was proved that $M_n(\mathbb{C})$ (then $\mathrm{End}(V)$) was simple, Prop. 10.1. The argument there seemed to be *ad hoc*, however, much more can be proved by a judiciously combination of our previous description of subrepresentations, Thm. 11.1, and Schur's Lemmas.

We will start by observing that the trivial statement that, given any non-zero vector v in a vector space and any other vector u, there is a linear map $L \colon V \to V$, such that $Lv = u$ (see Sect. A.2), holds if V is the support of an irreducible representation of the algebra A and we restrict ourselves to consider the linear maps provided by the representations itself.

Proposition 11.1
Let (ρ, V) be an irreducible representation of the algebra A, then given a non-zero vector $v \in V$, for any other vector $u \in V$ there is an element $a \in A$ such that $u = a \cdot v$.

Proof 11.2 Consider the linear subspace $A \cdot v = \{\rho(a)(v) = a \cdot v \mid a \in A\}$. Clearly, $A \cdot v$ is a subrepresentation but, because V is irreducible and $A \cdot v \neq \mathbf{0}$, then $A \cdot v = V$ and there exists $a \in A$, such that $a \cdot v = u$ for any u.

We can push the previous argument forward and show that the statement holds when we consider a family of linearly independent vectors and not just a single vector v, that is:

Theorem 11.2
Let (ρ, V) be a finite-dimensional irreducible representation of the algebra A, and $v_1, \ldots, v_r$ a finite family of linearly independent vectors in V (hence, $\dim V \geq r$), then, for any family $u_1, \ldots, u_r$, there exist $a \in A$, such that $u_k = a \cdot v_k$, $k = 1, \ldots, r$.

Proof 11.3 Consider the linear subspace $W = A \cdot (v_1 \oplus \cdots \oplus v_r)$ of $rV = V^{\oplus r}$ which defines a subrepresentation of $\rho^{\oplus r}$. Now, because of the description of subrepresentations discussed above, Thm. 11.1, W is isomorphic to $sV = V^{\oplus s}$ for some $s \leq r$. Moreover, the canonical inclusion $i\colon W \to V^{\oplus r}$ has the form $i(w_1 \oplus \cdots \oplus w_s) = A_{1j_1} w_{j_1} \oplus \cdots \oplus A_{rj_r} w_{j_r}$. Then, choosing $v_1 \oplus \cdots \oplus v_r \in W$ (recall that A has a unit), $v_1 = A_{1j_1} w_{j_1}, \ldots, v_r = A_{rj_r} w_{j_r}$, for some vectors $w_1, \ldots, w_s$. Then, $s = r$ because the vectors $v_1, \ldots, v_r$ are linearly independent, and this implies that W is isomorphic to $V^{\oplus r}$. Thus, we conclude that for any $u_1, \ldots, u_r$ vectors in V, there is $a \in A$, such that $u_1 \oplus \cdots \oplus u_r = a \cdot (v_1 \oplus \cdots \oplus v_r)$.

Remark 11.3 The property expressed in Thm. 11.2, can be called n-transitivity, $n = \dim V$, as it expresses the fact that the action of algebras on linear spaces supporting irreducible representations, are simultaneously transitive on families of independent vectors. ∎

As a direct consequence of the previous proposition, we get:

Corollary 11.1
Let (ρ, V) be a finite-dimensional irreducible representation of the algebra A, then the map $\rho\colon A \to \mathrm{End}(V)$ is surjective, i.e., ρ is an epimorphism of algebras.

Proof 11.4 Let $L \in \mathrm{End}(V)$ and $v_1, \ldots, v_n$ a linear basis of V. Consider the vectors $u_k = L(v_k)$. Then, because of the n-transitivity of V, Thm. 11.2, there is $a \in A$ such that $u_k = a \cdot v_k$, $k = 1, \ldots, n$. In consequence, $L = \rho(a)$ and the map ρ is surjective.

Again, we can push a bit further the previous argument to get as a consequence of the n-transitivity of irreducible representations the following 'closure' theorem:

Theorem 11.3
Let (ρ, V) be the direct sum of pairwise non-equivalent irreducible representations (ρ_k, V_k), i.e., $V = \bigoplus_{k=1}^r V_k$, then the map $\rho\colon A \to \bigoplus_{k=1}^r \mathrm{End}(V_k)$, $\rho(a)(v_1 \oplus \cdots \oplus v_r) = \rho_1(a)(v_1) \oplus \cdots \oplus \rho_r(a)(v_r)$, is surjective.

Proof 11.5 Consider the range $\rho(A) \subset \bigoplus_{k=1}^r \mathrm{End}(V_k)$ of the map ρ. Recall that we can consider $\mathrm{End}(V_k)$ as an A-module in a natural way and, by choosing a basis, identify it with $V_k^{\oplus n_k}$, $n_k = \dim V_k$. The range $\rho(A)$ is a submodule of $\bigoplus_{k=1}^r \mathrm{End}(V_k)$, because:

$$\begin{aligned}(a\cdot\rho(b))(v_1\oplus\cdots\oplus v_r) &= a\cdot(\rho(b)(v_1\oplus\cdots\oplus v_r))\\ &= a\cdot(\rho_1(b)(v_1)\oplus\cdots\oplus\rho_r(b)(v_r))\\ &= a\cdot(\rho_1(b)(v_1))\oplus\cdots\oplus a\cdot(\rho_r(b)(v_r))\\ &= (a\cdot\rho_1)(b)(v_1)\oplus\cdots\oplus(a\cdot\rho_r)(b)(v_r)\\ &= \oplus_{k=1}^r(a\cdot\rho_k)(b)(v_1\oplus\cdots\oplus v_r),\end{aligned}$$

and $a\cdot\rho(b)\in\bigoplus_{k=1}^r\mathrm{End}(V_k)$. Then, because of Thm. 11.1, the subrepresentation $\rho(A)$ will have the form $\rho(A)=\bigoplus_{k=1}^r d_kV_k$ and the inclusion $i\colon\rho(A)\to\bigoplus_{k=1}^r\mathrm{End}(V_k)$ is a direct sum of inclusions $i_k\colon d_kV_k\to\mathrm{End}(V_k)$. But then the map $i_k\circ\rho\colon A\to\mathrm{End}(V_k)$ is surjective because of Cor. 11.1, then we get that $d_k=n_k$ and ρ is surjective.

11.2 Semi-simple modules

Very often it happens in Mathematics that the presentation of a simple idea is convoluted, or obscure, or both. In fact, this happens more often in mature fields of Mathematics. The reason is that, as the understanding of a field of Mathematics deepens, new relations between the different notions that gave birth to the field are found, in many occasions unexpected, and some of the more relevant results of the theory happen to be consequences of some unifying property that was hidden in the beginning.

Mathematicians can hardly resist the temptation to present their fields of expertise starting from the most fundamental unifying principle and arriving, in many cases, to the motivating ideas after long and difficult arguments. 'Semi-simplicity' is just one of these notions. The idea behind the name is that a mathematical object of some class is 'semi-simple' if it is the 'sum' of 'simple' ones, where 'simple' here refers to objects without inner structure in that class ('elementary' would be an appropriate alternative terminology like the use of 'elementary particles' in Physics).

For instance, in linear algebra, we may say that a linear space is semi-simple if it is the direct sum of 'simple' ones, and a 'simple' linear space would be a linear space that has no linear subspaces (apart from the trivial ones, $\{\mathbf{0}\}$ and the space itself). Then, simple linear spaces are one-dimensional ones and any linear space V is semi-simple because it can be written as a direct sum of one-dimensional ones. In particular, if V is finite-dimensional, choose a linear basis $u_1,\ldots,u_n$, and $V=V_1\oplus\cdots\oplus V_n$ with V_k is the one-dimensional subspace generated by u_k, that is $V_k=\mathbb{C}u_k$. So far, everything is 'simple' enough.

The same idea can be used to define semi-simple modules (we will do this just after this introduction). However, we will face a problem when deciding what a semi-simple algebra is (see the discussion at the beginning of Sect. 11.4). In this section, we will present, as straightforward as possible, the notion and characterization of semi-simple modules, so that we will prepare the road for these ideas to be used in the coming sections when discussing the structure of the algebra of a groupoid or the algebra generated by a quiver.

Thus, we will proceed, as previously suggested, by discussing first the notion of semi-simple modules that will be always assumed to be left-A-modules (recall Sect. 10.4.1, Def. 10.9 and comments afterwards).

Definition 11.1 Let A be an algebra with unit. The A-module M is simple if it has no proper submodules, that is, submodules different from $\{\mathbf{0}\}$ and M. In other words, a left-A-module M is simple if it defines an irreducible representation of A.

Definition 11.2 An A-module M is semi-simple if it is the direct sum of simple modules: $M = M_1 \oplus \cdots \oplus M_r$, M_k, $k = 1, \dots, r$, simple A-modules.

In other words, using the identification between linear representations of algebras and left-A-modules, Def. 10.9, we may say that a representation (ρ, V) of an algebra A is semi-simple, if it is the direct sum of irreducible representations of A, $\rho = \bigoplus_{k=1}^{r} \rho_k$, with (ρ_k, V_k) irreducible representations of the algebra A. This notion agrees with the notion of complete reducibility used before in the case of groups, Thm. 7.1 and comments afterwards, and groupoids, Def. 9.16.

Example 11.2 *Consider the linear space $\mathbb{C}^n$ as an $\mathcal{A}$-module with $\mathcal{A}$ the algebra of $n \times n$ complex matrices $M_n(\mathbb{C})$. Then, $\mathbb{C}^n$ is simple as $\mathcal{A}$-module. The reason is that, if $N \subset \mathbb{C}^n$ is a non-zero linear subspace, then, given any vector $x \in \mathbb{C}^n$ and any vector $v \neq 0$ in N, there is a matrix A, such that $Av = x$, then $N = \mathbb{C}^n$.*

On the other hand, we may consider the subalgebra $\mathcal{A} \subset M_n(\mathbb{C})$, consisting of block diagonal matrices of sizes $n_1, \dots, n_r$ such that $n_1 + \cdots + n_r = n$. Note that $\mathcal{A} = \bigoplus_{k=1}^{r} M_{n_k}(\mathbb{C})$. Then, $\mathbb{C}^n$ is no longer simple as an $\mathcal{A}$-module. In fact, the subspaces V_k of vectors whose entries are zero, except for those located in positions running from $n_1 + n_2 + \cdots + n_{k-1} + 1$ to $n_1 + n_2 + \cdots + n_{k-1} + n_k$, are proper $\mathcal{A}$-submodules, but $\mathbb{C}^n = \bigoplus_{k=1}^{r} V_k$ and, because each submodule is simple, the representation of $\mathcal{A}$ in $\mathbb{C}^n$ is $\mathcal{A}$-semi-simple.

Example 11.3 *Consider the algebra $A = \mathbb{C}[x]$ of polynomials on the variable x and the representation $\rho_L \colon A \to \mathrm{End}(V)$, $\rho_L(p)(v) = p(L)v$, $p(x)$ a polynomial, $v \in V$ and $L \colon V \to V$ is a given linear map. Then, $V = \bigoplus_{k=1}^{r} V_k$, where V_k is a generalized eigenspace corresponding to a Jordan block of L. However, the representation ρ_L is not semi-simple (unless the linear map is diagonalizable) because the subspaces V_k are not simple (provided that they have dimension larger than 1).*

Example 11.4 *Consider an irreducible representation $\rho \colon A \to \mathrm{End}(V)$ of the algebra A. As it was discussed in Example 10.36, the linear space $\mathrm{End}(V)$ can be considered as an A-module, where multiplication by elements $a \in A$ is given by $(a \cdot L)(v) = a \cdot (L(v))$. Then the A-module $\mathrm{End}(V)$ is semi-simple. We just need to realize that $\mathrm{End}(V) \cong V^{\oplus n}$ and that each factor in the decomposition, that is, V, is irreducible by definition.*

Remark 11.4 The last example brings to our attention an interesting fact. The linear space $\text{End}(V)$ is semi-simple as a left-A-module, provided that V supports an irreducible representation of A while, at the same time, $\text{End}(V)$ is itself a simple algebra. ▮

Example 11.5 *An extremely interesting example is provided by the group algebra of a finite group. Consider the algebra $\mathbb{C}[G]$ of the finite group G. The regular representation of G decomposes as the direct sum of irreducible representations of the group G each one with multiplicity its own dimension. This is exactly the content of Burnside's theorem, Thms. 8.9 and 8.7. Then, $\mathbb{C}[G] = \sum_{k=1}^{r} d_k V_k$, where (μ_k, V_k) denote the irreducible representations of G. Hence, we conclude that the regular representation of a finite group is semi-simple.*

11.3 The Jordan-Hölder theorem

We have already obtained a particular form of the Jordan-Hölder theorem when discussing the first properties of representations of groupoids, Thm. 9.6. The same theorem holds for representations of finite-dimensional algebras, i.e, for left-A-modules. Let us see why. As in the case of groupoids, we start by introducing the notion of filtrations of a representation or, in other words, filtrations of A-modules.

Definition 11.3 Let A be a finite-dimensional algebra and (ρ, V) a representation. A (finite) filtration of V is a family of subrepresentations $\mathbf{0} = V_0 \subset V_1 \subset \cdots \subset V_n = V$ (compare with Def. 9.17).

Again, as in the case of groupoids, there are filtrations that are as fine as possible (compare with Lemma 9.3).

Lemma 11.1
Any finite dimensional representation (ρ, V) of a finite dimensional algebra A admits a finite filtration $\mathbf{0} = V_0 \subset V_1 \subset \cdots \subset V_n = V$, such that the successive quotients V_i/V_{i-1} are irreducible.

Exercise 11.6 *Prove the previous Lemma, using the proof of Lemma 9.3 as a guide.*

Definition 11.4 A filtration $\mathbf{0} = V_0 \subset V_1 \subset \cdots \subset V_n = V$ of a given A-module V satisfying the property of Lemma 11.1, that is, such that the quotients V_i/V_{i-1}, $i = 1, \ldots, n$, are irreducible, is called a composition series.

That the factors that appear in a composition series are characteristic of the representation is the content of the Jordan-Hölder theorem (compare with the Jordan-Hölder theorem for groupoids, Thm. 9.6).

Theorem 11.4
(Jordan-Hölder theorem for algebras) Let (ρ,V) be a finite-dimensional representation of the finite-dimensional algebra A. Let $\mathbf{0} = V_0 \subset V_1 \subset V_2 \subset \cdots \subset V_{n-1} \subset V_n = V$, and $\mathbf{0} = V_0' \subset V_1' \subset V_2' \subset \cdots \subset V_{m-1}' \subset V_m' = V$, be two composition series. Then $n = m$ and there exists a permutation σ such that $V_{\sigma(i)}$ is equivalent to V_i'.

Exercise 11.7 *Prove the previous theorem using the insight provided by the proof of Thm. 9.6.*

Thus, the Jordan-Hölder theorem shows that the number of terms in a composition series does not depend on the filtration and depends only on the representation. Such number is sometimes called the length of (ρ,V). Thus, for instance, an irreducible representation has length 1.

Theorem 11.5
Let M be a finite-dimensional semi-simple A-module. If $\mathbf{0} = M_0 \subset M_1 \subset \cdots \subset M_n = V$ is a composition series of (ρ,V), then:

$$V \cong M_1 \oplus M_2/M_1 \oplus \cdots \oplus V/M_{n-1}.$$

Proof 11.6 Clearly, if (ρ,V) is semi-simple, $V \cong \bigoplus_{k=1}^{n} V_n$, where (π_k, V_k) are irreducible representations, i.e., simple A-modules. Consider the filtration given by $M_0 = \mathbf{0}$ and $M_k = \bigoplus_{l=1}^{k} V_l$, $k = 1,\ldots n$. Then, $\mathbf{0} = M_0 \subset M_1 \subset \cdots \subset M_{n-1} \subset M_n = V$ is a composition series because $M_i/M_{i-1} = V_i$, which is irreducible.

If $\mathbf{0} = M_0' \subset M_1' \subset \cdots \subset M_{n-1}' \subset M_m' = V$ is another composition series, because of the Jordan-Hölder theorem, $n = m$ and the quotients are the same up to a permutation, then $V \cong M_1' \oplus M_2'/M_1' \oplus \cdots \oplus V/M_{n-1}'$.

11.4 Semi-simple algebras: The Jacobson radical

11.4.1 On the many roads to semi-simplicity

We would like to extend the notion of semi-simplicity to algebras. Thus, when considering the notion of 'semi-simple' algebras, we can use the principle stated in the introduction to this chapter: "a semi-simple thing is the sum of simple ones". Thus, a semi-simple algebra would be a sum of simple algebras, that is, a direct sum of algebras without proper *bilateral* ideals.

However, any algebra A is also a left A-module, that is, any algebra is the support of its left regular representation $\lambda\colon A \to \mathrm{End}(A)$, $\lambda(a)(x) = ax$, $a,x \in A$. Hence, considering A as a left A-module, we may consider whether if it is semi-simple or, we may declare that A is semi-simple if it is semi-simple as a left-A-module, or if it is the direct sum of simple A-modules.

The two notions of semi-simplicity introduced so far seem to be different because, a submodule W of A, considered as a left-A-module, is just a left ideal,

$A \cdot W \subset W$, then a simple A-module which is a submodule of A is just a left ideal without proper *left ideals*, which apparently is quite different from the perhaps more natural, but more restrictive, notion of semi-simplicity introduced at the beginning of this section. We insist on these concepts, note that the role played by left and bilateral ideals in the previous comments that, in the case of Abelian algebras becomes irrelevant, is fundamental with non-Abelian ones.

Things become even more involved as the reader may check that the standard definition given in many textbooks (although not always) is that "a semi-simple algebra is an algebra A such that its *radical* is zero". Of course, there are very good reasons to proceed in this indirect way as it will be clarified later on but, for the moment, we can anticipate that no matter from where we start, we arrive at the so called Wedderburn's theorem: Any semi-simple algebra is the direct sum of algebras of matrices $M_{d_k}(\mathbb{C})$, thus identifying the algebras $M_{d_k}(\mathbb{C})$ as the 'simple' bricks of the theory.

We could proceed straight away, declaring that a semi-simple algebra is an algebra satisfying Wedderburn's theorem. This would be alright, but proceeding in this way will rob the student of not only the beauty and the thrill of discovering the subtle relations existing between the ideas involved but, more importantly, some basic ideas and results that capture the many facets of semi-simple algebras and that will be used later on.

Then, we will start by using our first proposal as the definition of a semi-simple algebra.

Definition 11.5 We will say that the algebra A is semi-simple if it is the direct sum of simple algebras.

Our first task will be then to describe, more precisely, the structure of simple algebras.

Theorem 11.6
Let A be a finite-dimensional simple algebra, then A is isomorphic to the algebra $\mathrm{End}(V)$ *for some linear space V.*

Proof 11.7 Choose a minimal proper left ideal I of A, that is, I is such that if I' is another left ideal strictly contained in I, then $I' = \mathbf{0}$. Then, I is a left-A-module and it defines an irreducible representation $\rho\colon A \to \mathrm{End}(I)$ (note that I defines an irreducible representation because the ideal I has no proper left ideals, that is it has no proper subrepresentations). Then, because of the fundamental transitive property of irreducible representations, Cor. 11.1, we have that ρ is surjective. Let us check that ρ is injective too. Consider the kernel of ρ. Because ρ is a homomorphism of algebras $\ker\rho$ is a bilateral ideal of A, but A is simple, then $\ker\rho$ is A or 0. However, $\rho \neq 0$, because ρ is surjective and $\mathrm{End}(I) \neq 0$, then $\ker\rho = 0$ and ρ is a monomorphism.

Then, the previous theorem establishes our first version of Weddeburn's theorem: A is semi-simple if it is isomorphic to the direct sum of algebras of the form $\mathrm{End}(V_k)$.

11.4.2 The radical of an algebra

We could ask ourselves 'how far' is a general (finite-dimensional) algebra A from being semi-simple. The way we choose to address this question would be to identify those elements on A that behave 'oddly' with respect to the representations of A.

What do we mean by 'odd' in this context? If A is semi-simple, because of Thm. 11.6, it is a direct sum of endomorphisms of spaces supporting irreducible representations of A, then its non-zero elements decompose as linear maps of spaces supporting irreducible representations, not all of them zero. Then 'odd' elements would we those that 'dissappear' in the irreducible representations of A.

Thus, let us consider an algebra A and $\rho\colon A \to \mathrm{End}(V)$ a representation.

Definition 11.6 We will say that an element $a \in A$ is nilpotent if there is $r > 0$, such that $a^r = 0$. If the non-negative integer r is such that $a^{r-1} \neq 0$, then we say that r is the order of nilpotency of a.

Example 11.8 *1. Recall the algebra $A = \mathbb{C}[x]/(x^2)$ discussed in Example 10.13.3, then $\mathbf{x}^2 = 0$, with $\mathbf{x} = x + (x^2)$ and $\mathbf{x}$ it is a nilpotent element.*

2. Consider the algebra of $n \times n$ matrices. Then, any strictly upper triangular matrix A is nilpotent. For instance, let A be the matrix $J_{0,n}$, Eq. (9.3), corresponding to a Jordan block with eigenvalue 0, then A is nilpotent of order n, that is $A^n = 0$.

Definition 11.7 Let $I \lhd A$ be an ideal. We say that I is nilpotent if $I^r = 0$ for some non-negative integer r.

Remark 11.5 Note that all elements $a \in I$ in a nilpotent ideal are nilpotent, however the ideal generated by a nilpotent element is not nilpotent, in general. ▮

Example 11.9 *1. Consider again Example 11.8.1. The ideal generated by $\mathbf{x}$ is nilpotent.*

2. Consider the algebra A of upper triangular matrices. Then the subalgebra of A consisting of all strictly upper triangular matrices, that is, matrices of the form:

$$\begin{bmatrix} 0 & * & * & \cdots & * \\ 0 & 0 & * & \cdots & * \\ & \ddots & \ddots & \ddots & \vdots \\ & & 0 & 0 & * \\ & & & 0 & 0 \end{bmatrix},$$

form a nilpotent ideal of A.

Example 11.10 *If A is an Abelian algebra, the ideal generated by a nilpotent element is nilpotent and all its elements are nilpotent.*

Exercise 11.11 *Prove that, if A is Abelian, the union of all nilpotent elements is a nilpotent ideal (Hint: Use Newton's binomial theorem).*

That elements of nilpotent ideals do not behave as expected with respect to irreducible representations is shown in the following Lemma.

Lemma 11.2
If $I \triangleleft A$ is a nilpotent ideal, and (ρ, V) is an irreducible representation of A, then $\rho(I) = 0$.

Proof 11.8 Consider a non-zero vector $v \in V$. Then $I \cdot v \subset V$ is a subrepresentation. Then, either it is $\mathbf{0}$ or V. It cannot be the total space V because, in such case, there will be an element $b \in I$, such that $b \cdot v = v$, but then $b^p \cdot v = v$ for all $p \geq 1$ and $b \in I$, which contradicts that b is nilpotent. Then, $I \cdot v = \mathbf{0}$.

The previous Lemma shows that nilpotent ideals are represented as zero by any irreducible representation. Hence, elements in nilpotent ideals are exactly those 'oddly' elements we were searching for. We can collect all of them under the name of the radical of the algebra.

Definition 11.8 Let A be an algebra. The set of all elements a in A, such that $\rho(a) = 0$ for any irreducible representation of A is called the radical, also the Jacobson radical, of A, and will be denoted by $\mathrm{Rad}(A)$ or $J(A)$.

Remark 11.6 The theory of nilpotent ideals is very rich and we will not try to cover any of the subtle issues that are raised immediately from the previous considerations. We will refer to the radical of the algebra A as the Jacobson radical in honour of N. Jacobson who discussed for the first time semi-simplicity in this setting [37]. There are 'inner' and 'outer' characterizations of the Jacobson radical of an algebra. The one we are giving here is outer, that is, it uses the representations of the algebra. Inner ones are given in terms of intersections of maximal ideals and their properties (see, for instance, [44]). ▮

Proposition 11.2
Let A be a finite-dimensional algebra, then the Jacobson radical of A is an ideal and is the largest nilpotent ideal in A.

Proof 11.9 It is easy to show that $J(A)$ is a two-sided ideal. In fact, if $a, b \in J(A)$, $\rho(a+b) = \rho(a) + \rho(b) = 0$ for any irreducible representation. On the other hand, if $a \in J(A)$, then $\rho(xay) = \rho(x) \circ \rho(a) \circ \rho(y) = 0$ and $xay \in J(A)$.

Let $0 \subset A_1 \subset A_2 \subset \cdots \subset A_n = A$ be a filtration of A, such that A_{k+1}/A_k is simple (such filtration exists because A is an A-module and the Jordan-Hölder theorem, Thm.

11.4). Let $a \in J(A)$, then $a \cdot (A_{k+1}/A_k) = 0$, hence, $a \cdot A_{k+1} \subset A_k$, and, consequently, $a^n \cdot A = 0$. Then, $J(A)$ is nilpotent. Moreover, because of Lemma 11.2, any nilpotent ideal I is contained in the Jacobson radical, so the Jacobson radical is the largest nilpotent ideal in A.

Now, we are ready to prove our first structure theorem for finite-dimensional algebras. It tells us that, given any finite-dimensional algebra, its quotient by the Jacobson radical is semi-simple or, in other words, that the Jacobson radical is a good way of measuring how far an algebra is from being semi-simple.

Theorem 11.7

Let A be a finite dimensional unital algebra. Then, the algebra A has only finitely many non-equivalent irreducible representations (π_k, V_k) which are all finite-dimensional. Moreover,

$$A/J(A) \cong \bigoplus_{k=1}^{r} \mathrm{End}\,(V_k)\,.$$

Proof 11.10 Let (π, V) be any irreducible representation of A. Consider a non-zero vector $v \in V$ and the subrepresentation $A \cdot v \subset V$. However, $A \cdot v \neq \mathbf{0}$ as $1 \cdot v = v$. Then, $A \cdot v = V$ and because A is finite-dimensional, V is finite-dimensional too.

Moreover, because of the closure theorem, Thm. 11.3, the map $\rho = \bigoplus_k \pi_k : A \to \bigoplus_k \mathrm{End}\,(V_k)$ is surjective for any finite family of non-equivalent irreducible representations. Then, the family of non-equivalent irreducible representations must be finite (and at most $\dim A$).

Finally, notice that the kernel of ρ is just the intersection of the kernels of the maps π_k, that is the set of elements that vanish under all irreducible representations, that is, the Jacobson radical of A.

As an immediate corollary of Thm. 11.7, we get:

Corollary 11.2

Let (π_k, V_k), $k = 1, \ldots, r$, be non-equivalent irreducible representations of the finite-dimensional algebra A. Then, $\sum_{k=1}^{r} d_k^2 \leq \dim A$, where $d_k = \dim V_k$.

In particular, we already know that the equality holds in the case of the algebra of a finite group (recall Burnside's theorem, Thm. 8.7).

Theorem 11.8

Let A be a finite dimensional algebra. Then, A is semi-simple if and only if $J(A) = 0$.

Proof 11.11 Clearly, if $J(A) = 0$, then by Thm. 11.7, $A \cong \bigoplus_{k=1}^{r} \mathrm{End}\,(V_k)$ and A is semi-simple.

Conversely, if A is semi-simple then A is the direct sum of algebras of endomorphisms and its Jacobson radical must be zero.

An immediate consequence of the previous results is the following characterization of semi-simple algebras.

Corollary 11.3
A finite-dimensional algebra A is semi-simple if and only if $\dim A = \sum_{k=1}^{r} d_k^2$ *where* d_k *are the dimensions of the irreducible representations of A.*

Proof 11.12 If A is semi-simple, by Theorem 11.8, $J(A) = 0$, and because of Theorem 11.7, $\dim A = \sum_{k=1}^{r} d_k^2$.

Conversely, if $\dim A = \sum_{k=1}^{r} d_k^2$, because $\dim J(A) = \dim A - \sum_{k=1}^{r} d_k^2$ (as a consequence of Thm. 11.7 again), we get that $\dim J(A) = 0$, then $J(A) = 0$ and A is semi-simple.

We will end this section by stating Wedderburn's theorem as an immediate corollary of the previous results.

Theorem 11.9
(Wedderburn's theorem) A semi-simple finite-dimensional algebra (over $\mathbb{C}$*) is isomorphic to a direct sum of simple algebras* $M_{d_k}(\mathbb{C})$*:*

$$A \cong M_{d_1}(\mathbb{C}) \oplus \cdots \oplus M_{d_r}(\mathbb{C}) .$$

11.5 Characterizations of semisimplicity

We will finish the discussion on semi-simple algebras by showing, rather surprisingly, that the alternative definitions sketched on Sect. 11.4.1 are equivalent.

First, we note that the algebra $\mathrm{End}(V)$, considered as an $\mathrm{End}(V)$-module, is just $nV = V^{\oplus n}$, with $n = \dim V$, recall Example 10.36. Then if A is semi-simple, because of Wedderburn Theorem Thm. 11.9, $A = \bigoplus_{k=1}^{r} \mathrm{End}(V_k)$, with V_k the support of its irreducible representations (π_k, V_k), but then $A = \bigoplus_{k=1}^{r} d_k V_k$, $d_k = \dim V_k$ and A is the direct sum of simple A-modules. Then we have:

Theorem 11.10
The finite-dimensional algebra A is semi-simple if and only if A is semi-simple as a left A-module.

Proof 11.13 The direct implication was proved in the paragraph previous to the statement of the theorem.

To prove the converse, let us assume that $A = \bigoplus_{k=1}^r n_k V_k$ with V_k simple non-equivalent A-modules. Consider the endomorphisms of the left-A-module A, that is, $\mathrm{End}_A(A)$. Let $\varphi\colon A \to A$, $\varphi \in \mathrm{End}_A(A)$, then $\varphi(n_k V_k) \subset n_k V_k$ for all k (note that, by Schur's lemma, the restriction of the homomorphism φ to any copy of V_k must be zero on any copy of V_j with $j \neq k$ because they are non-equivalent representations of A). Then we get that $\mathrm{End}_A(A) = \bigoplus_{k=1}^r \mathrm{End}_A(n_k V_k)$.

On the other hand, by the same token, $\mathrm{End}_A(V_k) = \mathbb{C}$ because intertwiners of an irreducible representation are multiplication by scalars.

Then, because of (10.6):

$$\mathrm{End}_A(n_k V_k) = (V_k^{\oplus n_k})^* \otimes_A V_k^{\oplus n_k} = (V_k^* \otimes_A V_k)^{\oplus n_k \times \oplus n_k} = M_{n_k}(\mathbb{C}),$$

where in the last identification we have used that $V_k^* \otimes_A V_k = \mathrm{End}_A(V_k) = \mathbb{C}$ (and the trivial identifications (A.3)). Thus, we conclude that $\mathrm{End}_A(A) = \bigoplus_{k=1}^r M_{n_k}(\mathbb{C})$, but because of Prop. 10.7, $\mathrm{End}_A(A) = A^{\mathrm{op}}$, then because of Example 10.16, $A = \sum_{k=1}^r M_{n_k}(\mathbb{C})^{\mathrm{op}} \cong \sum_{k=1}^r M_{n_k}(\mathbb{C})$, and A is semi-simple.

We conclude this section by observing that, if A is semi-simple, any finite-dimensional representation of A is semi-simple, that is, any left-A-module is semi-simple (not just A itself). Actually, to prove this, it suffices to state that any representation is decomposable, that is, that for any subrepresentation $W \subset V$ there exists a supplementary subrepresentation U, $V = W \oplus U$ with W and U left A-submodules of V.

Theorem 11.11

Let A be a finite-dimensional algebra. Then a finite-dimensional representation (ρ, V) of A is completely reducible or semi-simple if and only if any A-submodule W has a supplementary A-submodule.

Proof 11.14 Let $V = \bigoplus_{k=1}^r n_k V_k$, with V_k irreducible representations, be a completely reducible representation. Thus, the direct implication is trivial because if $W \subset V$ is a subrepresentation, because of Thm. 11.1, $W = \bigoplus_k m_k V_k$, $m_k \leq n_n$ with natural inclusion $i = \bigoplus_k i_k$ on each factor $n_k V_k$ of V, and then it suffices to consider the supplementary space of the form $U = \mathbf{0} \oplus \bigoplus_{k=1}^r (n_k - m_k) V_k$.

The converse is proved as follows. Consider a vector $v \in V$ and the smallest irreducible invariant subspace V_1 containing v (if $V_1 = V$ we are done). Then, choose an invariant supplementary space U and repeat the argument. Because V is finite-dimensional, the process will end after a finite number of steps and we have found a decompositon of V in irreducible representations of A.

The result we were searching for follows rather easily:

Theorem 11.12

Let A be a finite-dimensional algebra, then A is semi-simple iff any finite-dimensional left-A-module M is semi-simple or, in other words, any finite-dimensional representation of A is completely reducible.

Proof 11.15 Clearly, if any representation of A is semi-simple, because A is an A-module, then it is a semi-simple A-module and, because of Thm. 11.10, A is a semi-simple algebra.

To prove the direct implication, we will proceed by induction on the dimension n of the representation. The case $n = 1$ is trivial. Let M be an A-module of dimension n. If it is not simple, then it is reducible as a linear representation. This means that there exists a proper submodule M' (i.e., a proper subrepresentation). Then, because of Thm. 11.11, there exists a supplementary submodule M'', that is $M = M' \oplus M''$. However, both submodules M' and M'' have dimensions smaller than n, then, by the induction hypothesis, they are completely reducible as well as their sum. Then, M is completely reducible, i.e., also semi-simple.

We can collect all the many different characterizations of semi-simple algebras we have found so far.

Theorem 11.13

(Characterization of semi-simple algebras) For a finite dimensional unital algebra A, they are equivalent:

1. *A is semi-simple.*
2. *The Jacobson radical of A is zero (i.e., the largest nilpotent two-sided ideal of A is zero).*
3. *A is a direct sum of matrix algebras* $A \cong M_{d_1}(\mathbb{C}) \oplus \cdots \oplus M_{d_r}(\mathbb{C})$.
4. $\dim A = \sum_k (\dim V_k)^2$ *with* V_k *the irreducible representations of A.*
5. *Any finite dimensional representation of A is completely reducible.*
6. *Any A-submodule U of an A-module V has a supplementary factor, that is, there is an A-submodule W, such that* $V = U \oplus W$.

Proof 11.16 The equivalence between (1) and (2) is the content of Thm. 11.8.

The equivalence between (2) and (3) is Weddeburn's theorem, Thm. 11.9.

The equivalence between (2) and (4) is established in Cor. 11.3.

The equivalence between (1) and (5) is provided by Thm. 11.12.

Finally, the equivalence between (5) and (6) was established in Thm. 11.11.

11.6 The algebra of a finite groupoid is semisimple

The main theorem concluding the discussion in this chapter is that the algebra of a finite groupoid is semi-simple.

Theorem 11.14
The groupoid algebra $\mathbb{C}[\mathbf{G}]$ of a finite groupoid $\mathbf{G}$ is semi-simple.

Proof 11.17 Let V be a $\mathbb{C}[\mathbf{G}]$-module and U an $\mathbb{C}[\mathbf{G}]$-submodule. Let U' be a supplementary subspace to U, that is, $V = U \oplus U'$. Let $P_U : V \to U$ be the projector onto U parallel to U', that is $P_U^2 = P_U$, $P_U\mid_U = \mathrm{id}_U$, $\ker P_U = U'$. Define:

$$P(v) = \frac{1}{|\mathbf{G}|}\sum_{\alpha\in\mathbf{G}} \alpha^{-1}\cdot(P_U(\alpha\cdot v)),$$

or, more compactly,

$$P = \frac{1}{|\mathbf{G}|}\sum_{\alpha\in\mathbf{G}} \alpha^{-1} P_U\,\alpha\,.$$

Clearly P is a projector $P^2 = P$, $P\mid_U = \mathrm{id}_U$.

Consider now the subspace $W = \ker P$, which is a supplementary subspace to U: $V = U \oplus W$.

Then, we notice that W is an $\mathbb{C}[\mathbf{G}]$-submodule. For any $v \in W = \ker P$:

$$\begin{aligned} P(\beta\cdot v) &= \frac{1}{|\mathbf{G}|}\sum_{\alpha\in\mathbf{G}} \alpha^{-1}\cdot P_U(\alpha\cdot\beta\cdot v) = \frac{1}{|\mathbf{G}|}\sum_{\alpha\in\mathbf{G}} \alpha^{-1}\cdot P_U((\alpha\cdot\beta)v) \\ &= \frac{1}{|\mathbf{G}|}\sum_{\alpha'\in\mathbf{G}} \beta\cdot\alpha'^{-1}\cdot P_U(\alpha'\cdot v) = \beta\cdot\left(\frac{1}{|\mathbf{G}|}\sum_{\alpha'\in\mathbf{G}} \alpha'^{-1}\cdot P_U(\alpha'\cdot v)\right) \\ &= \beta\cdot(Pv) = 0\,. \end{aligned}$$

Then because of the characterization of semi-simple algebras provided by Thm. 11.13.6, the algebra $\mathbb{C}[\mathbf{G}]$ of the groupoid $\mathbf{G}$ is semi-simple.

As an immediate consequence of Thm. 11.14, we obtain:

Corollary 11.4
Let $\mathbf{G}$ be a finite groupoid. Then:

$$\mathbb{C}[\mathbf{G}] \cong M_{n_1}(\mathbb{C})\cdots\oplus M_{n_r}(\mathbb{C}),$$

and $|\mathbf{G}| = \sum_{k=1}^{r} n_k^2$.

Proof 11.18 The proof is a direct consequence of Wedderburn theorem, Thm. 11.13.3,4: Any semi-simple finite-dimensional algebra A is the direct sum of matrix algebras: $A \cong M_{n_1}(\mathbb{C}) \oplus \cdots \oplus M_{n_r}(\mathbb{C})$. Hence, if we consider the algebra $\mathbb{C}[\mathbf{G}]$ of the finite groupoid $\mathbf{G}$, because it is semi-simple, it will be the direct sum of matrix algebras.

If the groupoid $\mathbf{G}$ is a finite group, the previous corollary and theorems reproduce Maschke's theorem, a central result in the theory of linear representations of groups.

Corollary 11.5

(Maschke's theorem) The group algebra $\mathbb{C}[G]$ of a finite group G is semi-simple.

Then,

$$\mathbb{C}[G] \cong M_{d_1}(\mathbb{C}) \oplus \cdots \oplus M_{d_r}(\mathbb{C}),$$

and,

$$|G| = \sum_{k=1}^{r} d_k^2 .$$

Actually, in that vein, we may refine the last corollary by stating that the factors appearing in the decomposition of the groupoid algebra are in one-to-one correspondence with the irreducible representations of the group and they provide the canonical decomposition of its regular representation, result that we would like to prove in the setting of groupoids. This will be the purpose of next chapter.

Chapter 12

The Structure of Linear Representations of Finite Groupoids

We will close the analysis of the theory of linear representations of finite groupoids started in the previous chapters: The study and description by elementary means of linear representations of finite groups, the theory of characters of linear representations of finite groups, the study of representations of categories and the relation of linear representations of groupoids with the theory of modules over the algebra of a groupoid.

We will study the regular representation of a finite groupoid and will be able to extend the theorem already obtained in the case of groups, Thm. 8.9, proving that all irreducible representations are factors of the regular representation and providing a decomposition formula for any finite-dimensional representation of a given finite groupoid. The analysis will be done again by using elementary methods. We will exploit the powerful notion of characters, but now in the setting of modules.

Finally, a structure theorem for the decomposition of the regular representation of a finite groupoid will be offered based on the structure theorem of groupoids. It will be shown that any representation of a connected groupoid can be written as a direct sum of tensorial products of the canonical representation of the pair groupoid and irreducible representations of the isotropy group. In this sense, the theory of representations of groupoids can be reconstructed from the theory of representations of groups.

12.1 Characters again

The last chapter was devoted to unveiling the notion of semi-simplicity, where Schur's Lemma played a paramount role in the various proofs (recall, for instance, the proofs of Thm. 11.1, Prop. 11.1, Thm. 11.6, or Lem. 11.2).

In the case of groups, it was clear that Schur's lemmas, Sect. 7.4, were the guide to arriving at the main theorems on the structure of their representations: Schur's lemmas were used to establish orthogonality properties (Cor. 8.1) that, in the particular instance of characters (Sect. 8.4), allowed us to prove that simple characters provide an orthonormal basis for the space of central functions, and the multiplicity of an irreducible representation could be obtained just by taking the inner product of the corresponding characters.

In the present chapter, we will follow a similar path in order to describe the representations of groupoids. A theory of characters of representations of groupoids, in the spirit of that of groups, will be developed, however, before embarking on that, we would like to stress that a substantial part of the results obtained for the theory of representations of groups are valid for representations of finite-dimensional algebras, of which, the algebra of a group is just an instance.

12.1.1 *Characters of representations of algebras and groupoids*

As in the case of groups (Sect. 8.2), an important tool in the description of representations of finite dimensional algebras is provided by their characters. The character χ of a finite-dimensional representation will be defined in the standard way, that is:

Definition 12.1 Let $R\colon \mathbb{C}[\mathbf{G}] \to \mathrm{End}(V)$ be a finite-dimensional representation of the algebra of the finite groupoid $\mathbf{G}$, the character of the representation R is defined as the trace of the linear operator $R(a)$ for any $a \in \mathbb{C}[\mathbf{G}]$:

$$\chi_R(a) = \mathrm{Tr}\,(R(a)),$$

Proposition 12.1
Let R be a finite-dimensional representation of the groupoid $\mathbf{G}$, then, if u is a unitary element in the algebra of the groupoid, $\chi_R(u^{-1}) = \chi_R(u^*) = \overline{\chi_R(u)}$.

Proof 12.1 If u is unitary, we have: $u^{-1} = u^*$, and we get $R(u \cdot u^*) = R(\mathbf{1}_\Omega) = \mathrm{id}_V$. On the other hand $R(u \cdot u^*) = R(u)R(u^*)$ and $R(u^*) = R(u)^{-1} = R(u^{-1})$, then $\chi_R(u^\star) = \chi_R(u^{-1})$.

Regarding the third identity, let us observe that it is always possible to equip our linear space with a Hilbert space structure $\langle \cdot, \cdot \rangle$ (recall Sect. 7.3.1). Even more, it is possible to choose one such that the unitaries of A will be represented by unitary operators. For that, we observe that the collection of unitary elements define a group and, because A is finite dimensional, this group is isomorphic to the group of unitary matrices and the conclusion follows.

It follows directly from the properties of the trace that characters are central functions on the groupoid algebra, that is:

$$\chi_R(a \cdot b) = \chi_R(b \cdot a) ,$$

for all a, b in $\mathbb{C}[\mathbf{G}]$.

Definition 12.2 Let A be a finite dimensional algebra, a central function on A is any function $C\colon A \to \mathbb{C}$, such that $C(ab) = C(ba)$ or, in other words, a function vanishing in the derived algebra of A which is the subalgebra of A, usually denoted by A' or $[A,A]$, generated by elements of the form $ab - ba$ (also denoted as $[a,b]$).

Denoting by $\mathbb{C}[\mathbf{G}]' = [\mathbb{C}[\mathbf{G}], \mathbb{C}[\mathbf{G}]]$ the derived algebra of the groupoid algebra, it is clear that a character χ induces a map (denoted with the same symbol by a small abuse of notation) $\chi\colon \mathbb{C}[\mathbf{G}]/\mathbb{C}[\mathbf{G}]' \to \mathbb{C}$.

Characters of different irreducible finite-dimensional representations of $\mathbb{C}[\mathbf{G}]$ are linearly independent, results that we already found by exploiting Schur's Lemmas in the case of groups.

Proposition 12.2
Characters corresponding to non-equivalent irreducible representations of the algebra A are linearly independent.

Proof 12.2 Consider non-equivalent irreducible representations ρ_1, ρ_2, then, because of the closure theorem, Thm. 11.3, the map $\rho_1 \oplus \rho_2\colon A \to \mathrm{End}(V_1) \oplus \mathrm{End}(V_2)$ is surjective.

Suppose that there are real numbers λ_1, λ_2, such that $\lambda_1 \chi_1 + \lambda_2 \chi_2 = 0$, where χ_1, χ_2 are the characters of ρ_1 and ρ_2, respectively. Then, $\mathrm{Tr}(\lambda_1 \rho_1(a) + \lambda_2 \rho_2(a)) = 0$ for all $a \in A$, hence:

$$\lambda_1 \mathrm{Tr}(L_1) + \lambda_2 \mathrm{Tr}(L_2) = 0 ,$$

with L_1, L_2 arbitrary endomorphisms in V_1 and V_2, respectively ($\rho_1 \oplus \rho_2$ is surjecive!). Then $\lambda_1 = \lambda_2 = 0$.

The previous result offers another proof (Thm. 11.7) of the fact that there are just a finite number of non-equivalent irreducible representations because their characters will define linearly independent maps on the algebra A/A' which is finite-dimensional (we already knew that they must be finite-dimensional, Prop. 9.7). Furthermore, for semi-simple algebras, it can be proved that characters of irreducible representations form a linear basis of the dual space of the derived algebra A', again, a result that we found in the case of groups. We will not continue this line of inquiry and we will concentrate our attention instead on the case where the algebra A is the algebra of a groupoid, trying to make contact with the beautiful results obtained for the theory of characters of representations of finite groups.

In the particular instance of the algebra of the groupoid $\mathbf{G}$, we may consider the characters as functions defined in the groupoid itself, in other words, given a representation (R,V), we consider the character χ_R as the function:

$$\chi_R : \mathbf{G} \to \mathbb{C}, \qquad \chi_R(\alpha) = \mathrm{Tr}(R(\alpha)).$$

In what follows, we will use characters in this sense.

Notice that, if χ_R and χ_S denote the characters of the representations R and S, respectively, then we would like to know what are the characters corresponding to the representations $R \oplus S$, $R \otimes S$ and R^*.

Proposition 12.3
Let (R,V), (S,W) be two finite-dimensional representations of the groupoid $\mathbf{G}$, then we get:

1. $\chi_{R\oplus S} = \chi_R + \chi_S$,
2. $\chi_{R\otimes S} = \chi_R \cdot \chi_S$,

Proof 12.3 Statements (1) and (2) follow inmediately from the properties of the trace (Prop. A.14.5,6).

12.1.2 Orthogonality of characters

The previous observations are valid for characters of representations of arbitrary finite-dimensional algebras, however, in the particular instance of the algebra of a finite groupoid, we can be more specific.

Let R be a representation of the groupoid algebra $\mathbb{C}[\mathbf{G}]$ on the linear space V, i.e., V is a $\mathbb{C}[\mathbf{G}]$-module, then the character χ_R of the representation R is given by:

$$\chi_R(a) = \sum_{x\in\Omega} \sum_{\gamma_x \in G_x} a_{\gamma_x} \chi_R(\gamma_x), \qquad \forall a = \sum_{\alpha} a_\alpha \, \alpha \in \mathbb{C}[\mathbf{G}],$$

Actually, we may state the previous observation as:

Proposition 12.4
Let χ_R be the character of a finite-dimensional representation R of the finite groupoid $\mathbf{G}$, then

$$\chi_R = \sum_{x\in\Omega} \chi_x, \tag{12.1}$$

where $\chi_x : G_x \to \mathbb{C}$ is the character of the restriction (μ_x, V_x) of the representation R to the isotropy group G_x. The r.h.s. of the previous equation (12.1) must be understood as the character of the natural restriction of the representation R to the fundamental subgroupoid $\mathbf{G}_0$ to $\mathbf{G}$, i.e., $\chi_x(\alpha) = \chi_R(\alpha)$ if $\alpha \in G_x$, and $\chi_x(\alpha) = 0$ if $\alpha \notin G_x$.

Proof 12.4 If R is a representation of the groupoid $\mathbf{G}$, then $R(\alpha)\colon V_x \to V_y$, $\alpha\colon x \to y$, where V_x is the linear space associated to x by the functor R or, equivalently, the subspace $V_x = R(1_x)V$ of the $\mathbb{C}[\mathbf{G}]$-module V defined by R.

Choosing a linear basis $\{e_i(x)\}$ for V_x, $\mathcal{B} = \cup_{x\in\Omega}\{e_i(x)\}$ is a basis of V. Then, it is clear that if $\alpha\colon x \to y$, $x \neq y$, $\mathrm{Tr}\,(R(\alpha)) = 0$, hence, $\chi_R(\alpha)$ is different from zero only if $\alpha \in G_x$ for some x. However, $\chi_R(\gamma_x) = \mathrm{Tr}\,(R(\gamma_x)\mid_{V_x}) = \mathrm{Tr}\,(\mu_x(\gamma_x))$ with $\mu_x\colon G_x \to \mathrm{End}\,(V_x)$ the restriction of the representation R to G_x and V_x. Then, $\chi_R(\gamma_x) = \chi_x(\gamma_x)$, with χ_x the character of μ_x.

In other words, the previous proposition shows that, when considering the character χ_R as a function defined on $\mathbf{G}$, its support, i.e., the subset where it does not vanish, is the fundamental isotropy subgroupoid $\mathbf{G}_0$. Moreover, we have already pointed out that, because all isotropy groups corresponding to objects in the same connected component of a groupoid are isomorphic, the restrictions of a given representation are all equivalent, Prop. 9.4. Then, the previous results show:

Proposition 12.5

If χ is the character of a finite-dimensional linear representation (R,V) of a connected groupoid $\mathbf{G}$, then the characters χ_x of the representations (μ_x, V_x) defined on each one of the isotropy groups G_x by restriction of the representation R, are all equal: $\chi_x = \chi_y$. Moreover, we get:

$$\chi_R(\mathbf{1}) = \sum_{x\in\Omega} \chi_x(1_x) = |\Omega| d\,.$$

where d denotes the local dimension of the representation R, that is, the dimension of the linear space V_x for any x, and $|\Omega|$ denotes the number of elements of Ω.

Note that $\chi_R(\mathbf{1}) = |\Omega|d$ is the dimension of the total space V of the representation, in full agreement with the results in the case of groups. In particular, if π_0 denotes the fundamental representation of the groupoid, we have that:

$$\chi_{\pi_0}(\alpha) = \begin{cases} 1 & \alpha \in G_x\,, x \in \Omega\,, \\ 0 & \text{otherwise.} \end{cases} \tag{12.2}$$

Then,

$$\chi_{\pi_0}(\mathbf{1}) = |\Omega|\,.$$

The previous observations suggest to define the following inner product on the space $\mathcal{F}(\mathbf{G})$ of complex-valued functions on $\mathbf{G}$:

$$\langle \psi, \phi \rangle_{\mathbf{G}} = \frac{|\Omega|}{|\mathbf{G}|} \sum_{\alpha\in\mathbf{G}} \overline{\psi(\alpha)}\phi(\alpha)\,. \tag{12.3}$$

Note that, in the particular instance of groups, the inner product $\langle\cdot,\cdot\rangle_{\mathbf{G}}$ becomes the standard normalized inner product, Eq. (8.4).

In particular, when we restrict the inner product to characters, we get (assuming that the groupoid is connected):

$$\begin{aligned}\langle \chi, \chi' \rangle_{\mathbf{G}} &= \frac{|\Omega|}{|\mathbf{G}|} \sum_{\alpha \in \mathbf{G}} \overline{\chi(\alpha)} \chi'(\alpha) = \frac{|\Omega|}{|\mathbf{G}|} \sum_{\gamma \in \mathbf{G}_0} \overline{\chi(\gamma)} \chi'(\gamma) \\ &= \frac{|\Omega|}{|\mathbf{G}|} \sum_{x \in \Omega} \sum_{\gamma_x \in G_x} \overline{\chi(\gamma_x)} \chi'(\gamma_x) = \frac{1}{|\Omega|} \sum_{x \in \Omega} \frac{1}{|G_x|} \sum_{\gamma_x \in G_x} \overline{\chi(\gamma_x)} \chi'(\gamma_x) \\ &= \frac{1}{|\Omega|} \sum_{x \in \Omega} \langle \chi_x, \chi'_x \rangle_{G_x},\end{aligned}$$

where we have used that $|\mathbf{G}| = |\Omega|^2 |G_x|$ (Cor. 6.1) and, as in Prop. 12.5, we have denoted by χ_x the character of the restriction of the representation R to G_x. Then, we obtain the following formula for the inner product of characters on a connected groupoid $\mathbf{G}$:

$$\langle \chi, \chi' \rangle_{\mathbf{G}} = \frac{1}{|\Omega|} \sum_{x \in \Omega} \langle \chi_x, \chi'_x \rangle_{G_x}, \tag{12.4}$$

where $\langle \chi_x, \chi'_x \rangle_{G_x}$ denotes the standard normalized inner product of characters on the finite group G_x.

We will denote, as in the case of groups, by $\mathrm{Hom}_{\mathbf{G}}(V_1, V_2)$ the linear space of intertwiners between two representations (R_1, V_1) and (R_2, V_2), that is, $\mathrm{Hom}_{\mathbf{G}}(V_1, V_2) = \mathrm{Hom}_{\mathbb{C}[\mathbf{G}]}(V_1, V_2) = \{\varphi \colon V_1 \to V_2 \mid \varphi \circ R_1 = R_2 \circ \varphi\}$.

We have already shown that any finite-dimensional representation (R, V) of a groupoid is completely reducible, Thm. 9.5. Then, in full analogy with the theory of characters for representations of groups, Thm. 8.6, we have:

Theorem 12.1

Let (R, V) be a finite dimensional linear representation of the finite connected groupoid $\mathbf{G}$. The decomposition of (R, V) as sum of irreducible representations, $R = \bigoplus_{\nu=1}^{r} n_\nu R_\nu$, $V = \bigoplus_{\nu=1}^{r} n_\nu V_\nu$, is essentially unique in the sense that the multiplicity n_ν of the factors V_ν is unique; its character is given by $\chi = \sum_{\nu=1}^{r} n_\nu \chi_\nu$ and the multiplicity n_ν is given by: $n_\nu = \langle \chi_\nu, \chi \rangle_{\mathbf{G}}$.

In order to prove the previous theorem, we will need to establish the corresponding orthogonality relations for the characters of a finite groupoid. The following theorem, the counterpart for groupoids of Thm. 8.7 for groups, shows that the characters of the groupoid $\mathbf{G}$ form an orthonormal basis of the space of central functions.

Theorem 12.2

Let $R_a \colon \mathbb{C}[\mathbf{G}] \to \mathrm{End}(V_a)$, $a = 1, 2$, be two representations of the finite connected groupoid $\mathbf{G}$, then:

$$\langle \chi_{R_1}, \chi_{R_2} \rangle_{\mathbf{G}} = \dim \mathrm{Hom}_{\mathbf{G}}(V_1, V_2),$$

and, $\langle \chi_{R_1}, \chi_{R_2} \rangle_{\mathbf{G}} = 1$ if R_1 is equivalent to R_2 and 0 otherwise, if and only if (R_a, V_a), $a = 1, 2$, are irreducible.

Proof 12.5 The theorem follows easily from Thm. 9.4. More explicitly, we have:

$$\langle \chi_{R_1}, \chi_{R_2} \rangle_{\mathbf{G}} = \frac{1}{|\Omega|} \sum_{x \in \Omega} \langle \chi_x^{(1)}, \chi_x^{(2)} \rangle_{G_x},$$

with $\chi_x^{(a)}$ the character of the restriction of the representation R_a, $a = 1, 2$, to the isotropy subgroup G_x. However, from the theory of representations of finite groups, Thm. 8.7, we have:

$$\langle \chi_x^{(1)}, \chi_x^{(2)} \rangle_{G_x} = \dim \mathrm{Hom}_{G_x}(V_x^{(1)}, V_x^{(2)}),$$

and

$$\langle \chi_x^{(1)}, \chi_x^{(2)} \rangle_{G_x} = 1,$$

if and only if the representations $R_x(a)$, $a = 1, 2$, are irreducible and equivalent, and zero otherwise. Then, we have

$$\langle \chi_{R_1}, \chi_{R_2} \rangle_{\mathbf{G}} = \frac{1}{|\Omega|} \sum_{x \in \Omega} \dim \mathrm{Hom}_{G_x}(V_x^{(1)}, V_x^{(2)}) = \dim \mathrm{Hom}_{\mathbf{G}}(V^{(1)}, V^{(2)}).$$

Finally, we observe that if R is an irreducible representation of the groupoid $\mathbf{G}$, then the restriction of R to any G_x is irreducible too, Thm. 9.3, and we get:

$$\langle \chi_{R_1}, \chi_{R_2} \rangle_{\mathbf{G}} = \frac{1}{|\Omega|} \sum_{x \in \Omega} \dim \mathrm{Hom}_{G_x}(V_x^{(1)}, V_x^{(2)}) = 1,$$

and zero otherwise.

Conversely, if $\langle \chi_{R_1}, \chi_{R_2} \rangle_{\mathbf{G}} = 1$, then because all restrictions μ_x are equivalent, $\langle \chi_x, \chi_x \rangle_{G_x} = \langle \chi_y, \chi_y \rangle_{G_y}$, we get:

$$1 = \langle \chi_{R_1}, \chi_{R_2} \rangle_{\mathbf{G}} = \frac{1}{|\Omega|} \sum_{x \in \Omega} \langle \chi_x, \chi_x \rangle_{G_x} = \langle \chi_x, \chi_x \rangle_{G_x},$$

and, because of the characterization of simple characters of group representations, the representation μ_x is irreducible, hence, R is irreducible.

Exercise 12.1 *Once we have established our previous theorem, the proof of Thm. 12.1 is an easy exercise.*

Example 12.2 *From Eq. (12.2) we get immediately that $\langle \chi_{\pi_0}, \chi_{\pi_0} \rangle_{\mathbf{G}} = 1$, hence the fundamental representation is irreducible and it restricts to the trivial representations on each isotropy subgroup G_x (recall the discussion in Sect. 10.5.1, Prop. 10.10).*

In what follows, we will denote equivalence classes of irreducible representations of a groupoid $\mathbf{G}$ by π_ν, $\nu = 1,\ldots,r$. Suppose that R is a finite-dimensional linear representation of the finite groupoid $\mathbf{G}$. It decomposes as a direct sum of irreducible representations π_ν, then: $R = \bigoplus n_\nu \pi_\nu$, where n_ν is the multiplicity of the representation π_ν in R. If we denote by χ_ν the character of the irreducible representation π_ν (with this notation, $\chi_0 = \chi_{\pi_0}$) and by χ_R the character of R, we will get again, as in Thm. 12.1:

$$\langle \chi_R, \chi_\nu \rangle_{\mathbf{G}} = n_\nu \,. \tag{12.5}$$

Moreover, from Eq. (12.4), it follows that:

$$\langle \chi_R, \chi_\nu \rangle_{\mathbf{G}} = \frac{1}{|\Omega|} \sum_{x\in\Omega} \langle \chi_x, \chi_x^{(\nu)} \rangle_{G_x} = \langle \chi_x, \chi_x^{(\nu)} \rangle_{G_x} \,,$$

where, as before, $\chi_x^{(\nu)}$ denotes the character of the restriction of the irreducible representation π_ν to the isotropy group G_x, and we have used that the restrictions of both representations R and π_ν to all isotropy groups on the same connected component are equivalent among them. Thus, the previous formula shows that n_ν is also the multiplicity of the irreducible representation of the isotropy group G_x with character χ_x^ν in the restriction of R to G_x. On the other hand we get immediately from the previous remarks that:

$$\chi_\nu(\mathbf{1}) = |\Omega| d_\nu \,, \tag{12.6}$$

with d_ν the dimension of the subspace V_x^ν supporting the irreducible representation of G_x with character $\chi_x^{(\nu)}$.

12.2 Operations with groupoids and representations

As in the case of groups, there is a tight relation between the direct sum and tensor product of irreducible representations of groupoids and the irreducible representations of the disjoint union and the direct product, respectively.

We will start with the discussion of the tensor product of two irreducible representations of two groupoids $\mathbf{G}_1$ and $\mathbf{G}_2$, respectively.

Proposition 12.6
Let $\mathbf{G}_a \rightrightarrows \Omega_a$, $a = 1,2$, be two finite connected groupoids and (R_1,V_1), (R_2,V_2) two finite-dimensional representations of $\mathbf{G}_1$ and $\mathbf{G}_2$, respectively. Then, the tensor product representation $(R_1 \otimes R_2, V_1 \otimes V_2)$ of the direct product groupoid $\mathbf{G}_1 \times \mathbf{G}_2$ is irreducible if and only if (R_1,V_1) and (R_2,V_2) are irreducible.

Proof 12.6 The proof follows easily from the observation that the character $\chi_{R_1\otimes R_2}$ of the tensor product of the representations (R_1,V_1) and (R_2,V_2) can be written as, Eq. (12.1):

$$\chi_{R_1\otimes R_2} = \sum_{(x_1,x_2)\in\Omega_1\times\Omega_2} \chi_{(x_1,x_2)} \,,$$

but, the restriction of the representation $R_1 \otimes R_2$ to the isotropy group $(\mathbf{G}_1 \otimes \mathbf{G}_2)_{(x_1,x_2)}$ is the tensor product of the restriction of the representations R_1 and R_2 to the isotropy groups $(G_1)_{x_1}$ and $(G_2)_{x_2}$, respectively. Then, because of Prop. (8.9), the character $\chi_{(x_1,x_2)}$ is the tensor product of the corresponding characters, i.e., $\chi_{(x_1,x_2)} = \chi_{x_1}^{(1)} \otimes \chi_{x_2}^{(2)}$ and, consequently, we get:

$$\sum_{(x_1,x_2)\in\Omega_1\times\Omega_2} \chi_{(x_1,x_2)} = \sum_{(x_1,x_2)\in\Omega_1\times\Omega_2} \chi_{x_1}^{(1)} \otimes \chi_{x_2}^{(2)} = \chi_{R_1} \otimes \chi_{R_2} \, .$$

Then:

$$\langle \chi_{R_1\otimes R_2}, \chi_{R_1\otimes R_2} \rangle_{\mathbf{G}_1\times\mathbf{G}_2} = \langle \chi_{R_1}, \chi_{R_1} \rangle_{\mathbf{G}_1} \langle \chi_{R_2}, \chi_{R_2} \rangle_{\mathbf{G}_2} \, ,$$

and the conclusion follows.

As in the case of groups, the tensor product of two representations of the same groupoid $\mathbf{G}$ can be restricted to be considered as a representation of the groupoid $\mathbf{G}$ considered as a subgroupoid of the direct product $\mathbf{G} \times \mathbf{G}$ by means of the diagonal map $x \mapsto (x,x)$, $x \in \Omega$, and $\alpha \mapsto (\alpha,\alpha)$. Thus, α acts on an element of $V_1 \otimes V_2$ as $R_1(\alpha)v_1 \otimes R_2(\alpha)v_2$, for any $v_1 \in V_1$, $v_2 \in V_2$.

Of course, the representation of the groupoid $\mathbf{G}$ on $V_1 \otimes V_2$ obtained in this way will not be irreducible anymore and it will decompose as a direct sum of the irreducible representations of $\mathbf{G}$, then we could write:

$$R_k \otimes R_l = \bigoplus_{m\in\widehat{\mathbf{G}}} n_{k,l,m} R_m \, ,$$

where R_k are the irreducible representations of $\mathbf{G}$ and k,l,m are indices labelling them. Now, we can compute the frequency $n_{k,l,m}$ by considering the characters of the corresponding representation and using Eq. (12.5). Then:

$$n_{k,l,m} = \langle \chi_{R_k\otimes R_l}, \chi_m \rangle_{\mathbf{G}} = \frac{1}{|\Omega|} \sum_{x\in\Omega} \langle \chi_x^k \cdot \chi_x^l, \chi_x^m \rangle_{G_x} = C(k,l,m) \, ,$$

which is exactly the Clebsch-Gordan coefficient of the decomposition of the tensor product of the irreducible representations μ_x^k, μ_x^l.

It is particularly interesting to observe that, apart from the tensor product, we may also construct the direct sum of the representations of two groupoids. Then, in perfect analogy with Prop. 12.6, we get:

Proposition 12.7
Let $\mathbf{G}_a \rightrightarrows \Omega_a$, $a = 1,2$, be two finite connected groupoids and (R_1,V_1), (R_2,V_2) two finite-dimensional representations of $\mathbf{G}_1$ and $\mathbf{G}_2$, respectively. Then the direct sum representation $(R_1 \oplus R_2, V_1 \oplus V_2)$ of the disjoint union groupoid $\mathbf{G}_1 \bigsqcup \mathbf{G}_2$ has a decomposition in irreducible representations $R_1 \oplus R_2$ if and only if (R_1,V_1) and (R_2,V_2) are irreducible.

Proof 12.7 Note that the disjoint union groupoid $\mathbf{G}_1 \bigsqcup \mathbf{G}_2$ is not connected and, if $\mathbf{G}_a \rightrightarrows \Omega_a$, $a = 1, 2$ are connected groupoids, its connected components are $\mathbf{G}_1$ and $\mathbf{G}_2$, respectively. Then the statement of the proposition follows immediately by restricting the representation $R_1 \oplus R_2$ to any of the connected components of $\mathbf{G}_1 \bigsqcup \mathbf{G}_2$.

Remark 12.1 Computing the norm of the character of the direct sum $R_1 \oplus R_2$ we get:

$$\begin{aligned}
\langle \chi_{R_1 \oplus R_2}, \chi_{R_1 \oplus R_2} \rangle_{\mathbf{G}_1 \sqcup \mathbf{G}_2} &= \frac{1}{|\Omega_1 \sqcup \Omega_2|} \left(\sum_{x_1 \in \Omega_1} \langle \chi^1_{x_1}, \chi^1_{x_1} \rangle_{G_{1,x_1}} + \sum_{x_2 \in \Omega_2} \langle \chi^2_{x_2}, \chi^2_{x_2} \rangle_{G_{2,x_2}} \right) \\
&= \frac{1}{|\Omega_1| + |\Omega_2|} \left(|\Omega_1| \langle \chi^1, \chi^1 \rangle_1 + |\Omega_2| \langle \chi^2, \chi^2 \rangle_2 \right).
\end{aligned}$$

but R_1 and R_2 are irreducible, then $\langle \chi_{R_1 \oplus R_2}, \chi_{R_1 \oplus R_2} \rangle_{\mathbf{G}_1 \sqcup \mathbf{G}_2} = 1$. However, $R_1 \oplus R_2$ is not irreducible as $V_1 \oplus \mathbf{0}$ and $\mathbf{0} \oplus V_2$ are invariant subspaces, then the irreducible components will be the corresponding representation R_a. Note that this is not contradictory with Thm. 12.2 characterizing the characters of irreducible representations because, the disjoint union groupoid $\mathbf{G}_1 \bigsqcup \mathbf{G}_2$ is not connected. ■

12.3 The left and right regular representations of a finite groupoid

12.3.1 Burnside's theorem for groupoids

The regular representations (left and right) of a finite groupoid $\mathbf{G}$ are supported on the linear space $\mathcal{F}(\mathbf{G})$ of complex-valued functions on it[1] which has dimension $|\mathbf{G}|$. The left regular representation is defined by the homomorphism of algebras, $\lambda : \mathbb{C}[\mathbf{G}] \to \mathrm{End}(\mathcal{F}(\mathbf{G}))$ induced by the linear maps $\lambda(\alpha)$ given by:

$$(\lambda(\alpha)f)(\beta) = f(\alpha^{-1} \circ \beta),$$

provided that $t(\alpha) = t(\beta)$, and zero otherwise. Then, note:

$$\begin{aligned}
(\lambda(\alpha \circ \alpha')f)(\beta) &= f(\alpha'^{-1} \circ (\alpha^{-1} \circ \beta)) \\
&= (\lambda(\alpha')f)(\alpha^{-1} \circ \beta) \\
&= (\lambda(\alpha)(\lambda(\alpha')f))(\beta), \qquad (12.7)
\end{aligned}$$

and $\lambda(\alpha \circ \alpha') = \lambda(\alpha) \circ \lambda(\alpha')$ for all composable α and α'.

Similarly, we define the right regular representation as the $\mathbb{C}[\mathbf{G}]$-module structure on $\mathcal{F}(\mathbf{G})$ defined by the homomorphism $\rho : \mathbb{C}[\mathbf{G}] \to \mathrm{End}(\mathcal{F}(\mathbf{G}))$:

$$(\rho(\alpha)f)(\beta) = f(\beta \circ \alpha), \quad t(\alpha) = s(\beta),$$

[1] That, because of the finiteness of $\mathbf{G}$, is identified with the Hilbert space of square integrable functions on $\mathbf{G}$ (with respect to the counting measure).

and we check, as in (12.7), that $\rho(\alpha \circ \alpha') = \rho(\alpha) \circ \rho(\alpha')$ whenever α and α' can be composed.

Both, the left and right, regular representations of the groupoid $\mathbf{G}$ are unitary, and because they commute with each other, that is $\rho(a)\lambda(b) = \lambda(b)\rho(a)$, they will have the same structure.

Exercise 12.3 *Show that $\rho(a)\lambda(b) = \lambda(b)\rho(a)$, for all $a, b \in \mathbb{C}[\mathbf{G}]$.*

We will concentrate on the right representation ρ.

Lemma 12.1

Given a unit 1_x, the projector $P_x^\rho = \rho(1_x)$ is given by the characteristic function on the subset $\mathbf{G}_+(x)$, that is:

$$P_x^\rho(f) = \chi_{\mathbf{G}_+(x)} f \,. \tag{12.8}$$

Proof 12.8 From the definition of the right regular representation, we get:

$$(P_x^\rho f)(\beta) = (\rho(1_x) f)(\beta) = f(\beta \circ 1_x) = f(\beta) \,,$$

provided that $s(\beta) = x$. If $s(\beta) \neq x$, we get $(P_x^\rho f)(\beta) = 0$, then P_x^ρ is just the characteristic function on the set of morphisms $\beta : x \to y$, that is on the set $\mathbf{G}_+(x)$.

In other words, the right regular representation ρ is the functor assigning to any object x, the linear subspace $W_x^+ = \mathcal{F}(\mathbf{G}_+(x))$ and to any morphism $\alpha : x \to y$ the linear map $\rho(\alpha) : W_x^+ \to W_y^+$, $(\rho(\alpha) f)(\beta) = f(\beta \circ \alpha)$.

Notice that, because $\mathbf{G} = \bigsqcup_{x \in \Omega} \mathbf{G}_+(x)$, the total space $\mathcal{F}(\mathbf{G})$ decomposes as the direct sum of the spaces of functions on each spray $\mathbf{G}_+(x)$, $x \in \Omega$:

$$\mathcal{F}(\mathbf{G}) = \bigoplus_{x \in \Omega} \mathcal{F}(\mathbf{G}_+(x)) = \bigoplus_{x \in \Omega} W_x^+ \,. \tag{12.9}$$

Similarly, the projector P_x^λ associated with the unit element 1_x by means of the left regular representation is the multiplier by the characteristic function of the set $\mathbf{G}_-(x)$, that is, the linear spaces associated to the objects $x \in \Omega$ by the left regular representation are given by the spaces W_x^- of functions on $\mathbf{G}_-(x)$.

The following observation provides a critical clue to understanding the structure of the regular representation.

Lemma 12.2

The subspaces W_x^+ (respect. W_x^-) defined by the right regular representation are invariant with respect to the left regular representation λ (respectively, the right regular representation ρ). This shows that the left-regular representation λ decomposes as the direct sum of the λ-invariant subspaces $W_x^+ = \operatorname{ran} P_x^\rho$ (respec. the right-regular representation ρ decomposes as the direct sum of the ρ-invariant subspaces $W_x^- = \operatorname{ran} P_x^\lambda$).

Proof 12.9 The proof is a direct check. Consider a function f on W_x^+, then it vanishes out of $\mathbf{G}_+(x)$. The function $(\lambda(\alpha)f)(\beta) = f(\alpha^{-1} \circ \beta)$ is defined on the same $\mathbf{G}_+(x)$, because the composition $\alpha^{-1} \circ \beta$ does not change the source of the morphism.

The same argument applies with W_x^- and the right regular representation ρ.

As a consequence of the previous lemmas, in order to understand the structure of the left-regular representation λ, all we have to do is to study the decomposition of the invariant subspaces W_x^+. For that, we will realize that the isotropy group G_x has a natural representation on W_x^+ that commutes with λ.

For each $x \in \Omega$, there is a natural right action of the isotropy group G_x on $\mathbf{G}_+(x)$ given by the groupoid composition law, that is $(\alpha, \gamma_x) \mapsto \alpha \circ \gamma_x$, for any $\alpha \colon x \to y \in \mathbf{G}_+(x)$ and $\gamma_x \in G_x$. We will denote by $\mathbf{G}_+(x)/G_x$ the space of orbits and by $\pi_x \colon \mathbf{G}_+(x) \to \mathbf{G}_+(x)/G_x$ the canonical projection, $\pi_x(\alpha) = \alpha \circ G_x$. Then, Prop. 4.4 shows that there is natural bijection between the space of orbits $\mathbf{G}_+(x)/G_x$ of the action of the group G_x on the set $\mathbf{G}_+(x)$ and the orbit $\mathcal{O}_x \subset \Omega$. If the groupoid $\mathbf{G}$ is connected, then:

$$\mathbf{G}_+(x)/G_x \cong \Omega\,. \tag{12.10}$$

Lemma 12.3

The right regular representation ρ, restricted to G_x, provides a linear representation μ_x of G_x on the space of functions $W_x^+ = \mathcal{F}(\mathbf{G}_+(x))$ that commutes with the left-regular representation λ.

Proof 12.10 Note that the restriction to G_x of the right regular representation of $\mathbf{G}$ defines a linear representation μ_x of G_x on $\mathcal{F}(\mathbf{G}_+(x))$, because $\gamma_x \colon x \to x$, $\gamma_x \in G_x$, that is $(\rho(\gamma_x)f)(\alpha) = f(\alpha \circ \gamma_x) = (\rho_x(\gamma_x)f)(\alpha)$, for any $\gamma_x \in G_x$.

Moreover, it is easy to compute the character of both the left and right regular representations. According to Prop. (12.4), we will have that (for the right-regular representation ρ):

$$\chi_\rho = \sum_{x \in \Omega} \chi_{\mu_x}\,,$$

where χ_{μ_x} denotes the character of the restriction μ_x of the right-regular representation to the isotropy group G_x acting on the subspace $\rho(1_x)(\mathcal{F}(\mathbf{G})) = W_x^+ = \mathcal{F}(\mathbf{G}_+(x))$. Then, we get:

$$\chi_\rho(\mathbf{1}) = \sum_{x \in \Omega} \chi_{\mu_x}(1_x) = \sum_{x \in \Omega} \dim(W_x^+) = \sum_{x \in \Omega} |\Omega||G_x|\,,$$

because, from Eq. 12.10, we get $|\mathbf{G}_+(x)| = |\Omega||G_x|$ and $\dim W_x^+ = |\mathbf{G}_+(x)| = |\Omega||G_x|$. Moreover, if the groupoid $\mathbf{G}$ is connected, we get $|G_x| = |G_y|$ for all $x, y \in \Omega$, then we conclude:

$$\chi_\rho(\mathbf{1}) = |\Omega|^2 |G_x| = |\mathbf{G}|\,,$$

in full agreement with the structure theorem of groupoids (recall formula (6.7) in the case of connected groupoids). Finally, notice that $\chi_\rho(\alpha) = 0$ for all $\alpha : x \to y$, $x \neq y$. Notice also that $\chi_\rho(1_x) = |\Omega||G_x|$.

Now, because of Eqs. (12.5), (12.6), the decomposition of the right-regular representation $\rho = \oplus_\nu n_\nu \pi_\nu$ into its irreducible components is such that:

$$\begin{aligned} n_\nu &= \langle \chi_\rho, \chi_\nu \rangle_{\mathbf{G}} = \frac{1}{|\Omega|} \sum_x \langle \chi_{\mu_x}, \chi_x^\nu \rangle_{G_x} \\ &= \frac{1}{|\Omega|} \sum_x \frac{1}{|G_x|} \bar{\chi}_{\mu_x}(1_x) \chi_x^\nu(1_x) = \sum_{x \in \Omega} \chi_x^\nu(1_x) \\ &= \chi_\nu(\mathbf{1}) = |\Omega| d_\nu . \end{aligned} \tag{12.11}$$

with d_ν the dimension of the restriction of the irreducible representation π_ν to G_x.

Note that we get the same result computing the inner product $\langle \chi_\rho, \chi_\nu \rangle_{\mathbf{G}}$ directly, that is:

$$\langle \chi_\rho, \chi_\nu \rangle_{\mathbf{G}} = \frac{|\Omega|}{|\mathbf{G}|} \sum_{\alpha \in \mathbf{G}} \overline{\chi_\rho}(\alpha) \chi_\nu(\alpha) = \frac{|\Omega|}{|\mathbf{G}|} \sum_{x \in \Omega} \overline{\chi_\rho}(1_x) \chi_\nu(1_x) = |\Omega| d_\nu .$$

Thus, we have proved the extension of Burnside's theorem to finite groupoids, in the sense that we have proved that each irreducible representation of a finite groupoid appears in its regular representation with frequency the dimension of the representation. In particular, for the fundamental representation, $d_0 = 1$, as the trivial representation of G_x is one-dimensional and it appears in the regular representation with frequency $|\Omega|$. We can collect all previous results in the following theorem.

Theorem 12.3

Let $\mathbf{G}$ *be a finite connected groupoid over the space of objects* Ω*. The family* π_ν *of inequivalent irreducible representations of* $\mathbf{G}$ *is finite and they are in one-to-one correspondence with the set of equivalence classes of irreducible representations* μ_α *of any isotropy group* G_x*. Each irreducible representation* π_ν *of* $\mathbf{G}$ *has dimension* $|\Omega| d_\nu$*, where* d_ν *is the dimension of the corresponding irreducible representation of* G_x*, and they appear in the right (left) regular representation with multiplicity* $|\Omega| d_\nu$*. Then:*

$$\rho = \bigoplus_{\nu \in \widehat{G}_x} |\Omega| d_\nu \pi_\nu = |\Omega| \pi_0 \oplus \bigoplus_{0 \neq \nu \in \widehat{G}_x} |\Omega| d_\nu \pi_\nu ,$$

with $\widehat{G}_x$ *denotes the space of equivalence classes of irreducible unitary representations of* G_x*, and*

$$|\mathbf{G}| = \sum_{\nu \in \widehat{G}_x} |\Omega|^2 d_\nu^2 .$$

12.3.2 The canonical decomposition of the regular representation of a finite groupoid

There is a further refinement of the decomposition of the left-regular representation λ in the invariant subspaces W_x^+ provided by (12.9).

Because each subspace W_x^+ is invariant under the left regular representation it will decompose into its irreducible components, each one of dimension $|\Omega| d_\nu$ and with multiplicity d_ν, in accordance with Burnside's Theorem, Thm. 12.3. Note that each subspace has dimension $|\Omega||G_x|$, and, because of Burnside theorem for groups, $|G_x| = \sum_\nu d_\nu^2$.

Then, we can decompose the space W_x^+ by using the projectors P^ν associated to the irreducible representations μ_ν of G_x (recall Thm. 8.12) as:

$$W_x^+ = \sum_{\nu \in \widehat{G}_x} W_\nu \,, \qquad W_\nu = P^\nu (W_x^+) \,,$$

with P^ν the canonical orthogonal projector given by Eq. (8.10).

Moreover, since $\mathbf{G}_+(x) = \bigsqcup_{y \in \Omega} \mathbf{G}(y,x)$, we get:

$$W_x^+ = \bigoplus_{y \in \Omega} W_{y,x} \,, \tag{12.12}$$

where

$$W_{y,x} = \mathcal{F}(\mathbf{G}(y,x)) \,,$$

is the space of functions defined on the set of morphisms connecting x to y. If we denote by p the order of the isotropy subgroup G_x, then it is clear that the subspaces $W_{y,x}$ all have dimension p (recall that $\mathbf{G}(y,x)$ has the same order as G_x). Hence, if $\alpha \colon y \to z$, then $\lambda(\alpha) \colon W_{y,x} \to W_{z,x}$. In particular, if $\gamma_y \in G_y$, then $\lambda(\gamma_y) \colon W_{y,x} \to W_{y,x}$ and $\lambda(\gamma_x) \colon W_{x,x} \to W_{x,x}$. Because of (12.12), we can introduce a block notation for the space W_x^+, indicating by y the block corresponding to the subspace $W_{y,x}$. With this understanding, the matrix representing $\lambda(\alpha)$ above (with respect to the canonical basis defined by the elements of $\mathbf{G}$ themselves) has the block structure (each block has dimension p and the number of blocks is the order of Ω) shown below:

$$[\lambda(\alpha)] = \begin{array}{c} \\ z \\ \\ \end{array} \overset{\qquad\quad y}{\left[\begin{array}{c|c|c} \ddots & \ddots & \\ \hline \ddots & A^\lambda(\alpha) & \ddots \\ \hline & \ddots & \ddots \end{array}\right]} ,$$

with the permutation $p \times p$ matrix $A^\lambda(\alpha)$ representing $\lambda(\alpha) \colon W_{y,x} \to W_{z,x}$ sitting in the block (z,y) and zeros everywhere. Using this notation, the right action of the isotropy group G_x is represented by the diagonal matrices:

$$[\rho(\gamma_x)] = \left[\begin{array}{c|c|c|c} \ddots & 0 & 0 & \cdots \\ \hline 0 & A^\rho(\gamma_x) & 0 & \cdots \\ \hline 0 & 0 & A^\rho(\gamma_x) & \cdots \\ \hline \vdots & \vdots & \vdots & \ddots \end{array}\right],$$

with $A^\rho(\gamma_x)$ the $p \times p$ matrix representing the action of γ_x in the (right) regular representation of G_x.

12.4 Some simple examples

12.4.1 Representations of Loyd's group $\mathfrak{L}_2$

Because of the general results obtained in Sect. 12.1.2 we conclude that the irreducible representations of $\mathfrak{L}_2$ are in one-to-one correspondence with the irreducible representations of the isotropy group $\mathcal{A}_3 \cong \mathbb{Z}_3$. Because the isotropy group is Abelian, its irreducible representations are one-dimensional and its table of characters is given in Table 12.1. There are three irreducible representations of the group $\mathcal{A}_3$ with characters χ^ν. The characters of the three irreducible representations π_ν of Loyd's groupoid will be denoted accordingly as $\chi^\nu_{\mathfrak{L}_2}$ and are listed in Fig. 12.1. All three irreducible representations of $\mathfrak{L}_2$ are four-dimensional and they will appear in the decomposition of the regular representation with multiplicity 4 (see below).

Table 12.1: Table of characters of the groupoid $\mathfrak{L}_2$.

$\mathfrak{L}_2$	$G_i = \mathbb{Z}_3, i = 1,2,3,4$	$\mathbf{G}(i,j), i \neq j$	Dimension
$\chi^0_{\mathfrak{L}_2}$	χ_0	0	4
$\chi^1_{\mathfrak{L}_2}$	χ_1	0	4
$\chi^2_{\mathfrak{L}_2}$	χ_2	0	4

The character table of the groupoid $\mathfrak{L}_2$, where χ_i are the characters of the irreducible representations of $\mathbb{Z}_3$ (Table 8.1).

Since the isotropy group of any object is isomorphic to the alternating group $\mathcal{A}_3$, the representations of the groupoid are induced from the representations of this group. In particular, we will consider the irreducible representations of $\mathcal{A}_3$ which, being one-dimensional, coincide with its characters.

For instance, we may describe the fundamental representation π_0. The space $\mathcal{H}_\Omega$ supporting the fundamental representation π_0 is the four dimensional linear space generated by Ω and it is naturally isomorphic to $\mathbb{C}^4$. As a functor, the fundamental representation assigns to each state the vector space $\mathbb{C}$. The groupoid acts on $\mathcal{H}_\Omega$ as:

$$\pi_0(\alpha)(a) = \sum_{i=1}^{4} c_i \delta_{s(\alpha),s_i} t(\alpha), \quad \alpha \in \mathfrak{L}_2, \; a = \sum_{i=1}^{4} c_i s_i \in \mathcal{H}_\Omega$$

The action of a morphism α_{ji} ($s(\alpha) = s_i$, $t(\alpha) = s_j$) over the basis $\{s_1, s_2, s_3, s_4\}$ is

$$\pi_0(\alpha_{ji})(s_k) = \sum_{l=1}^{4} \delta_{kl}\delta_{il} s_j = \delta_{ik} s_j$$

that is, in this basis,

$$\pi_0(\alpha_{ji}) = E_{ji}\,.$$

This representation is irreducible and not faithful. In fact, it is a faithful representation of the quotient groupoid $\mathbf{G}/\mathbf{G}_0$. The endomorphisms $\pi_0(\alpha_{ji})$ depend only on the indices i, j which denote the cosets. Note that, as matrices, we can multiply two any of them (that is "compose" the corresponding morphisms), but when the morphisms cannot be composed, the result is the zero matrix.

The unitary elements in the groupoid algebra correspond to unitary matrices in this representation. For instance, the unit $\mathbf{1}$ is represented by identity matrix:

$$\mathbf{1} \to \pi_0(e_{11}) + \pi_0(e_{22}) + \pi_0(e_{33}) + \pi_0(e_{44}) = \mathbf{I}_4$$

In fact, the set of unitary elements in the groupoid algebra is the unitary group[2] $U(4)$, and the character of this representation is: $\chi_0(\alpha_{ij}) = \delta_{ij}$, $i, j = 1, 2, 3, 4$.

Let us discuss now the structure of the regular representation. As discussed in Sect. 12.3, the left (right) regular representation is defined on the space of complex valued functions over $\mathbf{G}$. Since $\mathfrak{L}_2$ is a finite groupoid of order 48, any function f can be characterized as a vector in a space of dimension 48, and the elements $\lambda(\alpha)$ as endomorphisms in this space, that is, 48×48 matrices in a given basis. However, as it was discussed above, this space decomposes as the direct sum of invariant subspaces $W_i^+ = \mathcal{F}(\mathbf{G}_+(i))$ of dimension 12 (see Sect. 12.3):

$$\rho(\alpha_{ji})\colon \mathcal{F}(\mathbf{G}_+(i)) \to \mathcal{F}(\mathbf{G}_+(j))\,.$$

Since

$$\mathbf{G}_+(i) = \mathbf{G}_{1i} \sqcup \mathbf{G}_{2i} \sqcup \mathbf{G}_{3i} \sqcup \mathbf{G}_{4i}$$

each $\mathbf{G}_{ji} = \mathbf{G}(j,i)$ having three elements, we can divide each 12×12 block into four blocks, each block being a square matrix of dimension 3.

Consider, for instance, the morphism $(132)_{32}\colon s_2 \to s_3$. According to the discussion above,

$$\lambda((132)_{32}) : \mathcal{F}(\mathbf{G}_+(1)) \to \mathcal{F}(\mathbf{G}_+(1))\,,$$

[2]The elements $\pi_0(\alpha_{ji}) = E_{ji}$ is a basis (over $\mathbb{C}$) of the 4×4 complex matrices. Then the matrices representing $\mathbb{C}[\mathbf{G}]$ are the whole space of 4×4 complex matrices. The unitary elements correspond to the unitary matrices (note that $\alpha_{ji} \to E_{ji}$ and $\alpha_{ji}^{-1} \to E_{ji}^T = E_{ij}$).

The morphisms in $\mathbf{G}_+(1)$ are:

$$\mathbf{G}_+(1) = \{(12)_{21}, (1342)_{21}, (1432)_{21}, (13)_{31}, (1243)_{31}, (1423)_{31},\\ e_{11}, (234)_{11}, (243)_{11}, (124)_{41}, (134)_{41}, (14)(23)_{41}\},$$

and the space $W_1^+ = \mathcal{F}(\mathbf{G}_+(1))$ decomposes as the direct sum of the 3-dimensional subspaces

$$\begin{aligned} W_{11} &= \mathcal{F}(\mathbf{G}(1,1)) = \mathcal{F}(\{e_{11}, (234)_{11}, (243)_{11}\}), \\ W_{21} &= \mathcal{F}(\mathbf{G}(2,1)) = \mathcal{F}(\{(12)_{21}, (1342)_{21}, (1432)_{21}\}), \text{etc.} \end{aligned}$$

Then, acting with $(123)_{23} = (132)_{32}^{-1}$ on the left we get a non-trivial map: $\lambda((132)_{32})\colon W_{31} \to W_{21}$, so that the matrix associated to it is given by (where all blank blocks being the zero matrix):

$$[\lambda((132)_{32})] = \left[\begin{array}{c|c|c|c} & & & \\ \hline & & \begin{matrix} 1 & 0 & 0 \\ 0 & 1 & 0 \\ 0 & 0 & 1 \end{matrix} & \\ \hline & & & \\ \hline & & & \end{array}\right].$$

Finally, notice that the isotropy group G_1 acts on the right on W_1^+ and with the notations before we have:

$$[\rho((234)_{11})] = \left[\begin{array}{c|c|c|c} \begin{matrix} 0 & 1 & 0 \\ 0 & 0 & 1 \\ 1 & 0 & 0 \end{matrix} & & & \\ \hline & \begin{matrix} 0 & 1 & 0 \\ 0 & 0 & 1 \\ 1 & 0 & 0 \end{matrix} & & \\ \hline & & \begin{matrix} 0 & 1 & 0 \\ 0 & 0 & 1 \\ 1 & 0 & 0 \end{matrix} & \\ \hline & & & \begin{matrix} 0 & 1 & 0 \\ 0 & 0 & 1 \\ 1 & 0 & 0 \end{matrix} \end{array}\right]$$

The matrices in the blocks are the regular representation of the alternating group A_3 in three elements, as expected. This representation can be decomposed in the sum of the three irreducible representations of dimension 1 of this abelian group. The scheme is repeated four times, which is the number of objects in the groupoid (that is, the dimension of the vector space $\mathcal{H}_\Omega$, which is the carrier space of the fundamental representation), given the multiplicity four of the irreducible representations.

12.4.2 Representations of cyclic groupoids

Cyclic groupoids are the natural extension to the category of groupoids of the notion of cyclic groups in the theory of groups, Sect. 3.2.3. The cyclic groupoid $\mathfrak{C}_n(m)$ on a space with n positions and recurrence m is generated by the moves $m_{1,0}$, $m_{2,1}, \ldots, m_{0,n-1}$ on a board labeled from 0 to $n-1$ and a permutation of order m acting on an auxiliary register R. If n and m are relatively prime, then $|\mathfrak{C}_n(m)| = n^2 m$ and the isotropy group is given by $\mathbb{Z}_m$.

Cyclic groupoids $\mathfrak{C}_n(m)$ are isomorphic to the direct product groupoid $\mathbf{G}(n) \times \mathbf{Z}_s$ with $l = ms$ the least common multiple of n and m. Then, the irreducible representations of $\mathfrak{C}_n(m)$ will be the tensor product of the (only) irreducible representation π_0 of $\mathbf{G}(n)$ and the irreducible representations μ_k of $\mathbb{Z}_s$ (see Sect. 8.7.1). The characters of the irreducible representations $\pi_0 \otimes \mu_k$ will be the product of the corresponding characters, $\chi_{\pi_0} \cdot \chi_k$, and are listed in Table 12.2.

Table 12.2: Table of characters of the cyclic groupoid $\mathfrak{C}_n(m)$.

$\mathfrak{C}_n(m)$	$G_x = \mathbb{Z}_s$	$\mathbf{G}(n)$	Dimension
$\chi^0_{\mathfrak{C}_n(m)}$	χ_0	χ_{π_0}	$\lvert K\rvert$
$\chi^1_{\mathfrak{C}_n(m)}$	χ_1	χ_{π_0}	$\lvert K\rvert$
$\vdots$	$\vdots$	$\vdots$	$\vdots$
$\chi^{s-1}_{\mathfrak{C}_n(m)}$	χ_{s-1}	χ_{π_0}	$\lvert K\rvert$

The character table of the cyclic groupoid $\mathfrak{C}_n(m)$, where χ_k are the characters of the irreducible representations of $\mathbb{Z}_s$ and $s = l/m$ with l the l.c.m. (n,m).

12.4.3 Representations of action groupoids

Let us consider now the action groupoid $\mathbf{G}(\Gamma, \Omega)$ corresponding to the action of a group Γ on a set Ω. We will start by considering that the set Ω is the quotient group K of the group Γ by a normal subgroup H, then $\Omega = K = \Gamma/H$ and the action is given by $gx = (gg_x)H$, where $g_x \in \Gamma$ is any element, such that $g_x H = x \in K$.

The action groupoid $\mathbf{G}(\Gamma, K)$ is defined as the set of triples (y, g, x), with $y = gx$, $x, y \in K = \Gamma/H$. Then, the isotropy groups Γ_x of $\mathbf{G}(\Gamma, K)$ are all H (because H is normal) and, in this particular instance, all of them are canonically isomorphic.

Irreducible representations will correspond to irreducible representations of H. We will denote the irreducible representations of H by $\mu_\alpha \in \widehat{H}$, thus, the character of the irreducible representations of $\mathbf{G}(\Gamma, K)$ will be given in Table 12.3. The irreducible representation corresponding to the trivial representation of the group H is the fundamental representation π_0 of the groupoid $\mathbf{G}(\Gamma, K) \rightrightarrows K$.

It is worth comparing with the irreducible representations of the group Γ and realizing that they are different. For instance, the group Γ has a natural representation on the space of functions $\psi \in \mathcal{H}_K$ on K by left or right translations (this representation is a generalization of the regular representation): $(\lambda(g)\psi)(x) = \psi(g^{-1}x), x \in K$, $g \in \Gamma$. Actually, a simple computation shows that the character of the representation λ is given by $\chi_\lambda(h) = |K|$ and zero otherwise. Then, $\langle \chi_\lambda, \chi_\lambda \rangle_\Gamma = \frac{1}{|\Gamma|} \sum_g |\chi_\lambda(g)|^2 = |K| > 1$, whenever $H \triangleleft \Gamma$ is a proper normal subgroup.

The character χ_λ is not simple and the representation λ of Γ on $\mathcal{H}_K$ is not irreducible while the fundamental representation π_0 of the action groupoid on the space $\mathcal{H}_K$ of functions on K is.

Table 12.3: Table of characters of the action groupoid $\mathbf{G}(\Gamma, K)$.

$\mathbf{G}(\Gamma, K)$	$G_x = H$	$\mathbf{G}(x, y), x \neq y$	Dimension
$\chi^0_{\mathbf{G}(\Gamma,K)}$	χ_0	0	$\|K\|$
$\chi^1_{\mathbf{G}(\Gamma,K)}$	χ_1	0	$\|K\| d_1$
$\vdots$	$\vdots$	$\vdots$	$\vdots$
$\chi^r_{\mathbf{G}(\Gamma,K)}$	χ_r	0	$\|K\| d_r$

The character table of the action groupoid $\mathbf{G}(\Gamma, K)$, where χ_i are the characters of the irreducible representations of H and d_k their dimension.

12.5 Discussion

The results obtained in this chapter show the close relation existing between the theory of representations groupoids and that of groups: There is a one-to-one correspondence between the irreducible representations of a finite groupoid and those of the isotropy groups of its connected components. Our work stops here, but it is obvious that these results raise a variety of significant questions that deserve a careful study.

The most prominent of it concerns the exact nature of such relation between irreducible representations. The restriction of the irreducible representations of the groupoid provide the irreducible representations of the isotropy groups. The converse mechanism is called induction and it is a significative construction that has been used widely in the setting of groups. The relation between restriction and induction of representations is called Frobenius theory and it would be relevant to understand it in the context of groupoids.

Another important aspect of the theory of representations that has not been addressed in this work is the extension of Tannaka-Krein duality for groupoids.

APPENDICES

Appendix A

Glossary of Linear Algebra

A succinct glossary of the main notions and notations from Linear Algebra that are used throughout the work is offered in this Appendix. The reader may enjoy the (unusual) presentation inspired by J.M. Souriau's 'analytical' approach to the subject [60].

A.1 Linear spaces and subspaces

All linear spaces considered in this work will be complex linear spaces unless explicitly stated, thus, the scalars will always be the field of complex numbers $\mathbb{C}$.

Definition A.1 A (complex) linear space V is a set equipped with two operations: An addition $+$, i.e., given two elements $u, v \in V$, called vectors, the addition of v and u is another vector denoted by $u + v$; and a multiplication by scalars $\cdot$, i.e., given a scalar $\lambda \in \mathbb{C}$ and a vector $v \in V$, λv is another vector. The addition and multiplication by scalar operations satisfy that if u, v are two different vectors in V, there exists a map $\alpha : V \to \mathbb{C}$, such that $\alpha(u) \neq \alpha(v)$ and:

1. $\alpha(u+v) = \alpha(u) + \alpha(v)$,
2. $\alpha(\lambda u) = \lambda \alpha(u)$ for all $\lambda \in \mathbb{C}$.

Maps $\alpha : V \to \mathbb{C}$ satisfying properties (1)-(2) will be called covectors.

Remark A.1 The previous definition, due to J.M. Souriau [60], provides a beautiful geometrical definition of linear spaces that is not so well known in contrast with the 'algebraic' presentation that emphasizes the properties of the operations (note that, so far, no algebraic properties of the operations, like associativity, commutativity, etc, have been postulated). ▮

Remark A.2 It is worth to point out that the family of covectors of a linear space separates vectors, i.e., for any two different vectors $u \neq v$, there is a covector α such that $\alpha(u) \neq \alpha(v)$. The existence of a family of maps separating points will have dramatic consequences for the structure of linear spaces. ▮

The main idea behind this presentation is to ascertain the existence of a 'large enough' family of 'linear maps', that is covectors, on V that will determine the algebraic properties of the given operations. Clearly, if $\alpha(u) = \alpha(v)$ for all linear maps $\alpha \colon V \to \mathbb{C}$, then $u = v$ (because, if not, will contradict the separation property above). A similar idea was exploited in [13], to provide a geometric definition of linear spaces.

The algebraic properties of the operations $+$ and $\cdot$ are obtained as follows. For instance, if we apply any covector α to $(u+v)+w$ and $u+(v+w)$, using property (1) repeatedly and the associativity of the addition of complex numbers, we get that $\alpha((u+v)+w) = \alpha(u+(v+w))$, hence, we conclude that $u+(v+w) = (u+v)+w$.

Similarly, we prove all other standard algebraic relations determining the structure of a linear space: $(V,+)$ is an Abelian group whose neutral element is denoted by $\mathbf{0}$ and is defined as $0v$ for any $v \in V$ (notice that $0u = 0v$ for any $u, v \in V$, because $\alpha(0u) = 0\alpha(u) = 0$ for all α, u); moreover, $\lambda(u+v) = \lambda u + \lambda v$, $(\lambda+\mu)u = \lambda u + \mu u$, $(\lambda\mu)u = \lambda(\mu u)$ and $1u = u$ (see, for instance, [35] for an standard introduction to the theory of linear spaces)[1].

Proposition A.1
The set of all covectors is a linear space denoted by V^* called the dual space of V.

The proof is immediate because, if we are given two different covectors α, β, then there must exist a vector u, such that $\alpha(u) \neq \beta(u)$, then, consider the map $e_u \colon V^* \to \mathbb{C}$, defined by $e_u(\alpha) = \alpha(u)$. Clearly, the family of maps e_u, $u \in V$ are covectors for V^*, satisfying the axioms in Def. A.1.

Definition A.2 A (linear) subspace of a linear space V is a subset $W \subset V$ such that $u+v \in W$ and $\lambda u \in W$ for all $u, v \in W$ and any $\lambda \in \mathbb{C}$.

Notice that any subspace is a linear space as, Def. A.1 is satisfied immediately for any subspace W (given two different vectors w and w' in W it suffices to take the restriction to W of a covector α in V separating them), hence, $\mathbf{0} \in W$ for any subspace. Clearly, $\mathbf{0}$ and V are subspaces of V. A subspace will be called proper if it

[1] To show that the converse holds, that is, that the standard definition of linear spaces implies our Def. A.1 is trivial in finite dimensions, however it is not, in general.

is different from $\mathbf{0}$ and V. It is also immediate from the definition that the intersection of two subspaces is again a subspace.

Definition A.3 Given a subset $S \subset V$ of the linear space V, the smallest subspace in V containing S will be called the subspace generated by S (or the span of S). It will be denoted by $\mathrm{span}(S)$.

The span of S is clearly the intersection of all subspaces containing S. Alternatively, it can be defined as the subspace of all finite linear combinations of vectors in S, that is, vectors of the form $w = \sum_{i=1}^{r} \lambda_i v_i$, where r is any finite number, $\lambda_i \in \mathbb{C}$ and $v_i \in S$ for all $i = 1, \ldots, r$, thus,

$$\mathrm{span}(S) = \left\{ w = \sum_{i=1}^{r} \lambda_i v_i \in V \mid r \in \mathbb{N}, \lambda_i \in \mathbb{C}, v_i \in S \right\}.$$

Notice that, if W is a subspace, $\mathrm{span}(W) = W$, but if W, U are subspaces, in general, $U \cup W \subsetneq \mathrm{span}(U \cup W)$. The linear subspace generated by the union of two subspaces is called the sum of U and W and is denoted by $U + W$.

Definition A.4 Given a subspace $W \subset V$ of the linear space V, a supplementary subspace of W in V is any subspace $U \subset V$, such that $W \cap U = \mathbf{0}$ and such that for any vector $v \in V$, there exist vectors $w \in W$ and $u \in U$, such that $v = w + u$. Notice that, in such case, the vectors u, w are unique.

If U is a supplementary subspace of W, then $V = W + U$ and we will denote it by $V = W \oplus U$. We will also say that V is the direct sum of the subspaces W and U.

Theorem A.1
Let V be a linear space, any subspace $W \subset V$ has a supplementary subspace.

Proof A.1 The proof exceeds the scope of this work, however we will sketch it for the sake of the interested reader. We use Zorn's Lemma. Consider only proper subspaces of V. Let $U_1 \subset U_2 \subset \ldots$ be a chain of subspaces in the class of subsets of V, whose intersection with W is $\mathbf{0}$. There exists such a chain because it must exist $u_1 \neq \mathbf{0}$ which is not in W. Then, we take $U_1 = \mathrm{span}(\{u_1\})$.

Consider now the linear subspace generated by $W + U_1$ and repeat the same argument to construct another subspace $U_2 = U_1 \cup \{u_2\}$, such that $U_1 \subsetneq U_2$ and $U_2 \cap W = \mathbf{0}$. The chain is bounded above by V, hence, by Zorn's Lemma, it has a maximal element, call it U. Then, it follows easily that U is the desired subspace.

Corollary A.1
Let W be a proper subspace of V, then, given any vector $u \in V$, $u \notin W$, there exists a supplementary subspace U to W, such that $u \in U$.

Remark A.3 When trying to prove a statement similar as the one in the previous theorem, for instance, proving the existence of linear basis, we are always going to find set theoretical difficulties that will involve the use of the axiom of choice or any equivalent formulation (like Zorn's Lemma used before).

We may restrict our linear spaces though to be finite-dimensional (see later) or simply add the previous theorem as a property in the definition of linear space. There is, however, the important class of (separable) Hilbert spaces, not necessarily finite-dimensional (see Sect. 7.3.1 for the basic definitions and properties), for which the previous result can be proved by direct methods and that are relevant for the purposes of this work. ∎

Proposition A.2

Given a subspace $W \subset V$, and a vector u in V not in W, there exists a covector $\alpha \in V^*$, such that $\alpha(u) \neq 0$ and $\alpha(w) = 0$ for all $w \in W$.

Proof A.2 Let $W \subset V$, then choose a supplementary subspace U to W containing u, i.e., $V = W \oplus U$. Because U is a linear space, there exists a covector $\beta \in U^*$, such that $\beta(u) \neq 0$. Define the covector $\alpha(v) = \beta(u')$, where $v = w + u'$ is the unique decomposition of v determined by the supplementary subspaces W and U.

A subspace W of the linear space V, induces a natural equivalence relation: $u \sim v$ iff $u - v \in W$. In other words, the vectors equivalent to a given one $v \in V$ are all those that can be obtained by adding vectors in W to it (that is, they are the 'hyperplanes' defined by W in V). The equivalence class defined by the subspace W containing v will be denoted by $v + W$ and the set of equivalence classes will be denoted by V/W.

Proposition A.3

The set of equivalence classes V/W defined by a subspace W of the linear space V is a linear space.

Proof A.3 The operations $+$ and $\cdot$ are naturally defined on V/W by:

$$(u+W)+(v+W) = (u+v)+W\,, \qquad \lambda(u+W) = \lambda u + W\,, \quad u, v \in V, \lambda \in \mathbb{C}.$$

Now, it is trivial to check that there exists a covector $\tilde{\alpha}$ in V/W separating two different elements $u + W \neq v + W$. Consider the vector $u - v \notin W$. Then, because Prop. A.2, there exists a covector α, such that $\alpha(u - v) \neq 0$, and $\alpha(w) = 0$ for all $w \in W$. Then, define the covector $\tilde{\alpha}(u + W) = \alpha(u)$.

Clearly, $\tilde{\alpha}$ is well defined and satisfies the required condition.

Remark A.4 It is also possible to prove that V/W is a linear space by directly checking the algebraic properties defining a linear space. ∎

A.2 Linear maps

Definition A.5 A linear map between two linear spaces V, W is a map $L\colon V \to W$, such that $L(u+v) = L(u) + L(v)$ and $L(\lambda u) = \lambda L(u)$ for all $u, v \in V$ and any $\lambda \in \mathbb{C}$.

A covector $\alpha \in V^*$ is a linear map $\alpha\colon V \to \mathbb{C}$.

Any linear map $L\colon V \to W$ induces a linear map $L^*: W^* \to V^*$, called its adjoint, as:

$$(L^*(\alpha))(v) = \alpha(L(v)), \qquad \forall v \in V, \quad \alpha \in W^*. \tag{A.1}$$

A basic property of linear maps is that they act 'transitively' on the linear space V, that is:

Proposition A.4
Let $v \neq \mathbf{0}$ be a non-zero vector in V and $u \in V$ any other vector, then there exists a linear map $L\colon V \to V$ such that $L(v) = u$.

Proof A.4 Let $\mathbb{C}v \neq \mathbf{0}$ the linear span of v and W be a supplementary subspace (Thm. A.1), then $V = \mathbb{C}v \oplus W$. Then, define the linear map $L(\lambda v \oplus w) = \lambda u$.

Definition A.6 Let $L\colon V \to W$ be a linear map. The set of vectors in V mapped to $\mathbf{0}$ is called the kernel of L, it defines a subspace of V and is denoted as $\ker L = \{v \in V \mid L(v) = 0\}$. The range of the linear map L is a subspace of W, denoted as $L(V)$ or $\mathrm{ran}(L)$. The quotient space $W/L(V)$ is called the cokernel of L and denoted by $\mathrm{coker} L$.

If we are given two linear maps $L\colon V \to W$ and $M\colon W \to U$, we denote by $M \circ L$ the composition of the maps L and M, that is $(M \circ L)(v) = M(L(v))$, $v \in V$. The map $M \circ L$ is linear. A family of linear maps $L_i\colon V_i \to V_{i+1}$, $i = 1, \ldots, r$ is called a sequence of linear maps (or just a sequence for short). We will say that the sequence of linear maps $L_i\colon V_i \to V_{i+1}$ is exact if $\ker L_{i+1} = \mathrm{ran} L_i$, for all $i = 1, \ldots, r$. If $r = 2$, we call the sequence $V_1 \to V_2 \to V_3$ short.

Proposition A.5
A linear map $L\colon V \to W$ is injective iff $\ker L = \mathbf{0}$ and it is surjective iff $\mathrm{coker} L = 0$. *A linear map which is both injective and surjective will be called a (linear) isomorphism. If $L\colon V \to W$ is an isomorphism we will often denote $V \cong W$ (and the isomorphism L is omitted unless there is risk of confusion or the isomorphism is canonical).*

Remark A.5

1. If $W \subset V$ is a subspace, the canonical inclusion map $i\colon W \to V$ is an injective linear map.

2. If we denote by $\pi\colon V \to V/W$ the canonical projection $\pi(v) = v + V$, π is a surjective linear map.

3. The linear map $L\colon V \to W$ is injective iff the sequence $\mathbf{0} \to L \to M$ is exact (the linear map $\mathbf{0} \to V$ is the canonical inclusion map of the trivial subspace $\mathbf{0}$.)

4. The linear map $L\colon V \to W$ is surjective iff the sequence $L \to M \to \mathbf{0}$ is exact (the linear map $M \to \mathbf{0}$ is the linear map mapping all vectors to zero).

5. The linear map $L\colon V \to W$ is an isomorphism iff the canonical sequence $\mathbf{0} \to L \to M \to \mathbf{0}$ is exact.

6. The inverse of a bijective linear map is linear.

■

Definition A.7 Given a surjective linear map $\pi\colon V \to W$, a (cross) section of π is a linear map $\sigma\colon W \to V$, such that $\pi \circ \sigma = \mathrm{id}_W$, i.e., a section of π is a right inverse of π which is also a linear map.

Proposition A.6
For any surjective linear map π there exists a section.

Proof A.5 Consider the subspace $\ker \pi \subset V$. Then, because of Thm. A.1 there exists a supplementary subspace U to $\ker \pi$, then $V = \ker \pi \oplus U$. Consider the restriction $\pi_U = \pi|_U$ of the map π to U. It follows immediately that $\pi\mid_U$ is injective (if not, there would be a nonzero vector $u \in U$, such that $\pi_U(u) = 0$, but then $\pi(u) = 0$, and $u \in U \cap \ker \pi = \mathbf{0}$). On the other hand, π_U is surjective too, then it is bijective, then there exists the inverse map $(\pi_U)^{-1}\colon W \to U$ which is linear. Then, define the map $\sigma\colon W \to V$ as $\sigma(w) = \mathbf{0} \oplus \pi_U^{-1}(w)$, $w \in W$. The linear map σ is a section of π.

A.2.1 First isomorphism theorem

Isomorphisms between two linear spaces V and W allow us to identify them because an isomorphism L is a bijection between them and, in addition, it transforms the linear structure of one space into the linear structure of the other. If an isomorphism $L\colon V \to W$ is selected, we may transport any structure on one space into the other. For instance, if α is a covector in V, then $\beta = \alpha \circ L^{-1}$ is a covector in W, or if $A\colon W \to W$ is an endomorphism of W, then $L^{-1} \circ A \circ L$ defines the corresponding endomorphism of V.

If there is a canonical isomorphism between two linear spaces V and W, meaning that it is defined directly in terms of the properties of the given spaces, then we may consider that both spaces are identified and the symbol $V \cong W$ will be used whenever we make use of this fact (later on we will encounter many examples of this). However, there will be occasions where we will be dealing with linear spaces V and W, for which there exists an isomorphism $L\colon V \to W$ (for instance, any pair of finite-dimensional spaces of the same dimension are isomorphic) but we do not want to identify them.

Theorem A.2
(First isomorphism theorem) If $L\colon V \to W$ is a linear map, there is a canonical isomorphism $\tilde{L}\colon V/\ker L \to \mathrm{ran}(L)$. The linear map L factorizes as $L = i \circ \tilde{L} \circ \pi$, where $\pi\colon V \to V/\ker L$ is the canonical projection and $i\colon \mathrm{ran}\, L \to W$ is the canonical inclusion.

Proof A.6 Define $\tilde{L}(v + \ker L) = L(v)$. Clearly $\tilde{L}$ is well defined. Checking the statements is a routine application of the definitions.

Let $\alpha \in V^*$ be a nonzero covector. Then, applying the first isomorpphism theorem to the linear map $\alpha\colon V \to \mathbb{C}$, Thm. A.2, we get $V/\ker\alpha \cong \mathbb{C}$.

A.2.2 Projectors and subspaces

Definition A.8 A projector on the linear space V is a linear map $P\colon V \to V$ such that $P^2 = P$. We will also say that P is a projector onto the subspace $U = \mathrm{ran}\, P$.

Definition A.9 Let $V = U \oplus W$ be a decomposition of the linear space V. The projector P onto the subspace U such that $\ker P = W$ is called the projector onto U along[2] the subspace W and is defined by: $P(u \oplus w) = u$.

Proposition A.7
Let $P\colon V \to V$ be a projector. Let $U = \mathrm{ran}\, P$ and $W = \ker P$. Then, $V = U \oplus W$ and P is the projector onto U along W.

Proof A.7 Let $v \in U \cap W$. Then, $Pw = 0$ $(w \in W)$ and $w = Pv$ $(w \in \mathrm{ran}\,(P))$, then $Pw = P^2 w = 0$. Now, because of the first isomorphism theorem, $V/W \cong U$. Hence, $V = U \oplus W$.

Corollary A.2
Let $0 \to U \to V \to W \to 0$ be a short exact sequence of linear maps, then $V \cong U \oplus W$. In addition, if $L\colon V \to W$ is a linear map, then we have the short exact sequence $0 \to \ker L \to V \to \mathrm{ran}\, L \to 0$ and $V \cong \ker L \oplus \mathrm{ran}\, \mathrm{L}$.

[2]Sometimes it is also called the projector onto U parallel to the subspace W.

A.2.3 *The linear space of linear maps:* $\mathrm{Hom}(V,W)$

Given two linear spaces V and W, the set of linear maps from V to W will be denoted by $\mathrm{Hom}(V,W)$. If $V=W$, we will denote $\mathrm{Hom}(V,W)=\mathrm{End}(V)$.

Two linear maps $L,M\in\mathrm{Hom}(V,W)$ can be added: $(L+M)(v)=L(v)+M(v)$, and multiplied by scalars: $(\lambda L)(v)=\lambda(L(v))$. Moreover, if $L\neq M$, then there exists a vector $v\in V$, such that $L(v)\neq M(v)$. Then there exists a covector $\beta\in W^*$, such that $\beta(L(v))\neq\beta(M(v))$. Now, consider the map $e_{v,\beta}\colon \mathrm{Hom}(V,W)\to\mathbb{C}$ defined as: $e_{v,\beta}(L)=\beta(L(v))$. Clearly, this map satisfies the requirements of Def. A.1, and $\mathrm{Hom}(V,W)$ is a linear space.

Notice that $\mathrm{Hom}(V,\mathbb{C})=V^*$ and $\mathrm{Hom}(\mathbb{C},V)\cong V$.

Linear maps $L\colon V\to V$ are also called endomorphisms of V. The collection of linear isomorphisms of V form a group called the general linear group of V and denoted by $GL(V)$.

The class of linear spaces together with linear maps form the objects and morphisms, respectively, of a category called the category of linear spaces and denoted by **Vect**. This category is not small (see also Sect. 1.2.2).

A.3 Operations with linear spaces and linear maps

There are a number of operations that can be performed with linear spaces and which are of great importance in this work. We will summarise their definitions and main properties.

A.3.1 *Direct sums*

Let V and W be two linear spaces, Let us consider the Cartesian product $V\times W$ with the operations $(v,w)+(v',w')=(v+v',w+w')$, $\lambda(v,w)=(\lambda v,\lambda w)$. Clearly, if $(v,w)\neq(v',w')$, then either $v\neq v'$, $w\neq w'$ or both. In the first case, let $\alpha\in V^*$ be a covector, such that $\alpha(v)\neq\alpha(v')$, or in the second, let $\beta\in W^*$ be a covector, such that $\beta(w)\neq\beta(w')$. Then, the map $\tilde{\alpha}\colon V\times W\to\mathbb{C}$, $\tilde{\alpha}(v,w)=\alpha(v)$, or the corresponding map $\tilde{\beta}$, separates points and $V\times W$ with the previous definitions is a linear space.

Definition A.10 Let V,W be two linear spaces, we will call the linear space defined on $V\times W$ by the canonical operations $+,\,\cdot$ induced from the corresponding ones in the factors V and W, the direct sum of V and W and we will denote it by $V\oplus W$. Vectors in $V\oplus W$ will be denoted by $v\oplus w$.

This notation is consistent with the notation for supplementary subspaces introduced before Def. A.4. In fact, given the direct sum $V_1\oplus V_2$ of the linear spaces V_a, $a=1,2$, there are natural maps $i_a\colon V_a\to V_1\oplus V_2$, $i_1(v_1)=v_1\oplus 0$, $i_2(v_2)=0\oplus v_2$, $v_a\in V_a$, $a=1,2$ that allows us to identify the original spaces as subspaces of $V_1\oplus V_2$. On the other hand, the subspaces $i_a(V_a)\subset V_1\oplus V_2$ are supplementary.

Let W, U be supplementary subspaces of the linear space V. We can form the direct sum $W \oplus U$ of the linear spaces W and U. Consider the canonical map $i\colon W \oplus U \to V$ defined as $i(w,u) = w+u$. This map is clearly a isomorphism and we can identify V with $W \oplus U$.

Proposition A.8

1. *Let $\pi\colon V \to W$ be a surjective linear map. Then, V is isomorphic, although not canonically to $W \oplus \ker \pi$.*
2. *If $U \subset V$ is a subspace, then V is isomorphic to $U \oplus V/U$, but the isomorphism is not canonical.*

Proof A.8 (1) Let σ be a section of π (Prop. A.6). Define the map $\varphi\colon W \oplus \ker \pi \to V$, as $\varphi(w \oplus u) = u + \sigma(w)$. It is obvious that the map φ is linear. It is also injective (if $\varphi(w \oplus u) = 0$, then $u + \sigma(w) = 0$, but applying π to both sides, we get $w = 0$, and then $u = 0$). To show that φ is surjective, consider $v \in V$, then the vector $u = v - \sigma(\pi(v))$ is in $\ker \pi$. The vector $u \oplus \pi(v)$ is such that $\varphi(u \oplus \pi(v)) = u + \sigma(\pi(v)) = v$.

(2) Similarly we prove that V is isomorphic to $U \oplus V/U$. Consider the map $\varphi\colon V \to U \oplus V/U$ given by $\varphi(v) = (v - \sigma(\pi(v))) \oplus \pi(v)$ where σ is a section of π. Then similar considerations to those in (1) lead to the conclusion.

Remark A.6 The previous proposition shows that, for the short exact sequence $0 \to U \to V \to V/U \to 0$ of linear spaces, the linear space V can be 'factorised' as $U \oplus V/U$ but such factorisation is not canonical (it depends on the choice of a section). In this sense, it would be proper to say that, whenever we have a short exact sequence of linear spaces $0 \to U \to V \to W \to 0$, then $V = U \oplus \sigma(W)$, where σ is a section. Sometimes, such choice of a factorisation of the linear space V is called a splitting of the short exact sequence. ■

We can apply the previous considerations to the linear map, defined by a covector $\alpha\colon V \to \mathbb{C}$, in order to show that the linear space V can always be factorised as $\ker \alpha \oplus \mathbb{C}$ or, even better, $V = \ker \alpha \oplus \sigma(\mathbb{C})$, where σ is a section of α. If we denote by v_σ the vector $\sigma(1)$, we have that any vector $v \in V$ can be written uniquely as $v = u + \lambda v_\sigma$, with $\alpha(u) = 0$.

Given two maps $f\colon X \to V$ and $g\colon X \to W$, we may define another map $f \oplus g\colon X \to V \oplus W$ as $(f \oplus g)(x) = f(x) \oplus g(x)$. We will call it the direct sum of the maps f and g. In particular if $L\colon U \to V$ and $M\colon U \to W$ are linear maps, then $L \oplus M\colon X \to V \oplus W$ is again a linear map.

In addition to that, there are also two more canonical maps $\pi_i\colon V_1 \oplus V_2 \to V_a$, defined as $\pi_a(v_1 \oplus v_2) = v_a$, $a = 1,2$. The linear maps are surjective and their direct sum $\pi_1 \oplus \pi_2\colon V_1 \oplus V_2 \to V_1 \oplus V_2$, is just the identity map $\pi_1 \oplus \pi_2 = \mathrm{id}_{V_1 \oplus V_2}$.

A.3.2 Tensor product

Definition A.11 We will define the tensor product $V \otimes W$ of the linear spaces V and W as the linear space $V \otimes W = \mathrm{Hom}(V^*, W)$. Vectors in $V \otimes W$ will be sometimes called tensors.

Notice that, because of the definition, we also get: $\mathrm{Hom}(V,W) = V^* \otimes W$.

There is a natural map $t \colon V \oplus W \to V \otimes W$ given by $t(v \oplus w) \colon V^* \to W$, $t(v \oplus w)(\alpha) = \alpha(v)w$ for any $v \oplus w \in V \oplus W$. The tensor $t(v \oplus w) \in V \otimes W$ will be denoted by $v \otimes w$ and will be called a separable tensor.

The 'tensorialization map' t is neither linear nor injective. For instance, $t(v \oplus \mathbf{0}) = \mathbf{0}$ for any $v \in V$. It also happens that $t(\lambda v \oplus w) = t(v \oplus \lambda w)$, but $\lambda v \oplus w \neq v \oplus \lambda w$.

Some properties of previous operations are given below. Given three linear spaces V, W, U, we get:

Proposition A.9

1. $V \oplus (W \oplus U) \cong (V \oplus W) \oplus U$.
2. $V \otimes (W \otimes U) \cong (V \otimes W) \otimes U$.
3. $V \oplus W \cong W \oplus V$.
4. $V \otimes W \cong W \otimes V$.
5. $V \oplus \mathbf{0} \cong V$.
6. $V \otimes \mathbf{0} \cong \mathbf{0}$.
7. $V \otimes (W \oplus U) \cong (V \otimes W) \oplus (V \otimes U)$.
8. $\mathrm{Hom}(V,W) \cong V^* \otimes W$.
9. $(V \oplus W)^* \cong V^* \oplus W^*$.
10. $\mathrm{Hom}(V, U \oplus W) \cong \mathrm{Hom}(V,U) \oplus \mathrm{Hom}(V,W)$.

Because of Prop. A.9.1, we will write the direct sum of the linear spaces $V_1, \ldots, V_n$ as $V_1 \oplus V_2 \oplus \cdots \oplus V_n = \bigoplus_{i=1}^n V_i$. As a direct consequence of Prop. A.9.10, we get that if $V = \bigoplus_{k=1}^r V_k$, then

$$\mathrm{Hom}(V,W) \cong \bigoplus_{k=1}^{r} \mathrm{Hom}(V_k, W), \tag{A.2}$$

and similarly in the second factor.

If the spaces $V_i = V$ in the direct sums above are all identical, then $\bigoplus_{i=1}^n V$ will be denoted by nV, except if $V = \mathbb{C}$, in which case, we will write $\bigoplus_{i=1}^n \mathbb{C}$ as $\mathbb{C}^n$. In fact, it would be more consistent to use a notation like $V^{\oplus n}$ for nV, but we will stick to the common usage.

In this sense, it is interesting to observe that $\mathrm{Hom}(\mathbb{C}^n, \mathbb{C}^m) \cong \bigoplus_{i=1}^n \bigoplus_{j=1}^m \mathrm{Hom}(\mathbb{C}, \mathbb{C})$, but because linear maps $L\colon \mathbb{C} \to \mathbb{C}$ are defined by multiplication by a complex number $L(z) = \lambda z$, $\mathrm{Hom}(\mathbb{C}, \mathbb{C}) \cong \mathbb{C}$ and

$$\mathrm{Hom}(\mathbb{C}^n, \mathbb{C}^m) \cong \bigoplus_{i=1}^{n} \bigoplus_{j=1}^{m} \mathbb{C}. \tag{A.3}$$

A.4 Finite-dimensional linear spaces and matrix algebra

A.4.1 Finite-dimensional linear spaces

Definition A.12 A family of vectors $\mathcal{S} \subset V$ is said to be linearly independent if for any finite family of vectors $v_1, \ldots, v_n \in \mathcal{S}$, such that $\lambda_1 v_1 + \cdots + \lambda_n v_n = 0$, then necessarily $\lambda_1 = \cdots = \lambda_n = 0$.

Definition A.13 A family of vectors $\mathcal{B} \subset V$ is called a basis of the linear space V if it is a linearly independent generating set for V, that is, the family of vectors $\mathcal{B}$ satisfies Def. A.12 and its span is V, Def. A.3 .

Theorem A.3
Any linear space V has a basis.

The proof of this theorem relies on similar arguments as the ones used in the proof of Thm. A.1 (see also Remark A.3).

Definition A.14 A linear space is finite-dimensional if it has a finite generating set.

Proposition A.10
Any finite-dimensional linear space has a finite basis and all of them have the same number of vectors.

Definition A.15 We call the dimension of a finite-dimensional linear space V the number of vectors in any of its bases and will be denoted by $\dim V$.

Example A.1 *The standard numerical linear space $\mathbb{C}^n$. Consider the linear space $\mathbb{C}^n$. Vectors X in $\mathbb{C}^n$ will be denoted as columns of complex numbers, thus*

$$X = \begin{bmatrix} x^1 \\ \vdots \\ x^i \\ \vdots \\ x^n \end{bmatrix} .$$

The numbers x_i will be called the components of the vector X. Vectors in $\mathbb{C}^n$ will also be called numerical vectors.

The standard basis of $\mathbb{C}^n$ is defined by the vectors e_i whose ith component is one and all the others zero. Any vector X in $\mathbb{C}^n$ can be written as $X = x^1 e_1 + \cdots + x^n e_n = \sum_{i=1}^n x^i e_i$.

If V is a finite-dimensional linear space of dimension n, the choice of a basis $\mathcal{B} = \{u_1, \ldots, u_n\}$ defines an isomorphism from V to the standard numerical linear space of dimension $\mathbb{C}^n$, $\varphi_{\mathcal{B}} \colon V \to \mathbb{C}^n$ given by $\varphi_{\mathcal{B}}(v) = X$ where the ith component x^i of X is the ith coordinate of v in the basis $\mathcal{B}$, that is, $v = \sum_{i=1}^n x^i u_i$.

Theorem A.4
All finite-dimensional linear spaces of the same dimension are isomorphic.

Proposition A.11
Let V, W be two linear spaces, then any tensor $t \in V \otimes W$ can be written as a linear combination of separable tensors, i.e., $t = v_1 \otimes w_1 + \cdots + v_r \otimes w_r$ albeit not in a unique form.

A.4.2 Linear maps and matrices

Given a finite dimensional linear space V, we may consider the set of all linear maps $L \colon V \to V$, we will denote it by $\mathrm{End}(V) = \mathrm{Hom}(V, V)$, and its most salient feature is that, apart from being a linear space itself with the standard operations of addition of matrices and multiplication by scalars, the composition of linear maps introduces an additional product structure. We reserve the name 'algebra' for such structure[3] and the space of linear maps $\mathrm{End}(V)$ will be called the algebra of endomorphisms of the linear space V. What lies behind such an imposing name is the familiar algebra of square matrices $M_n(\mathbb{C})$ that we will succinctly review in what follows.

Let $L \colon V \to W$ be a linear map between finite-dimensional spaces V and W of dimensions m and n and let $\mathcal{B}_V = \{v_i \mid i = 1, \ldots, m\}$ and $\mathcal{B}_W = \{w_j \mid j = 1, \ldots, n\}$ be linear bases of V and W, respectively.

[3] Algebras will be discussed at length in the main text, Sect. 10.1.1.

The linear map L is completely characterised by the family of vectors $L(v_i) \in W$. In fact, given any vector $v \in V$, $L(v) = \sum_{i=1}^{m} x^i L(v_i)$, where $v = \sum_{i=1}^{m} x^i v_i$. Conversely, if we are given a family of vectors $u_i \in W$, $i = 1, \ldots, m$ in W, they define uniquely a linear map $L\colon V \to W$ by means of $L(v) = \sum_{i=1}^{m} x^i u_i$. Finally, any of the vectors $u_i = L(v_i)$ have associated unique coordinates in the basis $\mathcal{B}_W = \{w_i\}$. We will denote such coordinates by $A(j,i)$ that is $L(v_i) = \sum_{j=1}^{n} A(j,i) w_j$.

Definition A.16 A $m \times n$ matrix A is an array of m rows and n columns of scalars (tipically delimited by brackets):

$$A = \begin{bmatrix} A(1,1) & \cdots & A(1,j) & \cdots & A(1,n) \\ \vdots & \ddots & \vdots & \ddots & \vdots \\ A(i,1) & \cdots & A(i,j) & \cdots & A(i,n) \\ \vdots & \ddots & \vdots & \ddots & \vdots \\ A(m,1) & \cdots & A(m,j) & \cdots & A(m,n) \end{bmatrix} .$$

The entries of the matrix A will be denoted[4] by $A(i,j)$ (or A_{ij}), $i = 1, \ldots, m$, $j = 1, \ldots, n$. The entries of the matrix A of the form $A(i,j)$, $i = 1, \ldots, m$ with j fixed define the jth column of A and the entries of the form $A(i,j)$, $j = 1, \ldots, m$, i fixed will define the ith row of A.

Thus, we conclude that, if we are given bases $\mathcal{B}_V$ and $\mathcal{B}_W$ of V and W, respectively, there is a $m \times n$ matrix A associated to any linear map $L\colon V \to W$. Conversely, given a $m \times n$ matrix A, there is a linear map $L\colon V \to W$ associated to it, defined by the formula:

$$L(v) = \sum_{i=1}^{m} \sum_{j=1}^{n} x^i A(j,i) w_j , \qquad v = \sum_{i=1}^{m} x^i v_i .$$

We will call the matrix A, defined above, the matrix associated to the linear map L in the bases $\mathcal{B}_V$ and $\mathcal{B}_W$.

If we denote the space of $m \times n$ matrices by $M_{m,n}(\mathbb{C})$, the previous correspondence $L \mapsto A$, establishes a bijection between the space $\mathrm{Hom}(V,W)$ of all linear maps between V and W and $M_{m,n}(\mathbb{C})$. Then, using (A.3), we get:

$$M_{m,n}(\mathbb{C}) = \mathrm{Hom}(\mathbb{C}^n, \mathbb{C}^m) \cong \bigoplus_{i=1}^{n} \bigoplus_{j=1}^{m} \mathbb{C} .$$

In particular, if V is a n-dimensional linear space:

$$M_n(\mathbb{C}) \cong \mathrm{End}(V) = \bigoplus_{i,j=1}^{n} \mathbb{C} .$$

[4]The standard usage in mathematical textbooks is to use the subindex/upperindex notation A_{ij} for the entries of a matrix, however it is more convenient to use the bracket notation $A(i,j)$, which is standard in programming languages, but also because it does not interferes with the standard usage for the indexes of components of tensor quantities.

A.4.3 Operations with matrices

All operations with linear maps discussed before define the corresponding operation with matrices. Thus, for instance, the sum of linear maps will define the sum of matrices. Then, it is immediate to check the following list of operations with matrices:

Proposition A.12

1. *Sum. Given two $m \times n$ matrices A, B, its sum $A + B$ is defined as the $m \times n$ matrix whose (i, j) entry is given by $A_{ij} + B_{ij}$, that is, $(A+B)(i, j) = A(i, j) + B(i, j)$.*

2. *Multiplication by scalars. If A is a $m \times n$ matrix and $\lambda \in \mathbb{C}$, we define λA as the matrix whose (i, j) entry is $\lambda(A(i, j))$.*

3. *Product of matrices. If A is a $m \times n$ matrix and B is a $n \times p$ matrix, then $A * B$ (or just AB) is a $m \times p$ matrix whose (i,k) entry is given by $(AB)(i,k) = \sum_{j=1}^{n} A(i, j)B(j,k)$.*

In addition to the inner algebraic operations with matrices listed above, there are some external operations associated to the operations with linear spaces discussed in Sect. A.2.3: Direct sums, tensor products and the adjoint. Remember that the direct sum of two linear maps $L \colon V \to W$, $L' \colon V' \to W'$, is the linear map $L \oplus L' \colon V \oplus V' \to W \oplus W'$ given by $(L \oplus L')(v \oplus v') = L(v) \oplus L'(v')$. Similarly, we define the tensor product of L and L' as the linear map $L \otimes L' \colon V \otimes V' \to W \oplus W'$, $(L \otimes L')(v \otimes v') = L(v) \otimes L'(v')$; and the adjoint map $L^* \colon W^* \to V^*$, given by $(L^*(\alpha))(v) = \alpha(L(v))$, $\alpha \in W^*$, $v \in V$, $v' \in V'$.

Then, choosing basis, we can identify matrices with linear maps and the previous operations have a matricial counterpart.

Proposition A.13

1. *Direct sum. Given two matrices A, B, A $n \times m$ and B $p \times q$ its direct sum $A \oplus B$ is the block diagonal $(n + p) \times (n + q)$ matrix:*

$$A \oplus B = \begin{bmatrix} A & 0 \\ 0 & B \end{bmatrix}$$

2. *Tensor product of matrices. If A is a $m \times n$ matrix and B is a $p \times q$ matrix we define tensor product $A \otimes B$ as the $np \times mq$ matrix whose $((i,k), (j,l))$ entry is $(A \otimes B)((i,k), (j,l)) = A(i, j)B(k,l)$, $i = 1 \ldots n$, $j = 1, \ldots, m$, $k = 1, \ldots, p$, $l = 1, \ldots, q$.*

3. *Transpose. If A is a $m \times n$ matrix, then its tranpose matrix A^T is the $n \times m$ matrix, $A^T(i, j) = A(j, i)$.*

The Trace

Definition A.17 Let $L: V \to V$ be a linear map. Then, we define the trace the of L and we will denote it by $\mathrm{Tr}(L)$, as $\mathrm{Tr}(L) = \sum_{i=1}^{n} D_{ii}$, where $n = \dim V$ and D is the matrix associated to L, with respect to some basis on V.

The previous definition is correct as the number $\sum_{i=1}^{n} D_{ii}$ does not depend on the basis we choose. In fact, if we had chosen a different basis, the corresponding matrices D and D' associated to L would have been related by $D' = MDM^{-1}$, but then $D'_{ii} = \sum_{kl} M_{ik} D_{kl} (M^{-1})_{li}$. Then, compute $\sum_{i=1}^{n} D'_{ii} = \sum_{kl} \sum_{i=1}^{n} M_{ik} D_{kl} (M^{-1})_{li} = \sum_{kl} (\sum_{i=1}^{n} (M^{-1})_{li} M_{ik}) D_{kl} = \sum_{kl} \delta_{kl} D_{kl} = \sum_{k=1}^{n} D_{kk}$.

Proposition A.14

1. *Circular property:* $\mathrm{Tr}(LM) = \mathrm{Tr}(ML)$, *for all* L, M.
2. *Linearity:* $\mathrm{Tr}(L+M) = \mathrm{Tr}(L) + \mathrm{Tr}(M)$, $\mathrm{Tr}(\lambda L) = \lambda \mathrm{Tr}(L)$.
3. *Hermitian:* $\mathrm{Tr}(L^T) = \mathrm{Tr}(L)$, $\mathrm{Tr}(\overline{L}) = \overline{\mathrm{Tr}(L)}$.
4. $\mathrm{Tr}(\mathrm{id}_V) = \dim V$.
5. *Additivity:* $\mathrm{Tr}(L \oplus M) = \mathrm{Tr}(L) + \mathrm{Tr}(M)$.
6. *Multiplicativity:* $\mathrm{Tr}(L \otimes M) = \mathrm{Tr}(L)\,\mathrm{Tr}(M)$

Appendix B

Generators and Relations

In this appendix, we will discuss the construction of groups, groupoids and categories using a set of generators and relations.

B.1 Free groups

We will devote some paragraphs to a discussion on the construction of groupoids out of generators and relations. This problem was addressed in the first Chapter under the name of the category generated by a quiver, that is, categories (or groupoids) that were obtained by abstracting the connecting dots games, Sect. 1.1.3. We are ready now to investigate the appropriate background for that.

We will start by discussing the construction of groups out of a set of generators and relations.

The possibility of constructing something as specific and tight as a group out of almost anything is part of the magic of abstract thinking (the very soul of Algebra[1]). In fact, it is convenient to play this game in various ways so that the reader gets convinced that the 'magic' if any, is just in the relations, everything else is superficial.

Thus, consider that we are given a list of symbols, sometimes called an alphabet. We will suppose that the list is finite, in order to keep with the spirit of the games proposed in this book, and we will denote its elements (or 'letters') by a, b, c, etc. Later on, in order to facilitate the writing of expressions (and also to stick to long stablished mathematical tradition) we will use also $a_1, a_2, \ldots$ to denote the letters of the alphabet. The alphabet itself will be denoted by A.

[1]Let us recall that the name "Algebra" comes from the begining of the title of the book written by Alworitzmi, the astronomer of the Sultan Al Mamum in the ancient Damasco Califat in 800, whose translation is "Treatise of the substitutions...".

We can form all possible 'words' using the letters of the alphabet, i.e., finite ordered sequences of letters $w = a_{i_1}a_{i_2}\ldots a_{i_s}$, obviously, there is no need that they make sense in any possible way, we just ask that they have a finite number of ordered letters. Examples of words will be $w_1 = a$, $w_2 = ab$, $w_3 = ba$, $w_4 = aaaa$, $w_5 = abab$, etc. We will also allow the empty word, that is, the word containing no letters. We use the special symbol e(mpty) for it, that is $e =$" ". We will denote this set by $M(A)$ and we introduce a composition law on it given simply by juxtaposition, $w \cdot w' = a_1 \ldots a_r a'_1 \ldots a'_s$ if the words w and w' were given by $a_1 \ldots a_r$ and $a'_1 \ldots a'_s$, respectively. The composition law $\cdot$ satisfies (obviously) the following properties:

1. Neutral element. $e \cdot w = w$, $w \cdot e = w$ for any $w \in M(A)$.

2. Associativity. $(w \cdot w') \cdot w'' = w \cdot (w' \cdot w'')$, for all $w, w', w'' \in M(A)$.

Hence, the set $M(A)$ becomes a monoid, called the free monoid generated by A. The letters used to construct the words are the generators of the monoid considered as words themselves. In general, given a monoid M, we will say that it is generated by a family $S = \{m_1, \ldots, m_n\}$ if any element $m \in M$ can be written as $m = m_{i_1} \cdot \ldots \cdot m_{i_s}$, $i_1, \ldots, i_s = 1, \ldots, n$, and the set S will be called a generating system for M.

The monoid $M(A)$ has the following important property. Let M be another monoid and $\varphi\colon A \to M$ be a map, that is $\varphi(a) \in M$ is an element in M for any letter a in the alphabet A. Then, there is a unique map $\Phi\colon M(A) \to M$ which preserves the composition law (called a homomorphism of monoids), that is, $\Phi(w \cdot w') = \Phi(w) \cdot \Phi(w')$ for all $w, w' \in M(A)$, and such that, when restricted to A, it is just φ. Actually, the map Φ is explicitly constructed as $\Phi(w) = \varphi(a_1) \cdot \ldots \cdot \varphi(a_r)$ if $w = a_1 \ldots a_r$. This property is called a universal property and it completely characterises $M(A)$ (provided that we have shown first that we know how to construct it).

Proposition B.1

Let M' be a monoid generated by A (suppose that there is an injective map $j\colon A \to M'$ and $j(A)$ is a generating system for M'), such that it satisfies the same universal property satisfied by $M(A)$, with respect to the set A, that is, for any map $\varphi\colon A \to M$, there is a monoid homomorphism $\Phi\colon M' \to M$, such that $\Phi\mid_{j(A)} = \varphi$; then M' is isomorphic to $M(A)$.

Proof B.1 The proof is simple. If M' satisfies the universal property defined by A, then, because there is a natural map $i\colon A \to M(A)$, defined simply by $i(a) = a$, then there is a homomorphism $\Psi\colon M' \to M(A)$, such that $\Psi \circ j = i$.

Moreover, because $M(A)$ satisfies the universal property by construction, it means that there is a homomorphism $\Phi\colon M(A) \to M'$, such that $\Phi \circ i = j$. Now, it is trivial to check that Ψ and Φ are the inverse of each other. Actually, if $\Phi \circ i = j$, then $\Phi(a) = j(a)$ for any element a in A, and any word $w = a_1 \ldots a_r$ is mapped into the element $\Phi(w) = j(a_1) \cdot \ldots j(a_r)$. Conversely, the element $\Phi(w) = j(a_1) \cdot \ldots j(a_r)$ is mapped into the word w by using the map Ψ, because $\Psi \circ j = i$.

As a simple example, we may consider the free monoid generated by a single letter "a". Clearly $M(\{a\})$, denoted in what follows as M_1, will consist of words $e, a, aa, aaa, \ldots, a^{(n)}, \ldots$, where $a^{(n)}$ denotes the word with n a's. The composition law will be $a^{(n)} \cdot a^{(m)} = a^{(n+m)}$, thus, again we find that M_1 is isomorphic to the natural numbers monoid $\mathbb{N}$ (which, on the other hand, is the category $\mathbf{C}(Q_{2,1})$ generated by the singleton). Another example is the monoid generated by the letters "a" and "b" and denoted by M_2. Clearly, it consists of all words $e, a, b, aa, ab, ba, bb, aaa, aab, \ldots$. Note that the monoid M_1 is commutative while M_2 is not.

Interesting as they are, monoids are not our target. We would like to be able to "erase" letters from words. In other words, we will introduce for each generator $a \in A$ another one denoted by a^{-1} with the property that $a \cdot a^{-1} = a^{-1} \cdot a = e$.

The formal way to do that will be to extend our alphabet A, incorporating the erasing letters a^{-1}, one for each letter a in A. Denoting the new alphabet by $A_c = A \cup A^{-1}$, now we construct the free monoid $M(A_c)$ generated by A_c and then we introduce an equivalence relation on it, described as follows: We will say that two words w and w' are equivalent if they differ only on sequences of letters aa^{-1} or $a^{-1}a$. This clearly defines an equivalence relation on $M(A_c)$ whose equivalence classes are denoted as $[w]$. For instance, $[a] = \{a, a^{-1}aa, aa^{-1}a, aaa^{-1}, \ldots\}$.

There is a way of selecting a unique representative on each equivalence class $[w]$. Let w be a word. Suppose that the word contains sequences of letters of the form aa^{-1} or $a^{-1}a$ for any generator $a \in A$. We will denote by $w_{\rm red}$ the word obtained by erasing the previous sequences of letters. The word thus obtained will be called the reduced word of w. For instance, if $w = baa^{-1}db^{-1}aaaa^{-1}$, we will get $w_{\rm red} = bdb^{-1}aa$. We will take $w_{\rm red}$ as the representative of the equivalence class containing w.

We observe that the set of equivalence classes $M(A_c)/\sim$ can be equipped with an associative composition law given by:

$$[w] \circ [w'] = [w \cdot w'], \tag{B.1}$$

for all $[w]$, $[w']$ in $M(A_c)/\sim$. It is an easy exercise to check that the previous definition is correct, that is, it does not depend on the representatives chosen for the equivalence classes $[w]$ and $[w']$ to perform the calculation on the r.h.s. of Eq. (B.1). Thus, we may use the reduced words to perform the computation and we get:

$$w_{\rm red} \circ w'_{\rm red} = (w \cdot w')_{\rm red} \,.$$

Note that we cannot just use the juxtaposition composition $\cdot$ for the reduced words because we may find a situation like $w = ab$ and $w' = b^{-1}a^{-1}$, then $w_{\rm red} \cdot w'_{\rm red} = abb^{-1}a^{-1}$ while $w \circ w' = e$ ($w = w_{\rm red}$ and $w' = w'_{\rm red}$).

We realise that we have obtained something else, the set $M(A_c)/\sim$ is not only a monoid with the composition law $\circ$ but a group. Actually, any reduced word has an inverse with respect to $\circ$; it is the word that 'erases' w, that is if $w = a_1 \ldots a_r$ is a reduced word, then $w^{-1} = a_r^{-1} \ldots a_1^{-1}$, then clearly $w \circ w^{-1} = w^{-1} \circ w = e$. The set $M(A_c)/\sim$ is called the free group generated by A and will be denoted by $F(A)$.

In the simple examples discussed before, we get that the free group generated by a single letter, denoted by F_1, is just the Abelian group of integer numbers $\mathbb{Z}$, any element there represented by the reduced word $a^{(n)}$ where $n \in \mathbb{Z}$. The free group generated by two symbols, denoted by F_2, is not Abelian (and much more interesting and paradoxical!).

As in the case of the free monoid generated by an alphabet, the free group generated by a set of symbols[2] (or generators) S satisfies a universal property.

Proposition B.2
The free group $F(S)$ generated by the set of generators S is the unique group (up to isomorphisms) that satisfies the following universal property: Let $\varphi\colon S \to G$ be a map from the set S to the group G, then there exists a unique homomorphism $\Phi\colon F(S) \to G$, such that $\Phi \circ i = \varphi$, where $i\colon S \to F(S)$ is the natural inclusion of S in $F(S)$.

Proof B.2 Let $w = a_1 \dots a_r$ be a reduced word in $F(S)$ with $a_k \in S$, $k = 1, \dots, r$. Then, we define $\Phi(w) = \varphi(a_1) \cdots \varphi(a_r) \in G$. Clearly, $\Phi\colon F(S) \to G$ is a group homomorphism and it is an easy exercise to check that is the unique map satisfying the required property.

B.2 Groups constructed out of generators and relations

Once we have learned how to construct groups almost out of nothing, that is, from any family of symbols, we would like to understand if it is possible to construct other groups, not just the free ones (that, except for F_1, are quite bizarre). The answer is easy: *Any group can be constructed as a quotient of a free group.* However, we will not try to prove this statement, but just provide a constructive way of 'designing' groups.

The way to do that will be by imposing conditions, called 'relations', among its generators. So, we will start choosing a family of generators (just plain symbols) and then we will require that some properties will have to be satisfied by them. For instance, we may start with two generators a and b and ask that they 'commute': $ab = ba$. How do we get a group where this condition is satisfied? Of course, we may proceed as we did in the previous section to pass from the monoid $M(A_c)$ to the free group $F(A)$, introducing an equivalence relation. In the example above, we will consider the free group F_2 generated by the two elements a, b and we will establish that two reduced words w and w' are equivalent if they differ only on pairs of letters ab, ba, or $a^{-1}b^{-1}$, $b^{-1}a^{-1}$, that is $w = baaaba^{-1}b^{-1}$ would be equivalent to $w' = ababab^{-1}a^{-1}$.

The same idea works when we consider any other formula, not just $ab = ba$, expressed in terms of the generators $a, b \ldots \in S$. That is, if we are given a finite family $\mathcal{R}$ of formulas $w_i = w_i'$, $i = 1, \dots, s$, each word w_i, w_i' written as a reduced word, we will say that two reduced words are equivalent if they differ only in sequences of letters listed in the formulas in $\mathcal{R}$. The quotient space $F(S)/\sim$ inherits a composition

[2]We will assume from now on the S always contains the symbols a and a^{-1}.

law. Denoting the equivalence classes again as $[w]$, we get:

$$[w][w'] = [w \cdot w'] ,$$

and it is easy to show that it is well defined.

This way of describing this 'natural' equivalence relation defined by the set of formulas $\mathcal{R}$ becomes very wordy and seems a bit involved (but easy to apply). Fortunately (at least from the point of view of proving theorems), there is a nice way of describing the space of equivalence classes defined by the previous relation, albeit a bit more abstract that the manipulations with words we were enjoying so much. It is based on the abstract notions of normal subgroups and quotient groups that we introduced in Sect. 2.1.3.

Recall that to quotient a group G under an equivalence relation and still have a group structure on the space of equivalence classes, such equivalence relation should be given by a normal subgroup H, that is, there must exist a subgroup H, such that $g \sim g'$ iff there is $h \in H$ such that $g' = gh$ (or $g' = hg$). In such a case, the space of equivalence classes, denoted as G/H, would be a group with the natural composition law if and only if H is normal, that is, for every $g \in G$, $ghg^{-1} \in H$ for any $h \in H$, recall Thm. 2.2.

We will take advantage of this realising that if $F(S)/\sim$ inherits a group structure there must exist a normal subgroup H of $F(S)$, such that $F(S)/H = F(S)/\sim$. Such a group is not difficult to find and is just the normalizer of the set of relations we have introduced in the set of generators.

Definition B.1 Let G be a group and S a subset of G. We will call the normalizer of S the smallest normal subgroup of G containing S and it will be denoted as $N(S)$.

Clearly, the normalizer of S exists because the intersection of normal subgroups is a normal subgroup. Then the normalizer of S is just the intersection of all normal subgroups containing it. There is even an explicit characterization of it.

Proposition B.3
Let S be a subset of the group G. Then $N(S) = \{gsg^{-1} \mid s \in S, g \in G\}$.

Proof B.3 Clearly the set $N(S)$ is a normal subgroup and it contains S.

On the other hand, if N' would be a normal subgroup containing S, then for any $s \in S$, we will get that $gsg^{-1} \in N'$, hence, $N(S) \subset N'$.

Then, if we are given a list of formulas $w_i = w_i'$, we can write them as $w_i^{-1} w_i' = e$ and consider the elements $r_i = w_i^{-1} w_i'$ defining the set S.

Thus, given $S = \{r_1, \ldots, r_p\}$, a finite family of reduced words defining the relations introduced in the family of generators S, we consider the normalizer $N(S)$ of $F(S)$. As shown in Prop. B.3, such groups can be constructed by taking finite products of elements of the form $wr_k w^{-1}$ where r_k is any relation in S and w any reduced word in $F(S)$. Then, we define the group G generated by the set of generators S and the set of relations R as the quotient group $F(S)/N(R)$.

It is not difficult to check that the subgroup $N(R)$ is just the normal subgroup defined by the equivalence relation $\sim$ introduced in the set of reduced words by the formulas in $\mathcal{R}$, however, the important thing to understand is that a judicious choice of the set of generators S and relations $\mathcal{R}$ allows us to 'customize' any group we wish.

For instance, if we wish to construct Abelian groups, all we have to do is to introduce relations $ab = ba$ for any pair of generators. If we want to guarantee that we have finite groups we have to introduce relations $a^r = e$ for any generator[3] a (the exponent or 'order' of the element can vary).

The previous construction can be used backwards to show that any group[4] can be obtained as a quotient of a free group. Let us consider a finite group G and the set of generators provided by the elements of the group itself $S = G$ and relations R given by the multiplication table of the group, that is, all formulas $gh = k$, $g,h,k \in G$. Then, clearly, $F(G)/N(R) = G$.

A selection of a set of generators $a_1, a_2, \ldots, a_n$ and a family of relations $r_1 = e, \ldots, r_s = e$, describing a group G will be called a presentation of the group G and will be denoted as:

$$G = \langle a_1, \ldots, a_n \mid r_1 = \cdots = r_s = e \rangle .$$

Example B.1 *The group S_3. We can describe the symmetric group S_3, which is isomorphic to the dihedral group D_3 by generators and relations as:*

$$S_3 = D_3 = \langle a, b \mid a^3 = b^2 = e, bab = a^{-1} \rangle .$$

Example B.2 *The dihedral group D_n. A presentation for the dihedral group of order $2n$ is given by:*

$$D_n = \langle a, b \mid a^n = b^2 = e, bab = a^{-1} \rangle .$$

Example B.3 *The symmetric group S_n. A presentation for the symmetric group S_n is given by (the generators a_k could be identified with transpositions $(k, k+1)$):*

$$S_n = \langle a_1, a_2, \ldots, a_{n-1} \mid a_k^2 = e, a_k a_l = a_l a_k, l \neq k \pm 1, (a_k a_{k+1})^3 = e, k = 1, \ldots, n-1 \rangle .$$

B.3 Other finite groups of small order defined by generators and relations

We will finish this section, listing the definition of a few finite groups of low order by means of generators and relations.

[3]This is just a necessary condition, not sufficient! Can you think on an example of an infinite group with two generators of finite order, for instance, 2?

[4]The idea works in the case of non-finite groups too.

1. The dicyclic group of order $4n$: $\mathrm{Dic}_n = Q_{4n}$. It is an extension of the cyclic group of order 2 by the cyclic group of order $2n$. In terms of generators and relations, the dicyclic group is the group given by the following presentation:

$$\mathrm{Dic}_n = \langle a, b \mid a^{2n} = 1, b^2 = a^n, b^{-1}ab = a^{-1} \rangle$$

   In particular $Q_8 = \mathrm{Dic}_2$, $Q_{12} = \mathrm{Dic}_3$, $Q_{16} = \mathrm{Dic}_4$, $Q_{20} = \mathrm{Dic}_5$.

2. The quasi-dihedral group of order 2^n: QD_n. A presentation for QD_n is given by:

$$QD_n = \langle a, b \mid a^{2^{n-1}} = b^2 = e, bab = a^{2^{n-2}-1} \rangle$$

A word of caution: The reader must be careful about the apparent innocence of the procedures to construct groups that we have just described. They essentially consist of manipulating a finite family of symbols using a finite family of rules or relations. We must mention the infamous 'word problem' for finitely generated groups that consists of determining whether or not two groups obtained from two sets of generators and relations are isomorphic. Such a problem is undecidable, [53], [5], that is, there is not a universal algorithm that would solve it, even more, in the previous references, presentations of finitely generated groups are given which are undecidable.

B.4 Free objects

Once we have discussed the problem of constructing groups out of generators and relations, we would like to study whether it is possible to proceed in a similar way to construct categories and groupoids.

The answer is yes and we have already witnessed the method in the case of categories generated by finite quivers. Actually the construction of the category $\mathbf{C}(Q)$ generated by a quiver $Q(\Omega, L)$ is similar to the construction of a free monoid from an alphabet, in this case the alphabet being the set of links L. The difference is that not all symbols (links) can be composed, the source and target maps impose constraints on the links that can be composed, thus, not all possible words for the alphabet defined by the links can be constructed.

The construction of a groupoid from a quiver and a set of relations will proceed by constructing first the 'free' category generated by the quiver, then constructing the free groupoid associated to it, adding the inverses of morphisms and then considering the quotient with respect to the equivalence relation defined by such relations.

We will proceed by presenting these ideas in a unified fashion.

The construction of free monoids, free groups, free categories and free groupoids all share the same categorical notion, that of a free object, that we will discuss succinctly here. For that, we will need the notion of adjoint functors.

The notion of adjoint functors is borrowed from the notion of adjoint operators in linear algebra. Let V and W be two linear spaces with inner products $\langle \cdot, \cdot \rangle_V$ and

$\langle \cdot, \cdot \rangle_W$, respectively. Let $A\colon V \to W$ and $B\colon W \to V$, we will say that A and B are adjoint of each other if:

$$\langle Av, w \rangle_W = \langle v, Bw \rangle_V$$

for all $v \in V$ and $w \in W$.

We may 'categorize' the previous definition by replacing the linear spaces V and W by categories $\mathbf{C}$ and $\mathbf{D}$, the linear maps A and B by functors $F\colon \mathbf{C} \to \mathbf{D}$ and $G\colon \mathbf{D} \to \mathbf{C}$, and the inner products, the most delicate decision, by the family of morphisms between two objects in the corresponding categories. Thus, we will say that the functors F and G are adjoint of each other (also we say that F is a left-adjoint of G, or G is a right-adjoint of F) if for any pair of objects x in $\mathbf{C}$ and y in $\mathbf{D}$, we have:

$$\mathrm{Mor}_{\mathbf{D}}(F(x), y) = \mathrm{Mor}_{\mathbf{C}}(x, G(y)).$$

Consider now a small category $\mathbf{D}$ and the forgetful functor $G\colon \mathbf{D} \to \mathbf{Sets}$, that takes any object $x \in \mathbf{D}$ into its corresponding set $\{x\}$, and any morphism $\alpha\colon x \to y$ into the corresponding map between sets. Let $F\colon \mathbf{Sets} \to \mathbf{D}$ be a left-adjoint functor to the functor G, that is a functor such that:

$$\mathrm{Mor}_{\mathbf{D}}(F(S), y) = \mathrm{Mor}_{\mathbf{Sets}}(S, G(y)).$$

The previous identification means that, for each map $f\colon S \to G(y)$, from the set S into the set $G(y)$ associated with the object y in the category $\mathbf{D}$, there is a unique morphism $\alpha_f\colon F(S) \to y$, from the object $F(S)$ in the object y. However, this is just a general way of writing the universal property used to characterise free monoids, groups, etc.

Example B.4 *For instance, consider the functor $M\colon$ **Sets** $\to$ **Mon**, from the category of sets to the category of monoids, given by assigning the monoid $M(A)$ (generated by the alphabet A) to the set A that was described at the beginning of this Appendix. Then, if we are given a map from $A \to M$, M a monoid considered just as a set, then there exists a map from $M(A) \to M$ which is a homomorphism of monoids. Then, we can say that the functor M that constructs out of any set a monoid is a left-adjoint to the forgetful functor G on the category of monoids.*

Hence, whenever we are able to construct a left-adjoint functor F to the forgetful functor G on a given small category, the object $F(S)$ will be called the free object generated by the set S.

Example B.5 *Another example of the construction of free objects is provided by the construction of the free group generated by a family of generators S. In this situation the category under consideration will be the category of groups* **Groups** *whose objects are groups G and whose morphisms are group homomorphisms $f\colon G \to G'$. The forgetful functor assigns to any group G the set defined by itself. We will not use a particular notation for that. Then a left-adjoint functor to the forgetful functor will be functor $F\colon$ **Sets** $\to$ **Groups**, such that:*

$$\mathrm{Mor}_{\mathbf{Groups}}(F(S), G) = \mathrm{Mor}_{\mathbf{Sets}}(S, G),$$

that is, that assigns to any map $\varphi\colon S \to G$ a unique map $\Phi\colon F(S) \to G$, but this is precisely what the free group $F(S)$ does for us.

B.5 Free categories

The idea of a constructing a free object can be applied to quivers, that is, consider the forgetful functor $G\colon \mathbf{SmallCat} \to \mathbf{Quiv}$ from the category of small categories into the category of quivers, that is, the objects of **Quiv** are quivers $Q(L,V)$ and morphisms are quiver maps, $\varphi\colon Q(L,V) \to (Q',V')$, satisfying $\varphi(l) = l' \in L'$ for any link $l \in L$, $\varphi(v) = v' \in V'$ for any vertex $v \in V$, and such that $s(\varphi(l)) = \varphi(s(l))$ and $t(\varphi(l)) = \varphi(t(l))$.

The forgetful functor G assigns to any small category the quiver defined by all its arrows and objects.

A left-adjoint functor C to G, would be a functor $C\colon \mathbf{Quiv} \to \mathbf{SmallCat}$ assigning a small category $\mathbf{C}(Q)$ to any quiver Q, such that:

$$\mathrm{Mor}_{\mathbf{Cat}}(\mathbf{C}(Q),\mathbf{C}) = \mathrm{Mor}_{\mathbf{Quiv}}(Q, G(\mathbf{C}))\,,$$

that is, such that, for each quiver map $\psi\colon Q \to G(\mathbf{C})$, there is a unique functor $\Psi\colon \mathbf{C}(Q) \to \mathbf{C}$.

At first, this seems very abstract and probably a bit intimidating, but the actual construction of $\mathbf{C}(Q)$ is easy, it is just the category generated by the quiver Q that we have been using already.

The objects of $\mathbf{C}(Q)$ are consistent histories $p = l_r l_{r-1} \ldots l_2 l_1$ on Q, that is $l_k \in L$, and $t(l_k) = s(l_{k+1})$, that is, the links l_k and l_{k+1} can be composed, for all $k = 1,\ldots,r-1$. For each $x \in \Omega$ we define the stationary path 1_x at x. The composition of paths p, p', is the obvious juxtaposition pp' whenever they can be composed, that is, when the end vertex of p' agrees with the first vertex of p. It is obvious that $\mathbf{C}(Q)$ is a small category.

Given a quiver map $\psi\colon Q \to \mathbf{C}$, we define the functor $\Psi\colon \mathbf{C}(Q) \to G(\mathbf{C})$ as follows: $\Psi(x) = \psi(x)$ which is an object of $\mathbf{C}$ and $\Psi(p) = \psi(l_r) \circ \ldots \circ \psi(l_1)$ which is a morphism of $\mathbf{C}$. Conversely, given a functor $\Psi\colon \mathbf{C}(Q) \to \mathbf{C}$, we define the map $\psi\colon Q \to G(\mathbf{C})$, $\psi(x) = \Psi(1_x)$ and $\psi(l) = \Psi(l)$. Clearly, ψ is a quiver map.

We conclude this paragraph by observing that the category $\mathbf{C}(Q)$ generated by the quiver Q can be properly called the free category generated by the quiver Q.

B.6 Free groupoids and groupoids constructed from quivers and relations

We will construct the free groupoid generated by a quiver $Q(\Omega,L)$ by ‘adding’ the inverses l^{-1} of generators $l \in L$ and using reduced consistent histories as we did in the case of the free group generated by a set (recall Sect. B.1).

Suppose that we have the free category $\mathbf{C}(Q)$ generated by the quiver $Q = (L,\Omega)$, then we may consider the quiver obtained by adding the inverse of the links in L, that is, $L_c = L \cup L^{-1}$. Then, we repeat the construction of free groups by constructing reduced consistent histories. The space of reduced consistent paths on the quiver $Q_c = (L_c,V)$ is by definition the free groupoid generated by $Q = (L,\Omega)$ and will be denoted by $\mathbf{G}(Q)$.

As it is easy to check, the free groupoid $\mathbf{G}(Q)$ generated by the quiver Q satisfies a universal property too:

Proposition B.4
Let $\mathbf{G}(Q)$ be the free groupoid generated by the quiver $Q(L,\Omega)$. Then, for any map $\psi\colon Q \to \mathbf{G}$, where $\mathbf{G} \rightrightarrows \Omega'$ is a groupoid, and $\psi(x) \in \Omega'$ and $\psi(l)\colon \psi(x) \to \psi(y)$, for any $l\colon x \to y \in L$, then there exists a unique homomorphism of groupoids $\Psi\colon \mathbf{G}(Q) \to \mathbf{G}$, such that $\Psi \circ i = \psi$, where $i\colon Q \to \mathbf{G}(Q)$, is the canonical inclusion map.

Again, as in the case of groups, once we have constructed a free groupoid $\mathbf{G}(Q)$ from a quiver Q, we may introduce relations on it and obtain new groupoids.

The set of relations $\mathcal{R}$ will be defined as a family of formulas $w_k = w_k'$, where w_k and w_k' are consistent histories, then we write the elements $r_k = w_k^{-1} \circ w_k'$, but notice that as elements in the groupoid $\mathbf{G}(Q)$, if $w_k, w_k'\colon x \to y$, then $r_k\colon x \to x$ and r_k is in the isotropy group at x of $\mathbf{G}(Q)$.

Then, the groupoid defined by the equivalence relation defined by the relations in $\mathcal{R}$ will be given as the quotient groupoid $\mathbf{G}(Q,\mathcal{R}) = \mathbf{G}(Q)/N(\mathcal{R})$ (notice that the argument here is exactly the same as in the case of groups because of the theorem that establishes the identification of equivalence relations defining quotient groupoids and normal subgroupoids, Thm. 6.1), where $N(\mathcal{R})$ is the normalizer of the set $\mathcal{R}$ (compare with the situation for groups discussed in Sect. B.2). As in the case of groups, the normalizer $N(\mathcal{R})$ is the smallest normal subgroupoid containing $\mathcal{R}$ and $N(\mathcal{R})$ is generated by the set of relations $\mathcal{R}$, in the sense that $N(\mathcal{R}) = \{\alpha \circ r_k \circ \alpha^{-1} \mid r_k\colon x \to x, \alpha\colon x \to y\}$.

As in the case of groups, we will call a quiver Q and a family of relations $\mathcal{R}$ a presentation of the groupoid $\mathbf{G}(Q,\mathcal{R})$.

Appendix C

An Application of Groupoids to Quantum Mechanics: Schwinger's Algebra

This appendix will be devoted to succinctly describing an interesting application of the theory of groupoids to Quantum Mechanics, elaborating on ideas introduced by J. Schwinger a long time ago when describing the basic algebraic structures involved in the basic description of quantum mechanical measurements [58].

C.1 Schwinger algebra and groupoids

In this appendix, we will discuss the structure of the Schwinger's algebra of selective measurements. It will be shown that Schwinger's algebra is just the regular representation of a groupoid that, in some simple instances, can be identified with the group algebra of a pair groupoid.

C.1.1 Introduction: Schwinger's picture of quantum mechanics

J. Schwinger's foundational approach to Quantum Mechanics described the algebraic relations between a family of symbols representing fundamental measurement processes, establishing a "mathematical language that constitutes a symbolic expression of the properties of microscopic measurements" [59]. Such a language couldn't consists of the classical representation of all physical quantities by numbers, they must be represented in terms of abstract symbols whose composition properties are postulated on the grounds of physical experiences. As it will be shown in what follows, the main algebraic structure behind such "mathematical language" is that of a groupoid. This fact was already anticipated by A. Connes [17]:

"The set of frequencies emitted by an atom does not form a group, and it is false that the sum of two frequencies of the spectrum is again one. What experiments dictate is the Ritz–Rydberg combination principle which permits indexing the spectral lines by the set Δ *of all pairs* (i,j) *of elements of a set* I *of indices. The frequencies* $\nu_{(ij)}$ *and* $\nu_{(kl)}$ *only combine when* $j=k$ *to yield* $\nu_{(il)} = \nu_{(ij)} + \nu_{(jl)}$ *(...). Due to the Ritz–Rydberg combination principle, one is not dealing with a group of frequencies but rather with a groupoid* $\Delta = \{(i,j); i,j \in I\}$ *having the composition rule* $(i,j)(j,k) = (i,k)$*. The convolution algebra still has a meaning when one passes from a group to a groupoid, and the convolution algebra of the groupoid* Δ *is none other that the algebra of matrices since the convolution product may be written* $(ab)_{(i,k)} = \sum_n a_{(i,n)} b_{(n,k)}$ *which is identical with the product rule of matrices."*

Thus, as Schwinger pretended, Connes relates the structure of certain measurements, in this case frequencies of the emission spectrum by atoms, to a specific algebraic structure, that of groupoids.

C.1.2 Categories in physics

It is quite convenient to think of groupoids as codifying processes, in the sense that the composition law determines the different ways that one can go from one base element to another one.

The process of abstracting properties of physical systems obtained by their observation, like the properties of measurements on microscopic systems discussed below, is extremely useful by itself, however, as it was pointed out by J. Baez, the mathematical language based on set theory is extremely restrictive and limited for many purposes [3].

Physics dealing with processes and relations both at the classical and quantum level, is particularly ill-suited to be described by set theory. On the contrary, category theory is exactly about that, the emphasis is not in the description of the elements of sets and their properties, but on the relations between objects, i.e., morphisms. Thus, "elements" in set theory correspond to "objects" of a category (then "elements" can be "sets" themselves without incurring in contradictions!) and "equations between elements" correspond to "isomorphisms between objects". "Sets" correspond, in this categorification of set theory, to "categories" and maps between sets to "functors" between categories. Finally, "equations between functions" will correspond to "natural transformations" between functors.

C.2 Schwinger's algebra of measurements symbols

Thus, we will start by analyzing the algebraic relations satisfied by the family of selective measurements under simplifying assumptions similar to those used by Schwinger himself [58, Chap. 1].

In its simplest form, a complete selective measurement can be defined as follows. Given a physical system, let us denote by $\mathcal{E}$ an ensemble associated to it, i.e., a large family of physical systems of the same type, satisfying the same specified conditions [22]. The elements S of the ensenble $\mathcal{E}$, that is, individual systems, are of course noninteracting.

It will be assumed that there is a family of observables $\mathcal{A}$, meaning by that, just a family of measurable physical quantities like position, momentum, etc., and such that the outcomes $a \in \mathbb{R}$ of their individuals $A \in \mathcal{A}$ can be measured on any system S of the ensemble $\mathcal{E}$.

Such measurement will be denoted as $a = \langle A : S \rangle$ and it clearly supposes an idealized simplification of a more elaborated statistical interpretation of an actual experiment.

It will be assumed that the ensemble $\mathcal{E}$ is large enough so that for any observable A and for each possible outcome a of the observable there are systems S, such that, when A is measured on them, the outcome is actually a. Under these conditions, we will say that the ensemble $\mathcal{E}$ is *sufficient* for the observable A.

Two observables $A, B \in \mathcal{A}$ will be said to be compatible if the outcomes of their measurements do not depend on the order in which they are performed. Thus, if we denote by $\langle A : B : S \rangle$ the process of measuring first the observable B and then, immediately after, the observable A on the system S, the outcomes (a, b) of such sequence of measurement will be the same as the outcomes of the sequence of measurements $\langle B : A : S \rangle$, that is, they will be (b, a).

A family of observables $\mathbf{A} = \{A_1, \ldots, A_N\} \subset \mathcal{A}$ is said to be compatible if they are compatible among them or, in other words, if the outcomes of their measurements do not depend on the order in which the measurements are performed.

By $M_{\mathbf{A}}(\mathbf{a}, \mathbf{a})$, or simply $M(\mathbf{a})$ for short, we will denote a procedure or device that rejects all systems S in the ensemble $\mathcal{E}$ whose outcomes $\mathbf{a}' = (a_1', \ldots, a_N')$ when measuring $\mathbf{A}$ are different from $\mathbf{a} = (a_1, \ldots, a_N)$, and that leaves the systems S, such that the measurement of $\mathbf{A}$ gives $\mathbf{a}$. Such procedure is called a selective measurement for the family of compatible observables $\mathbf{A} = \{A_1, \ldots, A_N\}$.

Denoting the outcomes of the measurement of $\mathbf{A}$ on the element S by $\langle \mathbf{A} : S \rangle$, s$M(\mathbf{a})S = S$ if $\mathbf{a} = \langle \mathbf{A} : S \rangle$ and $M(\mathbf{a})S = \emptyset$ if $\mathbf{a} \neq \langle \mathbf{A} : S \rangle$.

Using a set-theoretical notation, we may define the subensemble of $\mathcal{E}$ determined by the selective measurement $M(\mathbf{a})$:

$$\mathcal{E}_{\mathbf{A}}(\mathbf{a}) = \{S \in \mathcal{E} \mid \langle \mathbf{A} : S \rangle = \mathbf{a}\},$$

(in what follows, we will just write $\mathcal{E}(\mathbf{a})$ unless risk of confussion). Then, $M(\mathbf{a})S = S$ if $S \in \mathcal{E}_{\mathbf{A}}(\mathbf{a})$ and $M(\mathbf{a})S = \emptyset$ otherwise. The empty set $\emptyset$ will be formally added to the ensemble $\mathcal{E}$ with the physical meaning of the absence of the physical system. Notice, however, that if the ensemble $\mathcal{E}$ is sufficient for $\mathbf{A}$, it cannot happen that $M(\mathbf{a})S = \emptyset$ for all S. Then, with this notation, $M(\mathbf{a})\mathcal{E} = \mathcal{E}_{\mathbf{A}}(\mathbf{a})$.

A selective measurement will be called complete if the family of compatible observables $\mathbf{A}$ is maximal. It will be assumed that there exist complete selective measurements and only those will be considered in what follows.

We will assume that the state of a quantum system is established by performing on it a complete selective measurement. Thus, in this simplified picture, the space of states $\mathcal{S}$ of the theory is provided by the outcomes of complete selective measurements. A more precise definition will be given at the end of this section.

A more general class of symbols $M(\mathbf{a}', \mathbf{a})$ can be introduced, those denoting the complete selective measurements that change the systems of the ensemble in a compatible way, meaning by that the complete selective measurement that rejects all systems of an ensemble whose outcomes are different from $\mathbf{a}$ and that those accepted emerge in such a way that the measurement of $\mathbf{A}$ on them give the outcome $\mathbf{a}'$. Using the set-theoretical notation introduced above, we have: $M(\mathbf{a}', \mathbf{a})S \in \mathcal{E}_{\mathbf{A}}(\mathbf{a}')$ if $S \in \mathcal{E}_{\mathbf{A}}(\mathbf{a})$ and $M(\mathbf{a}', \mathbf{a})S = \emptyset$ otherwise.

It is clear now that the following composition law prevails:

$$M(\mathbf{a}'', \mathbf{a}') \circ M(\mathbf{a}', \mathbf{a}) = M(\mathbf{a}'', \mathbf{a}), \tag{C.1}$$

for all possible outcomes $\mathbf{a}, \mathbf{a}'$, and $\mathbf{a}''$, of $\mathbf{A}$. Moreover, it is also obvious that:

$$M(\mathbf{a}') \circ M(\mathbf{a}', \mathbf{a}) = M(\mathbf{a}', \mathbf{a}), \qquad M(\mathbf{a}', \mathbf{a}) \circ M(\mathbf{a}) = M(\mathbf{a}', \mathbf{a}), \tag{C.2}$$

where the symbol $\circ$ means performing the selective measurements $M(\mathbf{a}', \mathbf{a})$, and $M(\mathbf{a}'', \mathbf{a}')$ one after the other. It is clear, that performing two selective measurements $M(\mathbf{a}', \mathbf{a})$, and $M(\mathbf{a}''', \mathbf{a}'')$ one after the other will produce a selective measurement only if $\mathbf{a}'' = \mathbf{a}'$. Otherwise, if $\mathbf{a}'' \neq \mathbf{a}'$, then $M(\mathbf{a}''', \mathbf{a}'') \circ M(\mathbf{a}', \mathbf{a})S = \emptyset$ for all S which, provided that the ensemble is $\mathcal{E}$ is sufficient, is not a selective measurement of the form $M(\mathbf{a}', \mathbf{a})$.

Notice that if we have three selective measurements $M(\mathbf{a}, \mathbf{a}')$, $M(\mathbf{a}', \mathbf{a}'')$ and $M(\mathbf{a}'', \mathbf{a}''')$, then because of the definition, it is immediate that:

$$M(\mathbf{a}, \mathbf{a}') \circ (M(\mathbf{a}', \mathbf{a}'') \circ M(\mathbf{a}'', \mathbf{a}''')) = (M(\mathbf{a}, \mathbf{a}') \circ M(\mathbf{a}', \mathbf{a}'')) \circ M(\mathbf{a}'', \mathbf{a}'''). \tag{C.3}$$

Finally, it is worth observing that, given a measurement symbol $M(\mathbf{a}', \mathbf{a})$, the measurement symbol $M(\mathbf{a}, \mathbf{a}')$ is such that:

$$M(\mathbf{a}', \mathbf{a}) \circ M(\mathbf{a}, \mathbf{a}') = M(\mathbf{a}'), \quad M(\mathbf{a}, \mathbf{a}') \circ M(\mathbf{a}', \mathbf{a}) = M(\mathbf{a}). \tag{C.4}$$

Thus, we conclude that the natural composition law $\circ$ defined on the family of selective measurements determines a groupoid law in the collection $\mathbf{G}_{\mathbf{A}}$ of all measurement symbols $M(\mathbf{a}', \mathbf{a})$ associated to the complete family of observables $\mathbf{A}$, whose units are the possible outcomes of the observables $\mathbf{A}$.

Thus, to any given physical system there is associated a family of groupoids $\mathbf{G}_{\mathbf{A}}$, one for each complete families of compatible observables $\mathbf{A}$. States are defined as outcomes of complete selective measurements, thus, for each maximal family of compatible observables $\mathbf{A}$ there is a family of states, denoted by $|\mathbf{a}\rangle$.

Given another different maximal family of compatible observables $\mathbf{B}$, it would determine another family of states $|\mathbf{b}\rangle$. It could happen that the systems selected by the selective measurement $M_{\mathbf{A}}(\mathbf{a})$ would lie inside the family of systems selected by $M_{\mathbf{B}}(\mathbf{b})$, that is, $\mathcal{E}_{\mathbf{A}}(\mathbf{a}) \subset \mathcal{E}_{\mathbf{B}}(\mathbf{b})$. In such case, we will say that the state $|\mathbf{b}\rangle$ is subordinate to $|\mathbf{a}\rangle$ and we will denote it by $|\mathbf{b}\rangle \prec |\mathbf{a}\rangle$, and, if this happens, the measurement of $\mathbf{B}$ will not modify the state defined by the sub-ensemble determining the state $|\mathbf{a}\rangle$. If it happens that $|\mathbf{a}\rangle$ is subordinate to $|\mathbf{b}\rangle$ and viceversa, i.e., $|\mathbf{b}\rangle \prec |\mathbf{a}\rangle$ and $|\mathbf{a}\rangle \prec |\mathbf{b}\rangle$, we will consider that both states are the same, $|\mathbf{b}\rangle \sim |\mathbf{a}\rangle$, and we will treat them as the same object. Thus, the space of states $\mathcal{S}$ of the system is the collection of equivalence classes of states determined by all complete selective measurement determined by the subordination relation among them.

Notice that the disjoint union $\bigsqcup_{\mathbf{A}} \mathbf{G}_{\mathbf{A}}$ of all groupoids $\mathbf{G}_{\mathbf{A}}$ associated to all complete selective measurements is a groupoid itself over the disjoint union of the spaces of states $\mathcal{S}_{\mathbf{A}}$ of all complete selective measurements. However, we would like to consider, as indicated above, that two states are equivalent if they are both subordinated to each other. Then, we would like to consider the natural composition law in $\bigsqcup_{\mathbf{A}} \mathbf{G}_{\mathbf{A}}$ induced from the composition laws of the groupoids $\mathbf{G}_{\mathbf{A}}$. Notice that, under such premises, two complete selective measurements $M_{\mathbf{A}}(\mathbf{a}',\mathbf{a})$ and $M_{\mathbf{B}}(\mathbf{b}',\mathbf{b})$ could be composed if $|\mathbf{a}\rangle \sim |\mathbf{b}'\rangle$, in which case, $M_{\mathbf{A}}(\mathbf{a}',\mathbf{a}) \circ M_{\mathbf{B}}(\mathbf{b}',\mathbf{b})$ will correspond to a physical device that will take as inputs elements in the sub-ensemble $\mathcal{E}_{\mathbf{B}}(\mathbf{b})$ and will return elements in the sub-ensemble $\mathcal{E}_{\mathbf{A}}(\mathbf{a})$. If we denote it by $M_{\mathbf{AB}}(\mathbf{a}',\mathbf{b})$ (or simply $M(\mathbf{a}',\mathbf{b})$ for short), then, provided that $|\mathbf{a}\rangle \sim |\mathbf{b}'\rangle$:

$$M(\mathbf{a}',\mathbf{a}) \circ M(\mathbf{b}',\mathbf{b}) = M(\mathbf{a}',\mathbf{b}) .$$

Notice that this new family of operations does not correspond, in general, to a complete selective measurement $\mathbf{C}$ constructed out of $\mathbf{A}$ and $\mathbf{B}$, however, it is important to introduce them because they correspond to the physical operations of composition of Stern-Gerlach devices (so they could be called *composite Stern-Gerlach measurements*).

All together, the space $\mathbf{G}$ consisting of all complete selective measurements and composite Stern-Gerlach measurements with the composition law above, has the structure of a groupoid whose space of objects is the space of states $\mathcal{S}$ of the system. Thus, we will denote by $\mathbf{G} \rightrightarrows \mathcal{S}$ such groupoid where the source and target maps are given as:

$$s(M(\mathbf{a}',\mathbf{a})) = \mathbf{a} \qquad t(M(\mathbf{a}',\mathbf{a})) = \mathbf{a}' .$$

C.3 2-groupoids and quantum systems

C.3.1 The inner groupoid structure: Events and transitions

Thus, using as inspiration and motivation the discussion in the previous section, Sect. C.2, we will assume that the starting point on the description of a quantum system is the selection of a family $\mathbf{A}$ of compatible observables. The outcomes of a measurement of such observables will be called, in what follows, 'events' or simply 'outcomes', and we will denote them as $a, a', a'', \ldots$,etc.

We will not try to make precise the meaning of 'measurement' or the nature of the outcomes as we will consider them primary notions, determined solely by the experimental setting used to study our system, neither will we require any particular algebraic structure for the family of observables used in such setting. For instance, the outcomes $a, b, \ldots$ could be just collections of real numbers.

We will only be concerned with the structural relations among the various notions that are introduced.

We postulate that, for a given physical system, there are transitions among the outcomes of measurements, that is, if after performing a measurement of $\mathbf{A}$ whose outcome has been a, we will perform another measurement whose outcome would have been a', we will say that there has been a 'transition' from the event a to the event a'.

Such transitions are determined completely by the intrinsic dynamic of the system and by the interaction of the experimental setting with it. That is, we may consider experimental devices that, when acting on the system, cause it to change in such a way that a new measurement of $\mathbf{A}$ will give a different outcome. Consistently with the notation introduced for groupoids, we will denote such transitions using a diagrammatic notation as $\alpha\colon a \to a'$ and we will say that the event a is the source of the transition α and the event a' is its target. We will also say that the transition α transforms the event a into the event a'.

The allowed physical transitions must satisfy a small number of natural requirements or axioms. The first one is that transitions can be composed, that is, if $\alpha\colon a \to a'$, and $\beta\colon a' \to a''$ denote two allowed transitions, there is a transition $\beta \circ \alpha\colon a \to a''$. Notice that not all transitions can be composed, we may only compose compatible transitions, that is, transitions α, β such that β transforms the target event a' of α. This composition law must be associative, that is, if α, β are transitions as before, and $\gamma\colon a'' \to a'''$, then

$$(\gamma \circ \beta) \circ \alpha = \gamma \circ (\beta \circ \alpha) .$$

Moreover, we will assume that there are trivial transitions, that is, transitions $1_a\colon a \to a$, such that $1_{a'} \circ \alpha = \alpha$ and $\alpha \circ 1_a = \alpha$ for any transition $\alpha\colon a \to a'$. The physical meaning of transitions 1_a is that of manipulations of the system which do not change the outcome of a measurement of $\mathbf{A}$.

Finally, we will assume a (local) reversibility property of physical systems, that is, transitions are invertible: Given any transition $\alpha\colon a \to a'$, there is another one $\alpha'\colon a' \to a$ such that $\alpha \circ \alpha' = 1_{a'}$ and $\alpha' \circ \alpha = 1_a$. We will denote such transition as α^{-1} and it clearly must be unique.

From the previous axioms, it is clear that the collection of transitions α and events a form a groupoid. Such groupoid will be denoted by $\mathbf{G_A}$, its objects are the events a provided by measurements of a family of compatible observables $\mathbf{A}$, and its morphisms are the allowed physical transitions among events. We will denote by $\mathbf{G_A}(a, a')$ the family of transitions $\alpha\colon a \to a'$ between the events a and a'.

We will denote by Ω the collection of events and by s, t the standard source and target maps. Notice that we are not assuming that the groupoid $\mathbf{G_A}$ has any particular additional structure, that is, for instance, it could be connected or not, it could possess an Abelian isotropy group or the isotropy group could be trivial.

We will associate to the system a Hilbert space $\mathcal{H}_A$. Such Hilbert space $\mathcal{H}_A$ is the support of the fundamental representation of the groupoid $\mathbf{G_A}$. In Sect. 10.5.1 the case of finite groupoids was discussed. There, the Hilbert space was finite-dimensional and given explicitly as: $\mathcal{H}_A = \bigoplus_{a\in\Omega} \mathbb{C}|a\rangle$. In a more general situation, we will assume that the space of events Ω is a standard Borel space with measure μ. In that case, $\mathcal{H}_A$ is the direct integral of the field of Hilbert spaces $\mathbb{C}|a\rangle$:

$$\mathcal{H}_A = \int_\Omega^\oplus \mathbb{C}\,|a\rangle\, d\mu(a) \cong L^2(\Omega,\mu)\,.$$

For instance, in the particular situation where Ω is discrete countable, such measure will be the standard counting measure and $\mathcal{H}_A \cong L^2(\mathbb{Z}) = l^2$. Notice that, if the space of objects is countable, the unit elements 1_a are represented in the fundamental representation as the orthogonal projectors P_a on the subspaces $\mathbb{C}|a\rangle$. Such projectors provide a resolution of the identity for the Hilbert space $\mathcal{H}_A$.

The Hilbert space $\mathcal{H}$ will allow us to relate the picture provided by the groupoid $\mathbf{G_A}$ with the standard Dirac-Schrödinger picture and, in the particular instance of a discrete, finite space of events considered by Schwinger, it becomes a finite dimensional space, corresponding to a finite-level quantum system.

All the previous arguments can be repeated, when considering another system of observables **B**, to describe the transitions of the system. The system **B** may be incompatible with **A**, however, they may have common events, that is, events that are outcomes of both **A** and **B**. Thus, in addition to the transitions $\alpha\colon a \to_A a'$, among outcomes of the family **A**, or $\beta : b \to_B b'$ corresponding to outcomes of the family **B**, new transitions could be added to the previous ones, those of the form $\gamma = \alpha \circ \beta : a \to_A a' = b \to_B b'$ where $a' = b$ is a common outcome for both **A** and **B**.

We may form the groupoid **G** consisting of all possible transitions over the total space of events S. Each groupoid $\mathbf{G_A}$ is a normal subgroupoid of the groupoid **G**. The construction of **G** depends on the possibility of giving a description of a physical systems in terms of two or more maximal families of compatible observables. This instance naturally carries with it the possibility of relating different descriptions, and this, in turn, will lead us to the construction of another layer on our abstract groupoid structure. Specifically, we will build another groupoid Γ over **G** obtaining, thus, what is known as a 2-groupoid.

If the groupoid **G** describing the system is connected or transitive, we will say that there are no superselection rules. The connected components of the groupoid will determine the different sectors of the theory. In what follows, we will restrict ourselves to considering only a connected component of the total groupoid.

C.3.2 Two examples: The qubit and the harmonic oscillator

The qubit

Let us start the discussion of some simple examples by considering what is arguably the simplest non-trivial groupoid structure: The groupoid $\mathbf{G}(2) = \mathbf{A}_2$ of pairs of two elements. We called it the extended singleton and is given by the Dynkin diagram A_2 (see Fig. 1.7).

This diagram will correspond to a physical system described by a family of observables $\mathbf{A}$ producing just two outputs, denoted respectively by $+$ and $-$, and with just one transition $\alpha\colon + \to -$ among them. Notice that the groupoid $\mathbf{G}(2)_{\mathbf{A}}$ associated with this diagram has 4 elements $\{1_+, 1_-, \alpha, \alpha^{-1}\}$ and the space of events is just $\Omega_A = \{+,-\}$. The corresponding (non-commutative) groupoid algebra $\mathbb{C}[\mathbf{A}_2]$ is a complex vector space of dimension 4 generated by $e_1 = 1_+$, $e_2 = 1_-$, $e_3 = \alpha$ and $e_4 = \alpha^{-1}$, with structure constants given by the relations:

$$
\begin{array}{llll}
e_1^2 = e_1\,, & e_2^2 = e_2\,, & e_1e_2 = e_2e_1 = 0\,, & e_3e_4 = e_1\,, \\
e_4e_3 = e_2\,, & e_3e_3 = e_4e_4 = 0\,, & e_1e_3 = e_3\,, & e_3e_1 = 0\,, \\
e_4e_1 = e_4\,, & e_1e_4 = 0\,, & e_3e_2 = e_3\,, & e_2e_3 = 0\,.
\end{array}
$$

The fundamental representation of the groupoid algebra is supported on the 2-dimensional complex space $\mathcal{H} = \mathbb{C}^2$ with canonical basis $|+\rangle$, $|-\rangle$. The groupoid elements are represented by operators acting on the canonical basis as:

$$
A_+|+\rangle = \pi(1_+)|+\rangle = |+\rangle\,, \qquad A_+|-\rangle = \pi(1_+)|-\rangle = 0\,,
$$

etc., that is, for instance, the operator A_+ has associated matrix:

$$
A_+ = \begin{bmatrix} 1 & 0 \\ 0 & 0 \end{bmatrix}.
$$

Similarly we get:

$$
A_- = \pi(1_-) = \begin{bmatrix} 0 & 0 \\ 0 & 1 \end{bmatrix}, \quad A_\alpha = \pi(\alpha) = \begin{bmatrix} 0 & 0 \\ 1 & 0 \end{bmatrix},
$$

and

$$
A_{\alpha^{-1}} = \pi(\alpha^{-1}) = \begin{bmatrix} 0 & 1 \\ 0 & 0 \end{bmatrix},
$$

Thus, the groupoid algebra can be naturally identified with the algebra of 2×2 complex matrices $M_2(\mathbb{C})$ and the fundamental representation is just provided by the matrix-vector product of matrices and 2-component column vectors of $\mathbb{C}^2$. The dynamical aspects of this system are be described extensively in [16].

Before discussing the next example, it is interesting to observe that, if we consider the system without the transition α, that is, now the groupoid will consist solely of the elements $\{1_+, 1_-\}$, its corresponding groupoid algebra will be just the 2-dimensional Abelian algebra defined by the relations: $e_1^2 = e_1$, $e_2^2 = e_2$ and $e_1e_2 = e_2e_1 = 0$, that is, the classical bit. According to the physical interpretation of transitions given at the beginning of this section, by disregarding the transitions α and α^{-1} we are implicitely assuming that there is no experimental device that, when acting on the singleton, cause it to change in such a way that a new measurement will yeld a different outcome. Put differently, we are assuming that experimental devices do not influence the system, which is an assumption of genuinely classical flavour.

The harmonic oscillator

We will now discuss a family of genuinely infinite-dimensional examples. Their kinematical description is as follows. The events are labelled by the symbols a_n $n = 0, 1, 2, \ldots$, and the groupoid structure is generated by a family of transitions $\alpha_n : a_n \to a_{n+1}$ for all n.

The assignment of physical meaning to the events a_n and the transitions α_n, that is, the identification of events with outcomes of a certain observable and the observation of physical transitions depends on the specific system under study. This, in turn, implies an assignment of physical meaning to the observables and the identification of the dynamics, and fixing the experimental setting chosen by the observers. For instance, the events can be identified with the energy levels of a given system, an atom for instance, or the number of photons of a given frequency on a cavity. In the case of atoms, the transitions will correspond to the physical transitions observed by measuring the photons emitted or absorbed by the system. In the case of an electromagnetic field on a cavity, the transitions will correspond to the change in the number of photons that could be determined by counting the photons emitted by the cavity or pumping a determined number of photons into it.

At this point, no specific values have been assigned to the events a_n yet, they just represent a sort of kinematical background for the theory. An assignment of numerical values to the events will be part of the dynamical prescription of the system. For instance, in the case of energy levels, we will be assigning a real number E_n to each event while, in the case of photons, it will be a certain collection of non-negative integers $n_1, n_2, \ldots$. In what follows, we will focus on the simplest non-trivial assignment of the number n to the event a_n. The diagram describing this situation is shown in Fig. C.1.

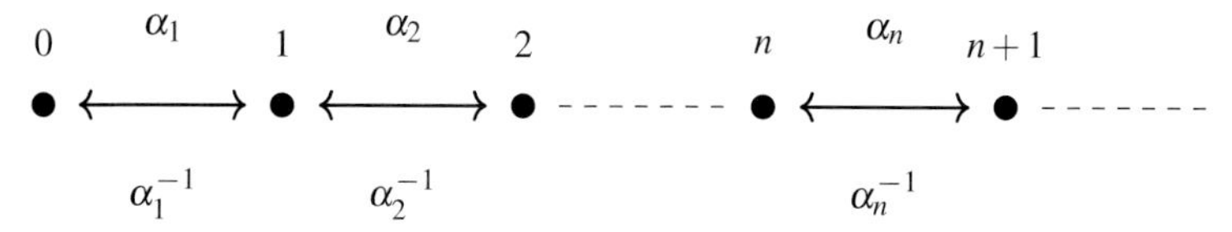

Figure C.1: The diagram K_∞ generating the quantum harmonic oscillator.

The groupoid of transitions generated by this system **G** is the groupoid of pairs of the natural numbers, that is, the complete graph with a countable number of vertices K_∞. Transitions $m \to n$ will be denoted by $\alpha_{n,m}$ or just (n,m) for short. The notation in the picture corresponds to $\alpha_n := \alpha_{n+1,n} = (n+1,n)$. With this notation, two transitions (n,m) and (j,k) are composable if and only if $m = j$, and their composition will be $(n,m) \circ (m,k) = (n,k)$. Notice that $(n,m)^{-1} = (m,n)$ and $1_n = (n,n)$ for all $n \in \mathbb{N}$.

The algebra of observables of the system will be given by functions on the groupoid $\mathbf{G}$ but, this time, in order to construct a C^*-algebra structure, we should start by considering first the set of functions which are zero except on a finite number of transitions and then take the closure with respect to an appropriate topology. Thus, denote by $\mathcal{F}_{\text{fin}}(\mathbf{G}) = \mathcal{F}_{\text{fin}}(K_\infty)$ the set of functions on K_∞ which are zero except on a finite number of pairs (n,m). We may write any one of these functions as:

$$f = \sum_{n,m=1}^{\infty} f(n,m)\delta_{(n,m)}, \tag{C.5}$$

where only a finite number of coefficients $f(n,m)$ are different from zero (the function $\delta_{(n,m)}$ is the obvious function $\delta_{(n,m)}(\alpha_{jk}) = \delta_{nj}\delta_{mk}$). We can define, as usual, the convolution product on $\mathcal{F}_{\text{fin}}(K_\infty)$:

$$(f \star g)(n,m) = \sum_{(n,j)\circ(j,m)=(n,m)} f(n,j)g(j,m) = \sum_j f(n,j)g(j,m).$$

Hence, using Heisenberg's interpretation of observables as (infinite) matrices, we may consider the coefficients $f(n,m)$, $n,m = 0,1,\ldots$, in the expansion (C.5) of the observable f as defining an infinite matrix F whose entries F_{nm} are $f(n,m)$, and the convolution product on the algebra $\mathcal{F}_{\text{fin}}(K_\infty)$ is just the matrix product of the matrices F and G corresponding to f and g, respectively (notice that the product is well defined as there are only finitely many non zero entries on both matrices). The involution $f \mapsto f^*$ is defined in the standard way $f^*(n,m) = \overline{f(m,n)}$ for all n,m.

The fundamental representation of the system will be supported on the Hilbert space $\mathcal{H}$ generated by vectors $|n\rangle$, $n = 0,1,\ldots$, that is, the family of vectors $\{|n\rangle\}$ define an orthonormal basis of $\mathcal{H}$. Thus, the Hilbert space $\mathcal{H}$ is the space $l^2(\mathbb{Z})$ of infinite sequences $z = (z_0, z_1, z_2, \ldots)$ of complex numbers with $||z||^2 = \sum_{n=0}^{\infty} |z_n|^2 < \infty$. The fundamental representation π of the algebra $\mathcal{F}_{\text{fin}}(K_\infty)$ is just given by[1]:

$$\pi(\alpha_{nm})|k\rangle = \delta_{mk}|n\rangle,$$

that is, $\pi(\alpha_{nm})$ is the operator in $\mathcal{H}$ that sends the vector $|m\rangle$ into the vector $|n\rangle$ and zero otherwise. Even in more concise terms: The fundamental representation of the transition α_n maps the vector $|n\rangle$ into the vector $|n+1\rangle$. In particular, $\pi(\alpha_1)|0\rangle = |1\rangle$. Notice that $\pi(\alpha_n^{-1})|n+1\rangle = |n\rangle$.

Using the fundamental representation π, we may define a norm on $\mathcal{F}_{\text{fin}}(K_\infty)$ as $||f|| = ||\pi(f)||_{\mathcal{H}}$ and consider the completion $\mathcal{F}(K_\infty)$ of $\mathcal{F}_{\text{fin}}(K_\infty)$ with respect to it. It is now clear that such completion is a C^*-algebra as $||f^* \star f|| = ||\pi(f^* \star f)||_{\mathcal{H}} = ||\pi(f^*)\pi(f)||_{\mathcal{H}} = ||\pi(f)^\dagger \pi(f)||_{\mathcal{H}} = ||\pi(f)||_{\mathcal{H}}^2 = ||f||^2$. Moreover, by construction, the representation π is continuous and has a continuous extension to the completed algebra $\mathcal{F}(K_\infty)$. By construction, the map π defines an isomorphism of algebras between the algebra $\mathcal{F}(K_\infty)$ and the algebra $\mathcal{B}(\mathcal{H})$ of bounded operators on the Hilbert

[1] With some abuse of notation as we are identifying the functions $\delta_{(n,m)}$ with the transition α_{nm}.

space $\mathcal{H}$. The elementary transitions α_n generating the graph K_∞ contain the relevant information of the system. Any transition α_{nm} can be obtained composing elementary transitions: $\alpha_{nm} = \alpha_n \alpha_{n+1} \cdots \alpha_{m-1}$ $(n < m)$.

Once we have chosen the assignment $a_n = n$, $n = 0, 1, 2, \ldots$, we may define the functions a and $a^\dagger$ in $\mathcal{F}(K_\infty)$ as follows:

$$a(\alpha_n) = \sqrt{n}\,, \qquad a^\dagger(\alpha_n) = \sqrt{n+1}\,, \tag{C.6}$$

or, in terms of the algebra of the groupoid K_∞, we will have that a and $a^\dagger$ are given as the formal series:

$$a = \sum_{n=0}^{\infty} \sqrt{n}\,\alpha_n^{-1}\,, \qquad a = \sum_{n=0}^{\infty} \sqrt{n+1}\,\alpha_n\,.$$

Strictly speaking, a, $a^\dagger$ are not elements in the groupoid algebra $\mathcal{F}(K_\infty)$, indeed, they define unbounded operators in the fundamental representation, however, they do define functions on K_∞ and we can manipulate them formally. A simple computation shows that:

$$[a, a^\dagger] = \mathbf{1}\,,$$

with $\mathbf{1} = \sum_{n=0}^{\infty} 1_n$, or in terms of functions in K_∞, $[a, a^\dagger](1_n) = 1$ for all n and zero otherwise. Hence, we may construct the Hamiltonian function:

$$h = a^\dagger a + \frac{1}{2} = \sum_{n=0}^{\infty} n\delta_n + \frac{1}{2}\,.$$

and the corresponding equations of motion:

$$\dot{a} = i[a, h] = -ia\,, \qquad \dot{a}^\dagger = i[a^\dagger, h] = ia^\dagger\,,$$

which constitute the standard equations of motion for the quantum harmonic oscillator.

Notice that the functions a, $a^\dagger$ on the groupoid K_∞ define densely defined unbounded operators on the Hilbert space $\mathcal{H} = l^2(\mathbb{Z})$ supporting the fundamental representation such that $\pi(a)^\dagger = \pi(a^\dagger)$. Moreover, the Hamiltonian operator $H = \pi(h)$ may be identified with the Hamiltonian operator of a harmonic oscillator with creation and annihilation operators $\pi(a^\dagger)$ and $\pi(a)$, respectively.

C.3.3 The 2-groupoid structure: Transformations

The dynamical behaviour of a system is described by a sequence of transitions $w = \alpha_1 \alpha_2 \cdots \alpha_r$ that will be called histories. In a certain limit, any history would define a one-parameter family α_t of transitions (but we may very well keep working with discrete sequences).

Typically, these sequences of transitions are generated by a given observable, promoted to be the infinitesimal generator of a family of automorphisms once the family of observables has an algebra structure itself. What we would like to stress here is that the explicit expression of such sequence of transitions depends on the complete measurements used to describe the behaviour of the system and it may look very different when observed using two different systems of observables **A** and **B**.

The existence of such alternative descriptions imply the existence of families of 'transformations' among transitions that would allow us to compare the descriptions of the dynamical behaviour of the system (and the kinematical structure as well) when using different references of measurements systems. We will also use a diagrammatic notation to denote transformations such as $\varphi\colon \alpha \Rightarrow \beta$, or as in Fig. C.2 below.

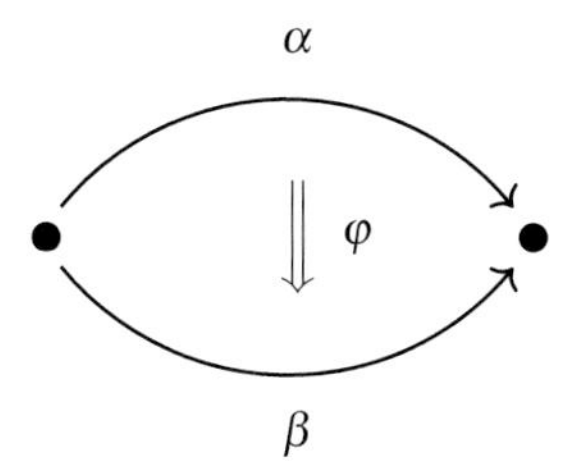

Figure C.2: A transformation φ between the transitions α and β.

The transformations φ must satisfy some obvious axioms. First, in order to make sense of the assignment $\alpha \Rightarrow \beta$, the transformation φ must be compatible with the source and target maps of the groupoid or, in other words, φ must map $\mathbf{G_A}(a,a')$ to $\mathbf{G_B}(b,b')$. Moreover, the transformations $\varphi\colon \alpha \Rightarrow \beta$ and $\psi\colon \beta \Rightarrow \gamma$ could be composed providing a new transformation $\varphi \circ_v \psi\colon \alpha \Rightarrow \gamma$, from transition α to γ. This composition law will be called the 'vertical' composition law and denoted accordingly by $\circ_v$ (see Fig. C.3(a) for a diagrammatic representation of the vertical composition of transformations). This vertical composition law of transformations must be associative as there is not a preferred role for the various arrangements of compositions between the various transformations involved. That is, we postulate:

$$\varphi \circ_v (\psi \circ_v \zeta) = (\varphi \circ_v \psi) \circ_v \zeta\,,$$

for any three transformations $\varphi\colon \alpha \Rightarrow \beta$, $\psi\colon \beta \Rightarrow \gamma$, $\zeta : \gamma \Rightarrow \delta$.

The transformation φ sends the identity transition 1_a into the identity transition 1_b, and there should be transformations 1_α, 1_β, such that $1_\alpha \circ_v \varphi = \varphi$ and $\varphi \circ_v 1_\beta = \varphi$.

Moreover, it will be assumed that transformations $\varphi\colon \alpha \Rightarrow \beta$ are reversible, provided that the physical information determined by using the family of observables **A**, that is, the groupoid $\mathbf{G_A}$, is equivalent to that provided by the groupoid $\mathbf{G_B}$, i.e., by the family of observables **B**. In other words, the transformations $\varphi\colon \alpha \Rightarrow \beta$ are invertible because there is no a natural precedence among the corresponding measurements systems. Thus, under such assumption, for any transformation $\varphi\colon \alpha \Rightarrow \beta$ there exists another one $\varphi^{-1}\colon \beta \Rightarrow \alpha$, such that $\varphi \circ_v \varphi^{-1} = 1_\alpha$ and $\varphi^{-1} \circ_v \varphi = 1_\beta$.

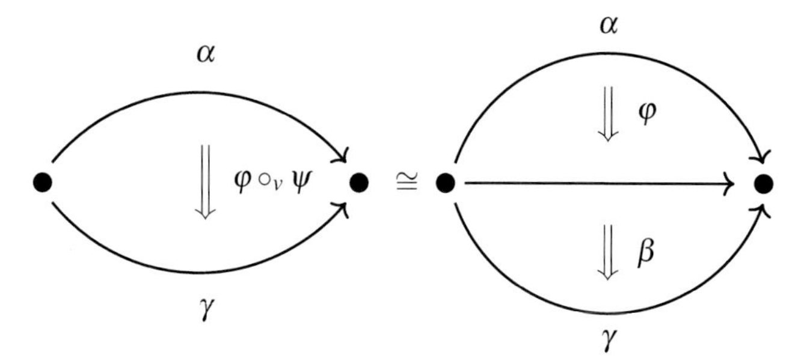

(a) Diagrammatic representation of the vertical composition law $\circ_v$.

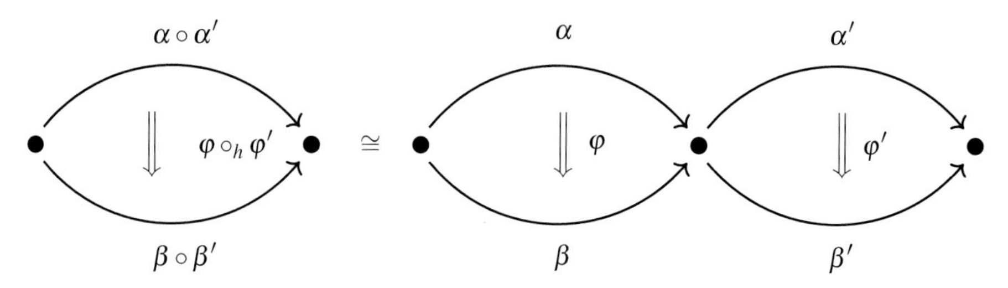

(b) Diagrammatic representation of the horizontal composition law $\circ_h$.

Figure C.3: A diagrammatic representation of vertical and horizontal composition of transformations.

Moreover, notice that if we have a transformation $\varphi\colon \alpha \Rightarrow \beta$ and another one $\varphi'\colon \alpha' \Rightarrow \beta'$, such that α and α' can be composed, then β and β' will be composable too because of the consistency condition for transformations, that is, given two transitions α, α' that can be composed, if the reference description for them is transformed, then the corresponding description of the transitions β and β' will be composable too and their composition must be the composition of the transformation of the original transitions (see Fig. C.3(b) for a diagrammatic description). In other words, there will be a natural transformation between the transition $\alpha \circ \alpha'$ to the transition $\beta \circ \beta'$ that will be denoted by $\varphi \circ_h \varphi'$ and called the 'horizontal' composition. Figure C.3 provides a diagrammatic representation of both operations $\circ_v$ and $\circ_h$.

It is clear that the horizontal composition law is associative too, i.e.,

$$\varphi \circ_h (\varphi' \circ_h \varphi'') = (\varphi \circ_h \varphi') \circ_h \varphi'' ,$$

for any three transformations $\varphi\colon \alpha \Rightarrow \beta$, $\varphi'\colon \alpha' \Rightarrow \beta'$, $\varphi''\colon \alpha'' \Rightarrow \beta''$, such that $\alpha\colon a \to a'$, $\beta\colon b \to b'$, $\alpha'\colon a' \to a''$, $\beta'\colon b' \to b''$ and $\alpha''\colon a'' \to a'''$, $\beta''\colon b'' \to b'''$. The horizontal composition rule has natural units, that is, if $\varphi\colon \alpha \Rightarrow \beta$, then the transformation $1_{a'b'}\colon 1_{a'} \Rightarrow 1_{b'}$, if $\alpha\colon a \to a'$ and $\beta\colon b \to b'$, is such that: $\varphi \circ_h 1_{a'b'} = \varphi$ and $1_{ab} \circ_h \varphi = \varphi$.

We observe that there must be a natural compatibility condition between the composition rules $\circ_v$ and $\circ_h$. That is, if we have two pairs of vertically composable trans-

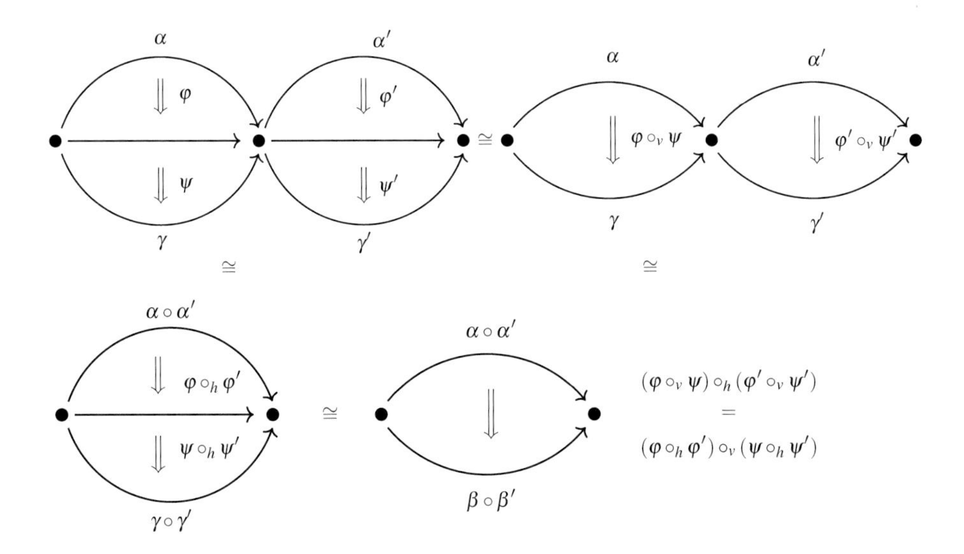

Figure C.4: A diagrammatic representation of the exchange identity: It is equivalent, first move right and then down, or first down and then right.

formations φ, ψ and φ', ψ' that can also be pairwise composed horizontally, then the horizontal composition of the previously vertically composed pairs must be the same as the vertical composition of the previously horizontally composed pairs. This consistency condition will be called the exchange identity. Formally, is written as follows and a diagramatic description is provided in Fig. C.4:

$$(\varphi \circ_v \psi) \circ_h (\varphi' \circ_v \psi') = (\varphi \circ_h \varphi') \circ_v (\psi \circ_h \psi') \,. \tag{C.7}$$

Notice that, given a transformation $\varphi\colon \alpha \Rightarrow \beta$, we can define two maps, similar to the source and target defined previously for transitions, that assign, respectively, α and β to φ. The family of invertible transformations Γ with the vertical composition law and the source and target maps defined in this way form again a groupoid over the space of transitions $\mathbf{G}$. Moreover, the source and target maps are morphisms of groupoids.

This 'double' groupoid structure, that is, a groupoid (the family of invertible transformations) whose objects (the family of all transitions) form again a groupoid (whose objects are the events) and such that the source and target maps are groupoid homomorphisms, is called a 2-groupoid.

The morphisms of the first groupoid structure Γ (or external groupoid structure) are sometimes called 2-morphism (or 2-cells). In our case, 2-morphisms correspond to what we have called transformations. The objects $\mathbf{G}$ of the first groupoid structure, which are the morphisms of the second groupoid structure (or inner groupoid structure), are called 1-morphisms (or 1-cells). In our setting, they will correspond to what we have called transitions. Finally, the objects S of the second groupoid structure are called 0-morphism (or 0-cells) and in the previous discussion they correspond to what we have called events or outcomes.

The set of axioms discussed above can, thus, be summarised by saying that the notions previously introduced to describe a physical system form a 2-groupoid with 0-, 1- and 2-cells being respectively events, transitions and transformations. Thus, if we denote the total 2-groupoid by Γ, we will denote its set of 1-morphisms by $\mathbf{G}$, and the outer groupoid structure will be provided by the source and target maps $\Gamma \rightrightarrows \mathbf{G}$. The groupoid composition law in Γ will be called the vertical composition law and denoted $\circ_v$. Since $\mathbf{G} \rightrightarrows S$ is itself a groupoid and the source and target maps of Γ are themselves groupoid homomorphisms, we can always define a horizontal composition law in a natural way denoted by $\circ_h$ and both composition laws must satisfy the exchange identity above, Eq. (C.7).

Thus, we conclude by postulating that the categorical description of a physical system is provided by a 2-groupoid $\Gamma \rightrightarrows \mathbf{G} \rightrightarrows S$. The 1-cells of the 2-groupoid will be interpreted as the transitions of the system and its 0-cells will be considered as the events or outcomes of measurements performed on the system. The 2-cells will be interpreted as the transformations among transitions of theory providing the basis for its dynamical interpretation.

The identification of these abstracts notions with the corresponding standard physical notions and their relations with other physical notions, like quantum states, unitary transformations, etc., will be provided by constructing representations of the given 2-groupoid (see [14], [15] for more details on this subject).

Appendix D

Solution to Rubik's Cube

In this appendix, we will graphically describe the solution to Rubik's pocket cube, discussed in Sect. 3.2.4.

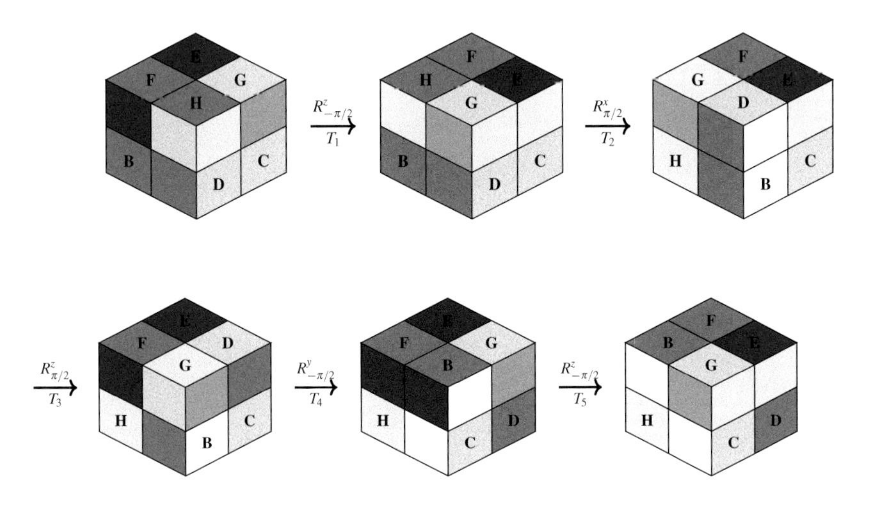

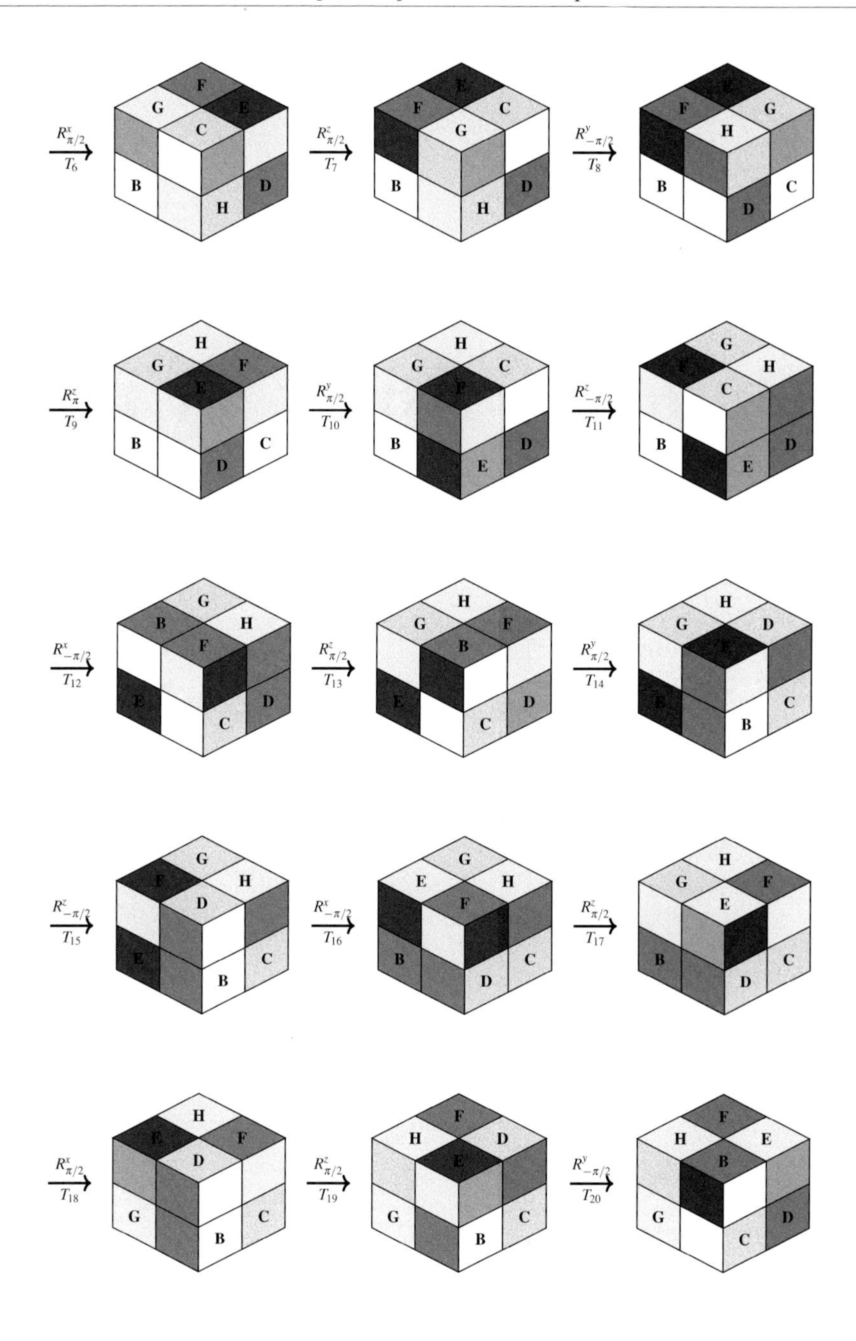
$R^x_{\pi/2}$
T_6
F
G
E
C
B
D
H
$R^z_{\pi/2}$
T_7
E
F
C
G
B
D
H
$R^y_{-\pi/2}$
T_8
F
G
H
B
C
D
R^z_{π}
T_9
H
G
F
E
B
C
D
$R^y_{\pi/2}$
T_{10}
H
G
C
F
B
D
E
$R^z_{-\pi/2}$
T_{11}
G
H
C
B
D
E
$R^x_{-\pi/2}$
T_{12}
G
B
H
F
E
D
C
$R^z_{\pi/2}$
T_{13}
H
G
F
B
E
D
C
$R^y_{\pi/2}$
T_{14}
H
G
D
C
B
$R^z_{-\pi/2}$
T_{15}
G
H
D
E
C
B
$R^x_{-\pi/2}$
T_{16}
G
E
H
F
B
C
D
$R^z_{\pi/2}$
T_{17}
H
G
F
E
B
C
D
$R^x_{\pi/2}$
T_{18}
H
E
F
D
G
C
B
$R^z_{\pi/2}$
T_{19}
F
H
D
E
G
C
B
$R^y_{-\pi/2}$
T_{20}
F
H
E
B
G
D
C

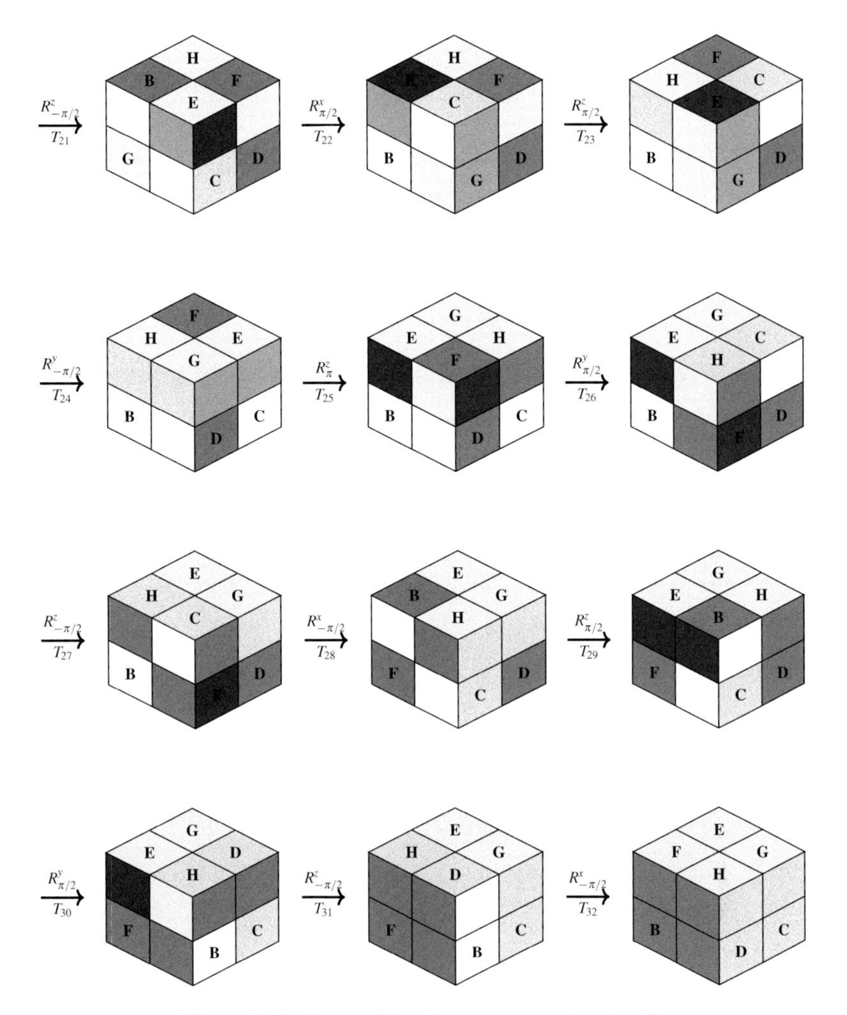

Figure D.1: A morphism from one state into itself.

References

[1] V. Arnold. Mathematics with a human face. *Priroda*, 3:117–119, 1988.

[2] R. Baer. Beiträge zur Galoisschen Theorie. *S.-B. Heidelberger Akad. Wiss. Math.-Verein. Kl*, 37:5–7, 1928.

[3] J. Baez. Higher-dimensional algebra and Planck-scale physics. In *Physics Meets Philosophy at the Planck Length*, pages 177–196. Cambridge University Press, Cambridge, 2001.

[4] S. Beckus, J. Bellisard, and G. De Nittis. Spectral continuity for aperiodic quantum systems I. General theory. *J. Funct. Anal.*, 275:2917–2977, 2018.

[5] W.W. Boone. The word problem. *Proc. Nat. Acad. Sci.*, 44:1061–1065, 1958.

[6] R. Bos. Continuous representations of groupoids. *Houston J. Math.*, 37:807–844, 2011.

[7] H. Brandt. Über eine Verallgemeinerung des Gruppenbegriffes. *Math. Ann.*, 96:360–366, 1927.

[8] H. Brandt. Idealtheorie in einer Dedekindsche Algebra. *Jber. Deutsch. Math. Verein.*, 37:5–7, 1928.

[9] R. Brown. Groupoids and Van Kampen's theorem. *Proc. London Math. Soc.*, 17:385–401, 1967.

[10] R. Brown. *Elements of Modern Topology*. McGraw-Hill, London, 1968.

[11] R. Brown. From groups to groupoids: a brief survey. *Bull. London Math. Soc.*, 19:113–134, 1987.

[12] W. Burnside. *Theory of groups of finite order*. Cambridge at The University Press, Cambridge, 1897.

[13] J.F. Cariñena, A. Ibort, G. Marmo, and G. Morandi. *Geometry from Dynamics, Classical and Quantum.* Theoretical, Mathematical and Computational Physics. Springer, Dordrecht, 2015.

[14] F.M. Ciaglia, A. Ibort, and G. Marmo. A gentle introduction to Schwinger's formulation of quantum mechanics: the groupoid picture. *Mod. Phys. Lett. A*, 33:1850122, 2018.

[15] F.M. Ciaglia, A. Ibort, and G. Marmo. Schwinger's picture of quantum mechanics I: Groupoids. *Int. J. Geom. Meth. Mod. Phys.*, 16:1950119, 2019.

[16] F.M. Ciaglia, A. Ibort, and G. Marmo. Schwinger's picture of quantum mechanics II: Algebras and observables. *Int. J. Geom. Meth. Mod. Phys.*, 16:1950136, 2019.

[17] A. Connes. *Noncommutative geometry.* Academic Press, San Diego, CA, 1994.

[18] D. Corfield. The importance of mathematical conceptualisation. *Stud. Hist. Phil. Sci.*, 32:507–533, 2001.

[19] J. Cortés, M. de León, J.C. Marrero, D.M. de Diego, and E. Martínez. A survey of Lagrangian mechanics and control on Lie algebroids and groupoids. *Int. J. Geom. Meth. Mod. Phys.*, 3:509–558, 2006.

[20] M. Crainic and R.L. Fernandes. Integrability of Lie brackets. *Ann. Math.*, 157:575–620, 2003.

[21] M. Crainic and R.L. Fernandes. Lectures on integrability of Lie brackets. *Geom. Topol. Monogr.*, 17:1–107, 2011.

[22] B. d'Espagnat. *Conceptual Foundations of Quantum Mechanics. 2nd ed.* Advanced Book Program, Perseus Books, Reading, MA, 1999.

[23] C. Ehresmann. *Catégories et structures.* Dunod, Paris, 1965.

[24] S. Eilenberg and S. MacLane. The general theory of natural equivalences. *Trans. Am. Math. Soc.*, 58:231–294, 1945.

[25] L. Esteban, R. Tapia-Rojo, A. Sancho, and L.J. Boya. Groups of order less than 32, revisited. *Rev. Real Academia de Ciencias de Zaragoza*, 66:63–104, 2011.

[26] P. Etingof, O. Golberg, S. Hensel, T. Liu, A. Schwender, D. Vaintrob, E. Yudovina, and S. Gerovitch. *Introduction to Representation Theory.* Student Mathematical Library, vol. 59. Am. Math. Soc., Providence, RI, 2011.

[27] P. Gabriel and A.V. Roiter. *Representations of finite-dimensional algebras.* Encyclopaedia of Mathematical Sciences, vol.73. Springer, Berlin, 1997.

[28] M. Gardner. *Mathematical Puzzles of Sam Loyd.* Dover Publications Inc., New York, NY, 1959.

[29] A. Giżycki and L. Pysiak. Multiciplicity formulas for representations of transformation groupoids. *Demonstratio Math.*, 50:42–50, 2017.

[30] A. Gracia-Saz and R.A. Mehta. $\mathcal{VB}$-groupoids and representation theory of Lie groupoids. *J. Symplectic Geom.*, 15:741–783, 2017.

[31] A. Grothendieck. Techniques de construction et théorèmes d'existence en géométrie algébrique III: préschémas quotients. *Séminaire Bourbaki*, 212:99–118, 1961.

[32] P.J. Higgins. Presentations of groupoids, with applications to groups. *Proc. Cam. Phil Soc.*, 60:7–20, 1964.

[33] P.J. Higgins. *Notes on Categories and Groupoids.* Mathematical Studies, vol.32. Van Nostrand Reinhold, London, 1971.

[34] H.-J. Hoehnke and M.-A. Knus. Algebras, their invariants and K-forms: A tribute to the work of Heinrich Brandt on the 50th anniversary of his death. `https://people.math.ethz.ch/~knus/papers/brandt04.pdf`.

[35] A. Ibort and M.A. Rodríguez. *Notas de Álgebra Lineal.* (in Spanish), 2001.

[36] A. Ibort and M.A. Rodríguez. On the structure of finite groupoids and their representations. *Symmetry*, 11:414, 2019.

[37] N. Jacobson. The radical and semi-simplicity for arbitrary rings. *Am. J. Math.*, 67:300–320, 1945.

[38] W. Johnson and W.E. Story. Notes on the "15" puzzle. *Am. J. Math.*, 2:397–404, 1879.

[39] A. Joyal and R. Street. An introduction to Tannaka duality and quantum groups. In *Part II of Category Theory, Proceedings, Como 1990, eds. A. Carboni, M. C. Pedicchio and G. Rosolini. Lectures Notes in Mathematics*, volume 1488, pages 413–492. Springer, 1991.

[40] I. Kleiner. The evolution of group theory: a brief survey. *Math. Magazine*, 59:195–215, 1986.

[41] A.I. Kostrikin. *Introduction to algebra.* Universitext. Springer, Berlin, 1982.

[42] A.I. Kostrikin and Yu.I. Manin. *Linear Algebra and Geometry.* Algebra, Logic and Applications Series, vol. 1. Taylor and Francis, Amsterdam, 1988.

[43] A. Kumjian, D. Pask, I. Raeburn, and J. Renault. Graphs, groupoids and Cuntz-Krieger algebras. *J. Funct. Anal.*, 144:505–541, 1997.

[44] T.Y. Lam. *A First Course in Noncommutative Rings.* Graduate Texts in Mathematics, vol. 131. Springer, New York, NY, 2001.

[45] N.P. Landsman. *Mathematical Topics between Classical and Quantum Mechanics*. Springer, New York, NY, 1998.

[46] N.P. Landsman. Lie groupoids and Lie algebroids in physics and noncommutative geometry. *J. Geom. Phys.*, 56:24–54, 2006.

[47] H. Loewy. Neue elementare Begründung und Eweiterung der Galoisschem Theorie. *S.-B. Heidelberger Akad. Wiss. Math. Nat. Kl, Abh*, 7, 1925.

[48] K. Mackenzie. *General theory of Lie groupoids and Lie algebroids*. London Math. Soc. Lecture Note Series, vol. 213. Cambridge Univ. Press, Cambridge, 2005.

[49] G.W. Mackey. Ergodic theory, group theory, and differential geometry. *Proc. Nat. Acad. Sci.*, 50:1184–1191, 1963.

[50] G.W. Mackey. Infinite-dimensional group representations. *Bull. Amer. Math. Soc.*, 69:628–686, 1963.

[51] G.W. Mackey. Ergodic theory and virtual group. *Math. Ann.*, 166:187–207, 1966.

[52] E. Martínez. Lagrangian mechanics on Lie algebroids. *Acta Appl. Math.*, 67:295–320, 2001.

[53] P.S. Novikov. On the algorithmic unsolvability of the word problem in group theory. (English transl.: AMS Transl. Series 2, vol. 9, pp. 1–122 1958). *Trudy Mat. Inst. Steklov*, 44:3–143, 1955.

[54] L. Pysiak. Groupoids, their representations and imprimitivity systems. *Demonstratio Math.*, 37:661–670, 2004.

[55] L. Pysiak. Imprimitive theorem for groupoid representations. *Demonstratio Math.*, 44:29–48, 2011.

[56] K. Reidemeister. *Einführung in die kombinatorische Topologie*. Braunschweig, Berlin. (See a translation into English, by J. Stillwell and W. Dicks in `arXiv:1402.3906v1 [math.HO] 2014`), 1932.

[57] J. Renault. *A Groupoid Approach to C^*-Algebras*. Lect. Notes Math., vol. 793. Springer, Berlin, 1980.

[58] J. Schwinger. *Quantum Kinematics and Dynamics*. Frontiers in Physics. W.A. Benjamin, Inc., New York, 1970.

[59] J Schwinger. *Quantum Mechanics Symbolism of Atomic measurements*. Ed. by B.-G. Englert. Springer, Berlin, 2001.

[60] J.M. Souriau. *Calcul Linéaire*. Presses Universitaires de France, Paris, 1959.

[61] J.M. Souriau. *Structure des Systèmes Dynamiques*. Dunod, Paris, 1970.

[62] G. Thompson. Classifying groups of small order. *Adv. Pure Math.*, 6:58–65, 2016.

[63] R.J. Trudeau. *Introduction to Graph Theory.* Dover Publ. Inc., New York, 1993.

[64] A. Vistoli. Groupoids: a local theory of symmetry. *Isonomia*, 6:1–12, `https://isonomia.uniurb.it/vecchiaserie/2011Vistoli.pdf` 2011.

[65] A. Weinstein. Coisotropic calculus and poisson groupoids. *J. Math. Soc. Japan*, 40:705–727, 1988.

[66] A. Weinstein. Groupoids: unifying internal and external symmetries. *Notices of the AMS*, 43:744–752, 1996.

[67] J.J. Westman. Harmonic analysis on groupoids. *Pacific J. Math.*, 27:621–632, 1968.

[68] H. Weyl. *The Classical Groups: Their Invariants and Representations.* Princeton University Press, 1916.

[69] H. Weyl. *The Theory of Groups and Quantum Mechanics.* Dover, New York, NY, 1950.

[70] H. Weyl. *Symmetry.* Princeton Univ. Press, Princeton, NJ, 1952.

[71] E. Wigner. On unitary representations of the inhomogenous Lorentz group. *Ann. Math.*, 40:149–204, 1939.

[72] H. Wussing. *The Genesis of the Abstract Group Concept.* Dover, New York, 2007.

Index